EPITHELIAL CANCER OF THE OVARY

EPITHELIAL CANCER OF THE OVARY

edited by

Frank G Lawton MD FRCOG
Gynaecological Oncologist
King's College Hospital, London, UK

Jan P Neijt MD PhD
Medical Oncologist
Department of Internal Medicine
University of Utrecht, The Netherlands

Kenneth D Swenerton MD FRCPC
Medical Oncologist
British Colombia Cancer Agency
Vancouver, Canada

BMJ
Publishing
Group

First published 1995
by the BMJ Publishing Group, BMA House, Tavistock Square,
London WC1H 9JR

British Library Cataloguing in Publication Data

A catalogue record for this book is available
from the British Library

ISBN 0-7279-0851-0

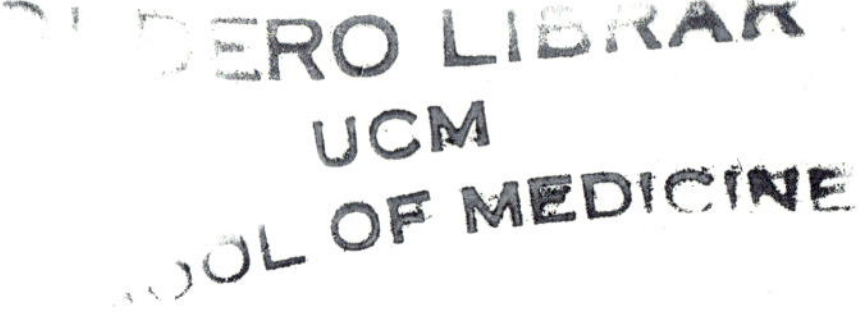

Typeset, printed and bound in Great Britain by
Latimer Trend & Company Ltd., Plymouth

Contents

Contributors

Hugh R K Barber MD
Director, Department of Obstetrics and Gynecology, Lenox Hill Hospital, 100 East 77th Street, New York, and Professor, Clinical Obstetrics and Gynecology, Cornell University Medical College, New York, NY 10021, USA

Robert C Bast Jr MD
Director, Division of Medicine, University of Texas M D Anderson Cancer Center, 1515 Holcombe Boulevard, Houston, TX 77030, USA

Philip J Beale BSc FRACP
Fellow, Department of Medical Oncology, Prince of Wales Hospital, High Street, Randwick, NSW 2031, Australia

Andrew Berchuck MD
Associate Professor, Department of Obstetrics and Gynecology, Duke University Medical Center, Durham, NC 27710, USA

Jonathan S Berek MD FACOG FACS
Professor and Vice-Chair, Chief, Gynecology and Gynecologic Oncology, UCLA School of Medicine, Center for the Health Sciences, Jonsson Comprehensive Cancer Center, 10833 Le Conte Avenue, Los Angeles, CA 90024, USA

Tom H Bourne PhD MRCOG
Visiting Professor, Department of Obstetrics and Gynaecology, University of Göteborg, Sahlgrenska Hospital, S-413 45, Göteborg, Sweden

Joel M Childers MD
Associate Professor, Department of Clinical Obstetrics and Gynecology, Division of Gynecologic Oncology, University of Arizona, Tuscon, AR 85712, USA

Joyce E Cornett MD
Prostate, Lung, Colorectal and Ovarian Cancer Screening Trial Coordinator, Early Detection Branch, Division of Cancer Prevention and Control, National Cancer Institute, National Institutes of Health, Bethesda, MD 20892, USA

William T Creasman MD
Sims-Hester Professor and Chairman, Department of Obstetrics and Gynecology, Medical University of South Carolina, 171 Ashley Avenue, Charleston, SC 29425, USA

Nina Einhorn MD PhD
Department of Gynecologic Oncology, Radiumhemmet, Karolinska Hospital, S-171 76 Stockholm, Sweden

Alan A Elbendary MD
Associate Professor, Division of Medicine, University of Texas MD Anderson Cancer Center, 1515 Holcombe Boulevard, Houston, TX 77030, USA

Frances Evans FRCS MRCOG
Senior Registrar, Department of Obstetrics and Gynaecology, King's College Hospital, Denmark Hill, London SE5 8RX, UK

Harold Fox MD FRCPath FRCOG
Emeritus Professor of Reproductive Pathology, Department of Pathological Sciences, Stopford Building, University of Manchester, Oxford Road, Manchester M13 9PT, UK

Michael L Friedlander PhD MRCP FRACP
Director of Medical Oncology, Prince of Wales Hospital, High Street, Randwick, NSW 2031, Australia

John K Gohagan PhD FACE
Chief, Early Detection Branch, Division of Cancer Prevention and Control, National Cancer Institute, National Institutes of Health, Bethesda, MD 20892, USA

Neville F Hacker MD FRCOG FRACOG FACOG FACS
Directory of Gynaecological Oncology, Royal Hospital for Women, 188 Oxford Street, Paddington, NSW 2021, Australia

A Peter M Heintz MD PhD
Professor, Department of Obstetrics and Gynaecology, Utrecht University Hospital, Heidelberglaan 100, 3584 CX Utrecht, The Netherlands

Paul J Hoskins BA MRCP FRCPC
Medical Oncologist, British Colombia Cancer Agency, Vancouver Clinic, West 10th Avenue, Vancouver, BC, Canada V5Z 4E6

Sean Kehoe MRCOG DCH
Lecturer, Academic Department of Obstetrics and Gynaecology, City Hospital, Dudley Road, Birmingham B18 7QH, UK

Barnett S Kramer MD MPH FACP
Associate Director, Early Detection and Community Oncology Program, Division of Cancer Prevention and Control, National Cancer Institute, National Institutes of Health, Bethesda, MD 20892, USA

Manfred Lahousen MD
Department of Obstetrics and Gynaecology, University Hospital, Graz, Austria

Frank G Lawton MD FRCOG
Consultant Gynaecological Oncologist, Department of Obstetrics and Gynaecology, King's College Hospital, Denmark Hill, London SE5 8RX, UK

Christoph Lees MB BS
Research Fellow, Department of Obstetrics and Gynaecology, King's College Hospital, Denmark Hill, London SE5 8RX, UK

Henry T Lynch MD
Professor and Head, Department of Preventive Medicine and Public Health, Creighton University, 2500 California Plaza, Omaha, NA 68178, USA

Susan E Mackey MD
Assistant Professor, Department of Obstetrics and Gynecology, Medical University of South Carolina, 171 Ashley Avenue, Charleston, SC 29425, USA

Amin Makar MD PhD
Department of Gynecologic Oncology, The Middelheim Hospital, Lindendreef 1, 2020 Antwerpen, Belgium

Wendy Makin MRCP FRCR
Macmillan Consultant in Palliative Care and Oncology, Christie Hospital, Wilmslow Road, Manchester M20 4BX, UK

Jonathan W F Mant MFPHM MSc MA
Clinical Lecturer, Department of Public Health and Primary Care, University of Oxford, Radcliffe Infirmary, Oxford OX2 6HE, UK

Otoniel Martinez-Maza PhD
Associate Professor, Departments of Obstetrics and Gynecology, Microbiology and Immunology, UCLA School of Medicine, Center for

the Health Sciences, 10833 Le Conte Avenue, Los Angeles CA 90024, USA

Jan P Neijt MD PhD
Associate Professor, Department of Internal Medicine, Section of Medical Oncology, Utrecht University Hospital, Heidelberglaan 100, 3584 CX, Utrecht, The Netherlands

Philip C Prorok PhD
Chief, Screening (Biometry), Division of Cancer Prevention and Control, National Cancer Institute, National Institutes of Health, Bethesda, MD 20892, USA

Kenneth D Swenerton MD FRCPC
Medical Oncologist, British Columbia Cancer Agency, Vancouver Clinic, West 10th Avenue, Vancouver, BC, Canada V5Z 4E6

Earl A Surwit MD
Professor, Department of Obstetrics and Gynecology, Division of Gynecologic Oncology, University of Arizona, Tucson, AZ 85712, USA

Claes G Tropé MD PhD
Professor, Chief of Department of Gynecologic Oncology, The Norwegian Radium Hospital, Oslo, Norway

Martin P Vessey MD FRCP FFPHM FRS
Professor of Public Health, Department of Public Health and Primary Care, University of Oxford, Radcliffe Infirmary, Oxford OX2 6HE, UK

Acknowledgements

Cover

The photograph was taken by Dr Michael Thomas, Department of Pathology, King's College Hospital, London.

Chapter 2

Figure 2.1a and b were provided by Professor Michael Wells, Department of Pathology, University of Leeds.

Chapter 4

Table 4.1 is reproduced from Hulka B, *Cancer* 1988;**62**:1776–80; Table 4.2 from Bast *et al*, *N Engl J Med* 1983;**309**:883–7; Table 4.3 from DePriest M *et al*, *Gynecol Oncol* 1993;**51**:205–9; and figures 4.1 and 4.2 from Boring C *et al*, *CA Cancer J Clin* 1994;**44**:7–26 all with permission of the publishers.

Chapter 7

Council for Tobacco Research grant no 1297E, the American Cancer Society grant no EDT-84, Department of the Army grant no DAMD 17-94-J-4340, and Nebraska Department of Health: Cancer and Smoking Disease Research Program-LB595. We thank the Aloka Co (Japan) for supplying ultrasound equipment for the ovarian screening clinic at King's College Hospital, London, and the Imperial Cancer Research Fund and the Cancer Research Campaign for financial support.

Chapter 11

Table 11.1 is reproduced from Eisenkop S *et al*, *Gynecol Oncol* 1992;**45**: 97; Table 11.3 is modified from Ferrier AJ, Petrillo AD. In: Sharp F, *et al*, eds. *Ovarian cancer*, New York: Chapman and Hall, 1992:388; Table 11.4 from Creasman WT. In: Sharp F, *et al*, eds. *Ovarian cancer*, New York: Chapman and Hall, 1992:375–83; figure 11.1 from Hacker NF *et al*, *Obstet Gynecol* 1983;**61**:413–20; figure 11.2 from Hacker NF *et al*. In: Sharp F, *et al*, eds. New York: Chapman and Hall 1992:351–6; figure 11.3 from Heintz APM, Berek JS. In: Piver MS, ed. *Ovarian malignancies*, Edinburgh:

Churchill Livingstone, 1987:134; figure 11.5 from Niejt JP *et al*, *Clin Oncol* 1987:5:1157–62; and figure 11.6 from van der Burg MEL, *et al*, *N Engl J Med* 1995;**332**:629–34, all with permission of the publishers.

Chapter 16

Parts are reproduced from Markman M, Hoskins WJ, eds. *Cancer of the ovary*. New York: Raven Press, 1993, with permission. The study was funded in part by the Romana Moscovitz Memorial Cancer Research Fund, the June Hill Ovarian Cancer Research Fund, and the Brindell and Milton Gottlieb Gynecologic Oncology Laboratory.

Chapter 17

Figures 17.1, 17.2, and 17.3 are reproduced from Childers GD. In: Adamson GD, ed. *Atlas of endoscopic management of gynecologic disease*. New York: Raven Press, in press, with permission of the publishers.

Foreword

Ovarian cancer was the first epithelial malignancy to show consistent responses to simple anti-cancer drug therapy and one might have expected that, by now, nearly 40 years after the initial observation, we would be doing much better than we are. However, treatment has been hampered by the silent nature of the disease, with symptoms usually apparent only when dissemination throughout the abdominal cavity has occurred. This means that surgeons, however skilled, cannot remove all tumour in the 50% of patients who present with stage III disease. Indeed, many patients may be left, after surgery, with a kilogram of tumour. Trying to cure such patients with drugs or radiotherapy is unlikely to succeed even when the tumours are exceedingly sensitive to present day cytotoxic drugs. Thus there is an urgent need for better methods to identify and diagnose ovarian cancer early. Screening programmes are being investigated and genetic markers are beginning to identify high risk patients.

Appropriate surgery remains the mainstay for cure in the 20% of patients who present with truly local tumours. Opinions differ as to whether the cure rate for later stage disease has improved. It might be appropriate at this point to remind younger readers of the situation in the years before modern treatment was available. In 1964 I reviewed the results of 251 cases treated at the Royal Marsden Hospital over the years 1945–58. The three year (not five year) survival figures by stage were: stage I, 55%; stage II, 32%; stage III, 32%; and stage IV, 12%; and this despite the fact that some borderline cases were, undoubtedly, included in the series. The equivalent figures for 1274 patients treated between 1972 and 1991, and with borderline tumours excluded, are 86%, 64%, 30% and 16%. Both sets of figures include all cases referred to the Royal Marsden Hospital irrespective of whether they had previously been treated elsewhere and whether they had been entered into clinical trials.

Unfortunately, such improvements in survival seen in cancer centres is not always reflected in regional or national figures. Partly this is due to the rather slow dissemination and acceptance of new practices by those who do not specialise in ovarian cancer care and who often work alone. It is books such as this one that can help raise standards generally by publishing up to date results and confirming the undoubted value of team work between family practitioner, surgeon, radiotherapist, and cancer physician—a collaboration that can only benefit the patient with epithelial cancer of the ovary.

EVE WILTSHAW MD, FRCP, FRCOG
President, International Gynecologic Cancer Society

Preface

Statistics that relate to epithelial cancer of the ovary make depressing reading. The disease will affect about 1% of the female population at some time during their lives and despite advances in many therapeutic areas over the last 10 to 20 years, fewer than 30% of those affected have a chance of long term survival or cure.

Nevertheless, with advances in cytotoxic treatment (particularly in those patients who have undergone "optimal" primary cytoreductive surgery) improvements in survival have been reported. It is probable, however, that even as our knowledge of the biology and immunology of the disease increases, as our ability to identify a genetic component in at least some cases is refined, and as further improvements in early diagnosis and chemotherapy occur, any substantial improvement in our management of the disease will be unlikely. Rather we are more likely to witness a continuing, gradual but real improvement in outcome over the next two decades.

It will be obvious to the reader that this book does not address all aspects of epithelial ovarian cancer, nor was that our intention. We were not seeking to produce a comprehensive tome, and in many instances we asked our authors not to give a global review, but to concentrate on particular and often quite precisely defined topics.

Screening remains controversial, not only because no benefit for such an expensive undertaking has yet been shown, but also because the concept of an "early" stage of the disease may be inappropriate. Prevention, particularly among those in the population who are at greater risk of the disease, may be a more useful therapeutic avenue to explore. While the prospect of cure for many patients remains elusive, increasingly we are able to schedule treatment more appropriately for some subgroups of patients.

Often this means confidence that surgery alone, sometimes surgery that spares fertility, can be curative, but for the foreseeable future most patients will present with advanced stage disease for which a combination of debulking surgery and combination chemotherapy will remain the cornerstone of treatment.

Controversy about the most appropriate cytotoxic schedules will continue, but as our knowledge of the biology and immunology of the disease increases, so might our therapeutic options. This plus the application of techniques of minimal access surgery may mean that our approach to many patients may change radically within a few years.

We should never lose sight of the fact that we are unable at present to cure most women with ovarian cancer. In this context the morbidity of our present treatment regimens and our ability to look after those who are dying of the disease needs to be examined.

Thirty four authors from eight countries have contributed to this book, a fact which acknowledges the importance of worldwide collaboration. We thank them and all their staff who have made their chapters possible. Many of them emphasise the need for, or lack of, prospective randomised studies large enough to answer many of the most pressing problems. Without such studies our ability to advance the treatment of epithelial ovarian cancer will be compromised and many of the potential advances documented in this book may remain untested with the result that our expectations may be unfulfilled.

FRANK G LAWTON

JAN P NEIJT

KENNETH D SWENERTON

August 1995

1 Mortality trends in ovarian cancer

JONATHAN WF MANT, MARTIN P VESSEY

Ovarian cancer is the commonest malignancy of the female genital tract and the fourth most common cancer among women in England and Wales (table 1.1).[1] It accounts for 5% of new cases of cancer and for 6% of deaths of cancer in women.[2] The prognosis of the disease is generally poor, as reflected in the relatively high proportion of deaths to incident cases, and the five year survival is just under 30%.[1] The incidence of ovarian cancer rises with age (fig. 1.1); it is relatively rare under the age of 30, with only 1.5 new cancers/100 000 women/year in the 20 to 29 age group, but over this age the incidence rises rapidly to a rate of 49/100 000 in women in the 60 to 69 age group. A total of 83% of ovarian cancers occur in women aged over 50.

Among developed countries there is a threefold difference in the incidence of ovarian cancer,[3] with high rates in the Scandinavian countries and low rates in Japan (fig. 1.2). Incidence tends to be low in developing countries and is generally higher in more affluent populations.[4] In England and Wales there is a social class gradient with the highest mortality in the professional classes.[5]

When interpreting the mortality trends presented in this chapter it is important to note that vital statistics cannot distinguish between the many different histological types of ovarian cancer, which may have quite different risk factors. Nevertheless it has been estimated that tumours of epithelial origin account for 66%–90% of all ovarian carcinomas.[6,7]

Table 1.1 Most common cancers among women in England and Wales

Site of cancer	No of new cases in England and Wales in 1987 (Proportion of total cancer registrations in women in parentheses.)	No of deaths in England and Wales in 1991 (Proportion of total cancer deaths in women in parentheses.)
1 Breast	23 740 (25)	13 786 (20)
2 Colon and rectum	12 777 (13)	8 619 (12)
3 Lung	10 836 (11)	10 882 (16)
4 Ovaries	**4 919 (5)**	**3 866 (6)**
5 Stomach	4 271 (4)	3 305 (5)
6 Cervix	3 972 (4)	1 668 (2)
7 Endometrium	3 308 (3)	943 (1)

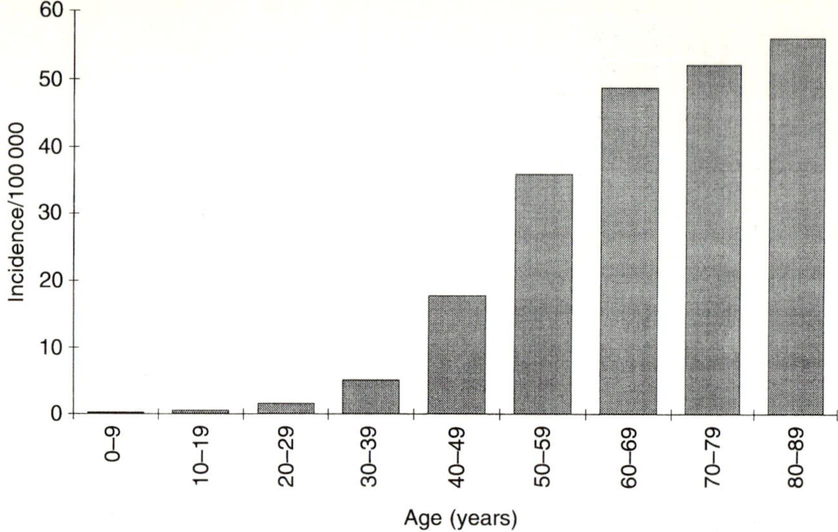

Fig. 1.1 Incidence of ovarian cancer by age group (based on data for England and Wales for 1987[1]).

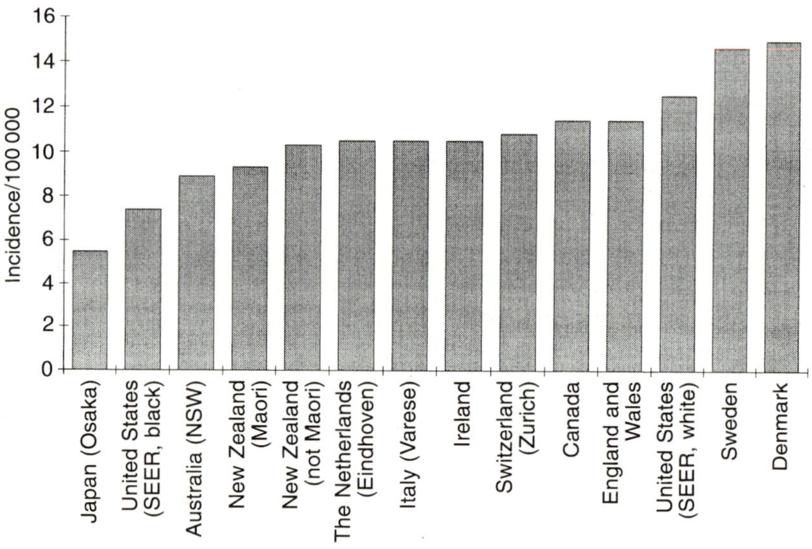

Fig. 1.2 International incidence of ovarian cancer in selected developed countries, 1983–87 (based on data from Parkin et al[3]). Rates have been standardised using the world standard population.

Mortality trends in England and Wales

Since the beginning of the century the age-standardised death rate from ovarian cancer in England and Wales has risen from 7 to 14/100 000.[8] This rise has been sustained in recent years: between 1950 and 1991, the death

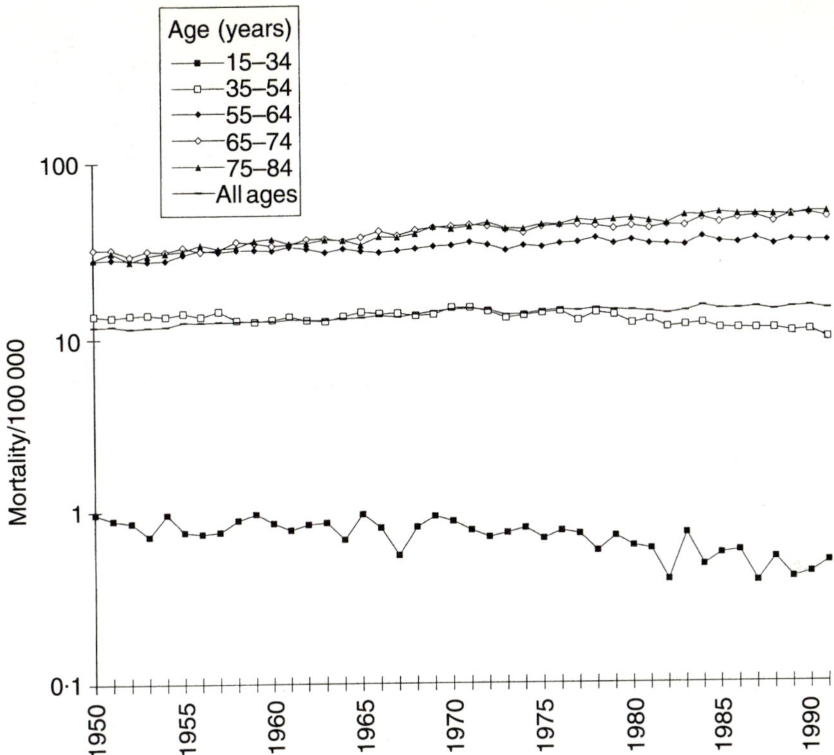

Fig. 1.3 Age specific mortality from ovarian cancer in England and Wales, 1950–91. Each age group is directly standardised by five-year age bands using the population of England and Wales in 1971.

rate rose from 11.8 to 14.2/100 000 (fig. 1.3). Consideration of all-ages rates can, however, obscure important differences in trends in age-specific mortality. Fig. 1.3 shows age-specific mortality rates for England and Wales from 1950–1991 based on national vital statistics.[9][10] There is a clear divergence in trend between the younger and older age groups. In the 15–34 age group there is no upward or downward trend between 1950 and 1970, during which time the rate fluctuated at around a mean of 0.8/ 100 000. From 1970 onwards there was a decline in mortality from 0.9/ 100 000 in 1970 to 0.5/100 000 in 1991. There is a similar pattern in the 35–54 age group. Between 1950 and 1970 the rate was stable at around a mean of 13.5/100 000, but thereafter it declined to 9.7/100 000 in 1991. In women aged 55 and over there was no decline in mortality in the 1970s—indeed, there was a sustained rise in the death rate from 1950–1991. This rise was more pronounced in the over 65s for whom there was more than a 50% increase in the death rate (from 31–48/100 000) compared with the 55–64 age group who experienced a 21% increase (from 29–35/ 100 000).

The pattern of mortality from ovarian cancer in England and Wales since 1950 has, therefore, shown two distinct trends. In women aged under 55 it was stable from 1950–1970, but has been falling since then. Among women aged over 55 the mortality rate has been rising for the last 40 years. The aggregation of these different trends produces an overall increase in the age-standardised death rate for all ages.

International mortality trends

To what extent is the experience of England and Wales representative of the international mortality from ovarian cancer? There is no homogeneous trend in the international mortality rates for ovarian cancer.[4 11] Figure 1.4 shows the age-standardised death rates for a selection of developed countries.[12] The countries have been divided into those such as Japan and Italy, which have shown a rise in mortality since 1955 and those such as Australia and The Netherlands, where there have not been sustained rises. Indeed, in countries such as the United States and Canada, the mortality rate has fallen over this period. It should be noted that the international rates shown are not directly comparable with the rates given for England and Wales because they were standardised against different populations. (If the England and Wales rates were standardised against the world standard population, which has a younger age structure than England and Wales, it would result in lower rates. Furthermore, by giving greater weight to the rates in the younger age groups, the rise in the all-age standardised rate would be abolished.)

The age standardised trends in mortality may conceal differences between age groups. Figure 1.5 divides women into two broad age groups (under 55 and 55 and over) and shows the mortality trends for six countries. The countries have been selected to represent both the mortality patterns shown in fig. 1.4. In the 15–54 age group, Canada, Denmark, and Australia show a similar pattern to England and Wales in that the mortality rate started falling around 1970. Italy and Japan show rising mortality over the period 1955–85, and there is no clear trend for Belgium. In the 55–84 age group most countries show rising mortality, but Denmark is an exception to this in that its rates started falling after 1970.

While many developed countries seem to have had similar experiences to England and Wales with regard to secular trends in mortality from ovarian cancer there are important exceptions. The decline in mortality in those aged under 55 in many countries since 1970 has not been mirrored in countries such as Japan, Belgium, and Italy.

Towards an explanation of the mortality trends

There are three key features of the secular trends in ovarian cancer mortality that must be explained:

4

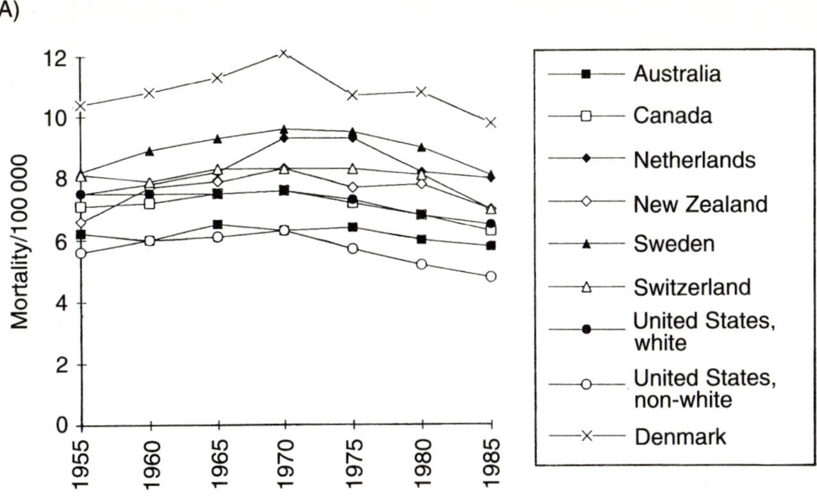

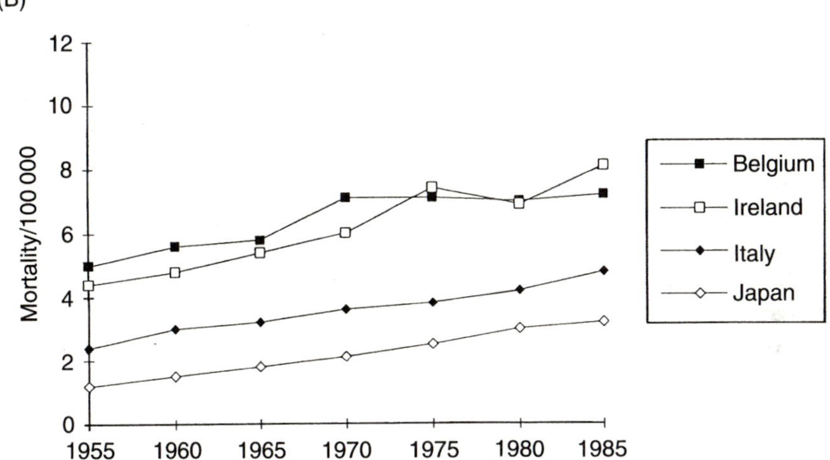

Fig. 1.4 International mortality from ovarian cancer in selected countries, 1953–87. (A) Countries that did not show a rise in mortality from ovarian cancer, and (B) countries that did show a rise. Rates are directly standardised using the world standard population. Points represent the mean of five years around the year shown.

- Why has mortality been rising in most countries in women aged over 55?
- Why has mortality been falling in some countries in women aged under 55?
- Why has this fall (in mortality in women aged under 55) not been mirrored in all countries?

There are three major categories of explanation that need to be considered:

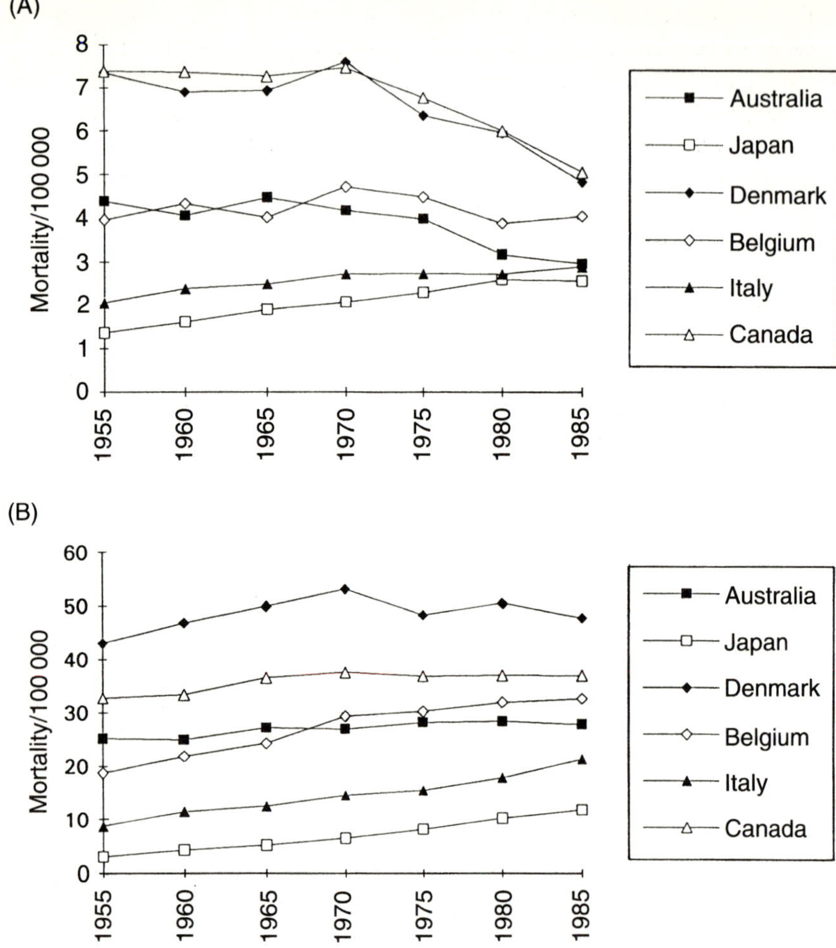

Fig. 1.5 Mortality trends from ovarian cancer by age group in six countries 1953–87. (A) Age group 15–54 years, and (B) 55–84 years. Rates are directly standardised using segments of the world standard population.

- Has there been a change in exposure to risk factors which promote ovarian cancer or to factors which protect against it?
- Has treatment had an impact on mortality?
- Are some of the trends simply artefacts of how the information has been collected?

Risk factors for ovarian cancer

A number of possible risk factors for, and protective factors against, cancer of the ovary have been identified. These include reproductive factors, genetic factors, lifestyle and environmental factors, and pelvic surgery.

The most important reproductive factor associated with ovarian cancer appears to be parity, with the relative risk reducing with the number of pregnancies.[8 13] Age at first birth is also associated; women who delay their first pregnancy to the age of 35 have a relative risk of ovarian cancer of 1.4 compared with those that have their first child under the age 25.[13] Other reproductive factors associated with increased risk in some studies include early age at menarche,[8] late age at menopause,[14] and infertility.[15 16] These reproductive factors are interrelated, and it is not clear whether they have independent effects on the risk of ovarian cancer. It has been reported that apparent effects of age at menarche, age at first birth, and age at menopause disappear if adjustments are made for parity.[17]

The higher incidence of ovarian cancer in whites compared with blacks in the United States (fig. 1.2) raises the possibility of differences in ethnic susceptibility to the disease. Conversely, while incidence rates are low in Japan, Japanese migrants living in the United States have a higher risk of ovarian cancer,[18] which suggests the importance of environmental factors. Family history is an important risk factor for ovarian cancer,[19 20] and a family history of breast cancer is also associated with an increased risk of the disease.[20 21] Because most women with ovarian cancer have no family history,[8] however, genetic factors are unlikely to be important determinants of the trends in mortality from ovarian cancer.

A number of factors associated with lifestyle and the environment have been identified as being possibly linked to ovarian cancer. These include the use of talc in the perineal area,[22] exposure to childhood infection,[6 23] various aspects of diet,[20 22 24] and exposure to asbestos.[8] The case for each of these factors is unconfirmed, so it is premature to speculate about what impact changes in each of them might have had on the mortality trends. One factor for which the evidence is strong is use of the combined oral contraceptive pill, which has been consistently shown in both case control and cohort studies to have a protective effect against ovarian cancer.[25]

Hysterectomy with conservation of ovaries and tubal sterilisation both seem to protect against ovarian cancer,[15] though the effect diminishes after two decades.[26] Most hysterectomies are carried out on women under the age of 44, and the operation rate has been rising over the last 20 years.[27] It has been estimated that about 20% of women in the United Kingdom will have had a hysterectomy by the age of 55,[27] and the rate is considerably higher in the United States.[28] Given these facts it is conceivable that this surgery might have had an impact on the incidence of ovarian cancer in some countries, but it would be difficult to explain the inflection point in mortality rates in the under 55 year olds around 1970 in terms of an operation that was introduced considerably earlier. The apparent effect of tubal sterilisation may provide some clues on the aetiology of ovarian cancer, but it is not an important factor in detecting trends in a population.

The two risk factors that stand out as having the potential to have had an important role in influencing ovarian cancer trends are parity and the

7

combined oral contraceptive pill.[4][29] It is therefore worth considering these factors in more detail.

Possible effect of parity

Beral *et al* analysed the risk of dying from ovarian cancer of successive cohorts of women born at five-year intervals between 1861 and 1931 in England and Wales and the United States and correlated this with average completed family size.[5] They concluded that the "increase in ovarian cancer . . . seems explicable . . . by the changes in the child bearing pattern." An analysis of the mortality experience of women born in Denmark between 1890 and 1934 reached similar conclusions.[30] The analysis by Beral *et al* was carried out before the reduction in the mortality rate in the under 55s in England and Wales had become apparent. A more recent review of the possible link between parity and ovarian cancer trend showed that changes in family size for women born since 1921 have been relatively modest and on the basis of the results of case control studies would not have been expected to have had an impact on vital statistics.[29] Furthermore, although completed family size did increase for women born between 1921 and 1940, it fell thereafter. The decrease in the death rate of women aged under 55 that was observed in the vital statistics largely reflects the experience of women born since 1936,[29] so, although changes in parity might explain the rise in ovarian cancer in the early part of the century, they are unlikely to be a major factor in explaining the recent decline in the under 55s.

Possible effect of the oral contraceptive pill

At least 15 case control studies have shown that the pill has a protective effect. An overview of the epidemiological studies of the link between ovarian cancer and the oral contraceptive pill estimates that ever using the pill results in a 36% reduction in ovarian cancer risk, and that five years' use reduces the risk by half.[32] As little as three months' use seems to give protection, and the benefit persists for at least 15 years.[33] There has been high usage of the oral contraceptive in many countries since the late 1960s,[34][35] to the extent that over 80% of women born in the UK in the 1950s have been exposed to it.[29] There are therefore strong grounds to anticipate a fall in ovarian cancer mortality in the generations of women exposed to the pill. To what extent does the mortality data reflect this? Recent changes in the secular trends in ovarian cancer in England and Wales and in Sweden have been attributed to a possible effect of the combined oral contraceptive pill.[29][36] The reduction in mortality in some countries in the under 55s is certainly consistent with an effect of the pill. Furthermore, in the countries that have not experienced a fall in mortality in this age group such as Italy, Japan, and Ireland (fig. 1.4) the pill is used relatively sparingly.[35] An exception to this general pattern is Belgium, which

8

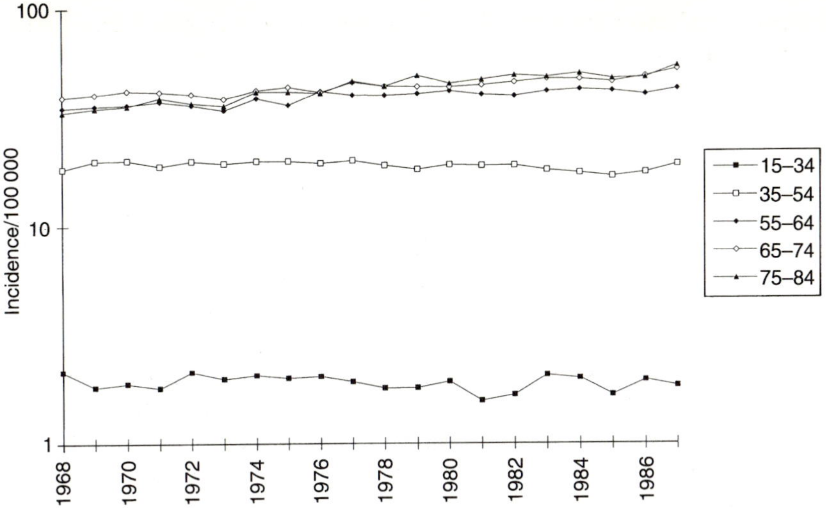

Fig. 1.6 Age specific incidence of ovarian cancer in England and Wales, 1968–87. Each group is directly standardised by five-year age bands using the population of England and Wales, 1971.

has a high use of the pill,[35] and rising mortality (fig. 1.4) though not in the under 55s (fig. 1.5).

Has improved treatment had an impact on mortality?

Figure 1.6 shows the age specific incidence of ovarian cancer in England and Wales between 1968 and 1987.[37 38] The decline in mortality in the younger age groups is not reflected by a decline in incidence. While it is possible that this is an artefact of improved registration over time, the difference in the incidence and mortality trends are also indicative of a possible effect of treatment. There is circumstantial evidence to support this in that some studies have suggested recent improvements in survival of patients with ovarian cancer.[39 40] A treatment effect might be the result of the introduction of more effective treatments, or of earlier diagnosis.

There have been major advances in the treatment of germ cell tumours of the ovary,[41] a type of cancer that predominantly affects young women, but advances in the treatment of epithelial ovarian cancer have been more modest.[41 42 43] While the introduction of more effective treatments may have had some impact on the mortality trends in the 15–34 age group on account of the advances in treating germ cell tumours, therefore, it is unlikely that they would have had a major impact on the mortality trends in older women.

The prognosis for women diagnosed early is considerably better than for those in whom the diagnosis is made late.[44] It is therefore conceivable that if there had been a trend toward earlier diagnosis, it would account for the difference between the incidence and mortality trends. The increasing use of ultrasound would be a plausible explanation for this, but in the group that is almost completely screened by ultrasound (pregnant women) few carcinomas are detected.[45] Furthermore, while there was an increase in the diagnosis of benign ovarian cysts between 1972 and 1986, there does not seem to have been any increase in the diagnosis of early ovarian cancer.[46]

Are the trends an artefact of the vital statistics?

Artificial rises in the incidence and mortality of ovarian cancer might result simply from changes in registration practices. While there have been no major changes in the International Classification of Disease codes with regard to ovarian cancer during its various revisions, it is feasible that there have been changes in certification of the disease over time. In this respect it is interesting that the countries that are shown in fig. 1.4 to have rising death rates from ovarian cancer all had a low baseline mortality rate in 1955 compared with those countries that do not show a rise.

Conclusion

The mortality trends for ovarian cancer in England and Wales indicate falling death rates in women under 55 since 1970 and rising death rates in women over 55. These trends are mirrored in some other developed countries but not in all. The reduction in the death rate in women under 55 is consistent with an effect of the oral contraceptive pill. It has not been observed in countries that have low use of the pill. The rise in the death rate in older women may reflect secular trends in parity, but in some countries it is possible that improved death certification has played a part.

References

1 Office of Population Censuses and Surveys. *Cancer statistics: registrations (England & Wales) 1987.* OPCS series MB1, no 20. London: HMSO 1993.
2 Office of Population Censuses and Surveys. *Mortality statistics: cause (England & Wales) 1991.* OPCS Series DH2, no 18. London: HMSO 1993.
3 Parkin DM, Muir CS, Whelan SL, Gao YT, Ferlay J, Powell J, eds. *Cancer incidence in five continents.* Vol VI. Lyons: International Agency for Research on Cancer Scientific Publications, no 120, 1992.
4 Coleman MP, Esteve J, Damiecki P, Arslan A, Renard H. Trends in cancer incidence and mortality. Lyons: International Agency for Research on Cancer Scientific Publications, no 121, 1993.
5 Beral V, Fraser P, Chilvers C. Does pregnancy protect against ovarian cancer? Lancet 1978; 1: 1083–7.
6 Baylis MS, Henderson WJ, Pierrepoint CG, Griffiths K. The aetiology of ovarian cancer. In: Morrow CP, Smart GE, eds. *Gynaecological oncology.* Berlin: Springer Verlag, 1986: 157–65.
7 Granal CO. Ovarian cancer: unrealistic expectations. *N Engl J Med* 1992; **327**: 197–200.
8 Beral V. The epidemiology of ovarian cancer. In: Sharp F, Soutter WP, eds. *Ovarian cancer: the way ahead.* Proceedings of the 17th study group of the Royal College of Obstetricians & Gynaecologists in conjunction with the Helene Harris Memorial Trust. London: Royal College of Obstetricians & Gynaecologists, 1987: 21–31.

9 Office of Population Censuses & Surveys. *Registrar General's statistical reviews of England & Wales 1950–73*. London: HMSO, 1952–1976.

10 Office of Population Censuses & Surveys. *Mortality statistics: cause 1974–91*. Series DH2, nos 1–18. London: HMSO, 1976–93.

11 Mant J, Vessey M. Trends in cancer incidence and mortality: endometrial and ovarian cancer. In: Doll R, Muir CS, Fraumeni JF, eds. *Cancer surveys Vol 19/20*. New York: Cold Spring Harbor Laboratory Press, 1994: 287–307.

12 Aoki K, Hayakawa N, Kurihara M, Suzuki S. *Death rates for malignant neoplasms for selected sites by sex and five year age group in 33 countries 1953–57 to 1983–87*. International Union against Cancer. Nagoya: University of Nagoya Cooperative Press, 1992.

13 Negri E, Franceschi S, Tzonou A, Booth M, La Vecchia C, Parazzini F, *et al*. Pooled analysis of 3 European case-control studies: 1. Reproductive factors and risk of epithelial ovarian cancer. *Int J Cancer* 1991; **49**: 50–6.

14 Franceschi S, La Vecchia C, Booth M, Tzonou A, Negri E, Parazzini F, *et al*. Pooled analysis of three European case-control studies of ovarian cancer: II. Age at menarche and at menopause. *Int J Cancer* 1991; **49**: 57–60.

15 Booth M, Beral V, Smith P. Risk factors for ovarian cancer: a case control study. *Br J Cancer* 1989; **60**: 592–8.

16 Whittemore AS, Wu ML, Paffenbarger RS Jr, Sarles DL, Kampert JB, Grosser S, *et al*. Epithelial ovarian cancer and the ability to conceive. *Cancer Res* 1989; **49**: 4047–52.

17 Kvale G, Heuch I, Nilssen S, Beral V. Reproductive factors and risk of ovarian cancer: a prospective study. *Int J Cancer* 1988; **42**: 246–51.

18 Haenszel W, Kurihara M. Studies of Japanese migrants 1. Mortality from cancer and other diseases among Japanese in the United States. *J Natl Cancer Inst* 1968; **40**: 43–68.

19 Schildkraut JM, Thompson WD. Familial ovarian cancer: a population based case control study. *Am J Epidemiol* 1988; **128**: 456–66.

20 Mori M, Harabuchi I, Miyake H, Casagrande JT, Henderson BE, Ross RK. Reproductive, genetic, and dietary risk factors for ovarian cancer. *Am J Epidemiol* 1988; **128**: 771–7.

21 Tulinius H, Egilsson V, Olafsdottir H, Sigvaldason H. Risk of prostate, ovarian, and endometrial cancer among relatives of women with breast cancer. *BMJ* 1992; **305**: 855–7.

22 Whittemore AS, Wu ML, Paffenbarger RS Jr, Sarles DL, Kampert JB, Grosser S, *et al*. Personal and environmental characteristics related to epithelial ovarian cancer II. Exposures to talcum powder, tobacco, alcohol, and coffee. *Am J Epidemiol* 1988; **128**: 1228–40.

23 Cramer DW, Welch WR, Cassalls S, Scully RE. Mumps, menarche, menopause and ovarian cancer. *Am J Obstet Gynecol* 1983; **147**: 1–6.

24 Shu XO, Gao YT, Yuan JM, Ziegler RG, Brinton LA. Dietary factors and epithelial ovarian cancer. *Br J Cancer* 1989; **59**: 92–6.

25 Vessey MP. The Jephcott Lecture: an overview of the benefits and risks of combined oral contraceptives. In: Mann RD, ed. *Oral contraceptives and breast cancer*. London: Parthenon Publishing Group, 1989.

26 Irwin KL, Weiss NS, Lee NC, Peterson HB. Tubal sterilisation, hysterectomy, and the subsequent occurrence of epithelial ovarian cancer. *Am J Epidemiol* 1991; **134**: 362–9.

27 Vessey MP, Villard-Mackintosh L, McPherson K, Coulter A, Yeates D. The epidemiology of hysterectomy: findings in a large cohort study. *Br J Obstet Gynaecol* 1992; **99**: 402–7.

28 McPherson K. Why do variations occur? In: Anderson TF, Mooney G, eds. *The challenges of medical practice variations*. London: MacMillan Press, 1990.

29 Villard-MacKintosh L, Vessey MP, Jones L. The effects of oral contraceptives and parity on ovarian cancer trends in women under 55 years of age. *Br J Obstet Gynaecol* 1989; **96**: 783–8.

30 Ewertz M, Kjaer SK. Ovarian cancer incidence and mortality in Denmark, 1943–1982. *Int J Cancer* 1988; **42**: 690–6.

31 Thorogood M, Villard-Mackintosh L. Combined oral contraceptives: risks and benefits. *Br Med Bull* 1993; **49**: 124–39.

32 Hankinson SE, Colditz GA, Hunter DJ, Spencer TL, Rosner B, Stampfer MJ. A quantitative assessment of oral contraceptive use and risk of ovarian cancer. *Obstet Gynecol* 1992; **80**: 708–14.

33 The Cancer and Steroid Hormone Study of the Centers for Disease Control and the National Institute of Child Health and Human Development. The reduction in risk of ovarian cancer associated with oral contraceptive use. *N Engl J Med* 1987; **316**: 650–5.

34 Thorogood M, Vessey MP. Trends in use of oral contraceptives in Britain. *British Journal of Family Planning* 1990; **16**: 41–53.

35 Wharton C, Blackburn R. Population reports: oral contraceptives. *Population Information Program, Johns Hopkins University* 1988; **16**: 1–31.

36 Adami H, Bergstrom R, Persson I, Sparen P. The incidence of ovarian cancer in Sweden, 1960–1984. *Am J Epidemiol* 1990; **132**: 446–52.

37 Office of Population Censuses & Surveys. *Registrar General's statistical reviews of England & Wales, 1968–1970, supplement on cancer.* London: HMSO 1975.

38 Office of Population Censuses & Surveys. *Cancer statistics: registrations 1971–87.* Series MB1, no 1–20. London: HMSO, 1979–93.

39 Balvert-Locht HR, Coebergh JW, Hop W, Brolman H, Crommelin M, van Wijck D *et al.* Improved prognosis of ovarian cancer in the Netherlands during the period 1975–85: a registry based study. *Gynecol Oncol* 1991; **42**: 3–8.

40 Haybittle JL, Kingsley-Pillers EM. Long term survival experience of female patients with genital cancer. *Br J Cancer* 1988; **56**: 322–5.

41 Granal CO. Ovarian cancer – unrealistic expectations. *N Engl J Med* 1992; **327**: 197–200.

42 Barker GH. Recent advances in the treatment of carcinoma of the ovary. *Br J Obstet Gynaecol* 1993; **100**: 803–5.

43 Cannistra SA. Cancer of the ovary. *N Engl J Med* 1993; **329**: 1550–9.

44 Campbell S, Bhan V, Royston P, Whithead M, Collins W. Transabdominal ultrasound screening for early ovarian cancer. *BMJ* 1989; **299**: 1363–7.

45 Thornton JG, Wells M. Ovarian cysts in pregnancy: does ultrasound make traditional management inappropriate? *Obstet Gynecol* 1987; **69**: 717–21.

46 Westhoff C, Clark C. Benign ovarian cysts in England and Wales and in the United States. *Br J Obstet Gynaecol* 1992; **99**: 329–32.

2 Histopathology of early ovarian cancer

HAROLD FOX

Definitions

Any consideration of the pathology of early ovarian cancer is bound to lead to semantic and conceptual difficulties. What, in this context, is meant by the term "early"? Gynaecologists usually consider stage 1A ovarian adenocarcinomas (particularly those less than 5 cm in diameter, or any tumours that can be completely removed) as being early lesions,[1] but this view conflicts with what is known about tumour cell kinetics and the natural history of malignant disease. Remarkably few data are available about cellular dynamics in ovarian cancer; in a comparable form of neoplasia, however, that of the gastrointestinal tract, it has been estimated that progression from a single cell that has undergone malignant change (accepting the monoclonal theory of the origin of cancer) to a tumour that is several millimetres in diameter takes between two and six years while the evolution of a clinically detectable lesion may take as long as 14 years.[2] By contrast, the time interval between clinical detection of a small tumour and death may be as short as two years. It is clear therefore that any tumour that can be seen with the naked eye, even one only 1 cm in diameter, is not an early lesion as such but is the early stage of a late lesion, one that already has a potential for metastatic spread and can lead to the death of the patient.

It is probable that an ovarian cancer should be classed as "early" only if it is not perceptible to the surgeon on examination of the ovary at operation and is not visible to the pathologist on examination of the outer and cut surfaces.[3] Even a lesion which falls into such a restrictive definition may be capable of metastasising, for microscopic carcinomas have been detected (retrospectively after development of peritoneal carcinomatosis), in ovaries previously removed and judged to be normal.[4,5]

Further, how does one define "malignant"? Transformation of a cell from "normal" to "malignant" is not a one-stage process but is characterised by a sequence of changes in the genetic structure of the cell. Where exactly in this chain of events a cell becomes "malignant" is currently an unanswered question that cannot be resolved in terms of morphology. It could, of course, be claimed that the acquisition of the ability to invade is clear

13

evidence of malignancy: acceptance of this view would, however, not only lead to the exclusion of intraepithelial neoplasia from the category of "early malignant lesion" but would also obviate any discussion of tumours of borderline malignancy.

Natural history of ovarian cancer

Clearly it is difficult to define, let alone discuss, the pathology of early malignant change in the ovary. Nevertheless there are recognised histological abnormalities in the ovary that are thought to indicate a high risk of progression to a frankly malignant neoplasm. Overgrowth of germ cells in a gonadoblastoma is clearly seen to be a precursor of a dysgerminoma despite the germ cells still being within the cell islands of the original benign lesion. Similarly, granulosa cell proliferation within the cell nests of a sex cord tumour with annular tubules is regarded as a starting point of some granulosa cell neoplasms.

These examples refer, however, to relatively rare tumours and of much greater importance is the delineation of the early stages of a malignant epithelial neoplasm of the ovary.

Role of inclusion cysts

An ovarian adenocarcinona could arise in an otherwise normal ovary, develop from a benign tumour, evolve from a neoplasm of borderline malignancy, or originate in a focus of heterotopic tissue. In the normal ovary it is widely thought that the first stage in the evolution of an epithelial neoplasm, particularly a serous one, is commonly an invagination of the surface epithelium of the ovary into the underlying cortical stroma to form an inclusion cyst, malignant change subsequently occurring within the cyst.[67] Credence has been lent to this concept by reports that the contralateral ovaries from women with unilateral ovarian adenocarcinoma contain an increased number of inclusion cysts,[8] that inclusion cysts have an unusually high incidence of metaplastic and hyperplastic changes,[59] and that the epithelium in these cysts may show overexpression of proto-oncogenes and epithelial tumour markers.[510] Not everyone has, however, been able to confirm the finding of an increased number of inclusion cysts in association with ovarian adenocarcinoma,[511] and while inclusion cysts are commonly seen in the ovary it has been the experience of most pathologists that it is unusual to see cytological atypia within their lining epithelium; certainly, there has been no systematic documentation of in situ neoplastic change within inclusion cysts.

Role of the epithelium

Carcinomas could, of course, arise directly from the surface epithelium of the ovary and Scully[5] has described a number of extremely small ovarian

14

adenocarcimonas which involved, and seemed to arise directly from, this epithelium. In recent years changes have been described within the ovarian surface epithelium which are thought to represent "ovarian intraepithelial neoplasia".[12 13] This histological abnormality was first recognised in a study of three pairs of identical twins, one of each pair having an ovarian adenocarcinoma: the other sister in each pair underwent prophylactic oophorectomy and in each case "dysplastic" abnormalities were noted in the ovarian surface epithelium. Subsequently, nuclear pleomorphism, irregular distribution of nuclear chromatin, stratification, and loss of nuclear polarity have been described as characteristic features of ovarian intraepithelial neoplasia, this characterisation being based on the findings in the surface epithelium adjacent to ovarian adenocarcinomas. Whether these abnormalities truly represent a preinvasive neoplastic lesion, and if so the magnitude of its potential for progression to frankly invasive adenocarcinoma, is currently unknown; indeed, it is inherently unlikely that the true nature and risk of this cytological abnormality will ever be found out. A high incidence of surface papillomatosis has also been described on the surface of contralateral ovaries of patients with unilateral ovarian epithelial tumours,[9] but this change was even more commonly found in the ovaries of women with endometrial adenocarcinoma and in polycystic ovaries and so seems to be a hormone-induced effect rather than a true precursor of malignant change.

Role of benign ovarian cystadenomas

Does malignant change occur with any frequency in benign ovarian cystadenomas? Can such neoplasms be regarded as important precursors of adenocarcinoma? It has been said that the evidence for overt adenocarcinomas arising in benign ovarian tumours is "unimpeachable",[14] but in reality any views expressed about this must be largely anecdotal and highly subjective. It is certainly true that the average age of patients with cystadenomas is about 12 years less than that of women with ovarian adenocarcinoma,[3 15] but this observation is no proof of progression from benign to malignant neoplasia and is more than counterbalanced by the rarity with which focal adenocarcinomatous change is seen in serous cystadenomas. Focal mural nodules of carcinoma have been noted in mucinous cystadenomas[16 17] but are nevertheless extremely uncommon. It can be argued that the relatively rapid growth of a malignant neoplasm will rapidly lead to partial or complete obliteration by an adenocarcinoma of any evidence of an antecedent benign cystadenoma. It has recently been claimed that residual evidence of a preceding cystadenoma is present in a high proportion of serous adenocarcinomas,[18] this view being based on the finding of areas of apparently benign epithelium in many malignant serous neoplasms. The strength of this observation is, however, greatly diluted by the knowledge that areas of benign-looking epithelium are present in a considerable proportion of metastases to the ovary from primary gastrointestinal adenocarcinomas.[19]

Tumours of borderline malignancy

Early malignant change is clearly recognised in ovarian epithelial tumours of borderline malignancy. Overt foci of malignant change in serous borderline neoplasms are uncommon in conventional terms but focal, limited, non-destructive invasion of the tumour stroma by epithelial cells has been described in a subset of such tumours.[20,21] The invasion takes the form of isolated cells, nests of neoplastic cells, or papillary clusters of cells; in some cases lymphatic invasion is also seen. Tumours that show this pattern of invasion seem to behave in exactly the same manner as do serous borderline neoplasms without focal stromal invasion, so although they are within the strict limits of the definition malignant, they can still be treated as borderline tumours rather than as adenocarcinomas. It is a moot point whether this is an example of early malignant change.

There is no doubt that focal evolution into an overt adenocarcinoma occurs much more often in mucinous tumours of borderline malignancy than in their serous counterparts. It is relatively common to encounter a mucinous neoplasm that shows a pattern of borderline malignancy in some areas and a clearly malignant pattern in others, while mucinous neoplasms often show a melange of benign, borderline, and malignant areas.[22] Mucinous tumours that show definite (albeit focal) adenocarcinomatous change are considered to be genuine adenocarcinomas but their natural history is not known. In particular, no information is available about whether the prognosis for patients with such tumours is the same, or better, than is that for women with conventional, well-differentiated stage 1A mucinous adenocarcinomas.

Endometriosis

Malignant epithelial ovarian neoplasms, usually either endometrioid or clear cell adenocarcinomas, can arise in foci of endometriosis.[23] Some endometrioid adenocarcinomas seem to develop in foci of ovarian endometriosis in a manner analogous to the development of many endometrial adenocarcinomas; there is a detectable transition from benign endometriosis to atypical hyperplasia and eventually to adenocarcinoma. Often the neoplasms develop in an endometriotic cyst. Such cysts commonly do not fulfill the stringent histological criteria for a diagnosis of endometriosis—that is the presence of both glandular epithelium and endometrial stroma—and there is much to suggest that such cysts are really endometrioid cystadenenomas rather than true foci of endometriosis.[24] This includes the recent evidence that a high percentage of endometriotic cysts are monoclonal.[25] Both the incidence of malignant change in ovarian endometriosis and the quantitative significance of endometriosis as a

16

precursor of ovarian adenocarcinoma are difficult to assess; often the diagnosis of adenocarcinoma arising in ovarian endometriosis has been based only on evidence of endometriosis in the contralateral ovary or elsewhere in the pelvis. Nevertheless it has been estimated that between 15% and 20% of ovarian endometrioid adenocarcinomas arise from endometriosis.[26] If true this suggests that endometrioid cystadenomas are important precursors of adenocarcinoma to an extent that does not seem to be the case for mucinous or serous cystadenomas. It must be borne in mind, however, that haemosiderin deposition is common in endometriotic cysts and that iron is a well recognised carcinogen; endometrioid cystadenomas cannot therefore be taken as typical benign ovarian neoplasms in terms of their malignant potential.

What we do not know

It will be clear from this that our knowledge of the morphology of early ovarian cancer is woefully inadequate. We do not have any real knowledge of the malignant potential of benign cystadenomas and we do not know the natural history of tumours characterised by focal malignant change in neoplasms of borderline malignancy. We do not know if the lesion classed as ovarian intraepithelial neoplasia has the potential for progression to an invasive neoplasm. It is difficult to see how any of the lacunae in our knowledge can be repaired by purely morphological techniques and it is probable that further progress in our understanding and definition of early malignant change in the ovary will come only from a more sophisticated approach involving the study within ovarian epithelial cells of changes such as altered glycoprotein synthesis,[27] abnormalities in the expression of oncogenes, and tumour suppression genes, or changes in the genetic makeup.

The place of screening

How relevant is this discussion of early ovarian cancer to the question of screening for ovarian neoplasia? Screening for ovarian cancer is based on the belief that if ovarian neoplasia can be detected at an early stage in its evolution the chance of cure must be increased. Any adenocarcinoma detected by conventional screening techniques is not in the true sense of the word an "early" lesion, but is rather the early stage of a late lesion, and there is no compelling evidence that detection of an adenocarcinoma will result in a better survival rate though there will be a longer survival time for women whose malignant neoplasms are detected at this stage.

Current screening techniques will detect tumours of borderline malignancy and there is no dispute that such neoplasms should be resected; it is debatable, however, if removal of clinically covert tumours of borderline malignancy that have been detected by screening will lead to any significant reduction in the incidence of ovarian adenocarcinoma; the risk of eventual malignancy is virtually confined to those of mucinous type and must be low in small, clinically undetected, neoplasms.

Asymptomatic benign epithelial tumours will also be detected by screening. While the risk of eventual malignancy in such neoplasms is unquantitated there are few, if any, grounds for believing it to be of considerable importance and the return for a considerable number of operations in terms of reduced mortality from ovarian cancer is likely to be small.

It has been suggested that patients at high risk for ovarian cancer could be subjected to repeated laparoscopic biopsy of the ovarian surface epithelium to detect the changes of "ovarian intraepithelial neoplasia":[13] it is, however, likely that genetic testing will prove to be of more value than repetitive biopsy for estimating the risk of cancer in such women.

Fig. 2.1 "Early" ovarian cancer showing atypical proliferation of surface epithelium described as "ovarian intraepithelial neoplasia".

References

1 Guthrie D, Davy MLJ, Philips PR. A study of 656 patients with "early" ovarian cancer. *Gynecol Oncol* 1984; **17**: 363–9.

2 Wright NR, Alison M. *The biology of epithelial cell populations*. Oxford: Oxford University Press, 1984.

3 Scully RE. Minimal cancer of the ovary. *Clin Oncol* 1982; **1**: 379–87.

4 Chen KTK, Schooley JL, Flam MS. Peritoneal carcinomatosis after prophylactic oophorectomy in familial ovarian cancer syndrome. *Obstet Gynecol* 1985; **66**: 93S–94S.

5 Scully RE. Early ovarian cancer. In: Sharp F, Mason WP, Creasman W, eds. *Ovarian cancer 2: biology, diagnosis and management*. London: Chapman & Hall, 1992: 199–205.

6 Cramer DW, Hutchinson GB, Welch WR, Scully RE, Ryan J. Determinants of ovarian cancer risk. I. Reproductive experience and family history. II. Inferences regarding pathogenesis. *J Natl Cancer Inst* 1983; **71**: 711–21.

7 Langley FA, Fox H. Ovarian tumours: classification, histogenesis and aetiology. In: Fox H, ed. *Haines and Taylor; obstetrical and gynaecological pathology.* Edinburgh: Churchill-Livingstone, 1987: 543–55.

8 Mittal KR, Zeleniuch-Jacquotte A, Cooper JL, Demoppoulos RI. Contralateral ovary in unilateral ovarian carcinoma: a search for preneoplastic lesions. *Int J Gynecol Pathol* 1993; **12**: 59–63.

9 Resta L, Russo S, Colucci GA, Prat J. Morphological precursors of ovarian epithelial tumors. *Obstet Gynecol* 1993; **82**: 181–6.

10 Wang DP, Knonishi I, Koshiyama M. *et al.* Immunohistochemical localization of c-erbB-2 protein and epidermal growth factor receptor in normal surface epithelium, surface inclusion cysts, and common epithelial tumours of the ovary. *Virchows Arch A Pathol Anat Histopathol* 1992; **421**: 393–400.

11 Westhoff C, Murphy P, Heller D, Halim A. Is ovarian cancer associated with an increased frequency of germinal inclusion cysts? *Am J Epidemiol* 1993; **138**: 90–3.

12 Deligdisch L, Gil J. Characterization of ovarian dysplasia by interactive morphometry. *Cancer* 1989; **63**: 748–55.

13 Plaxe SC, Deligdisch L, Dottino PR, Cohen CJ. Ovarian intraepithelial neoplasia demonstrated in patients with stage 1 ovarian carcinoma. *Gynecol Oncol* 1990; **38**: 367–72.

14 Novak ER, Woodruff JD. *Novak's gynecologic and obstetric pathology.* Philadelphia: Saunders, 1977.

15 Anderson MC. Malignant potential of benign ovarian cysts; the case "for". In: Sharp F, Mason WP, Leake RE, eds. *Ovarian cancer biological and therapeutic challenges.* London: Chapman & Hall, 1990: 187–90.

16 Prat J, Young RH, Scully RE. Ovarian mucinous tumors with foci of anaplastic carcinoma. *Cancer* 1982; **50**: 300–4.

17 Sondergaard G, Kaspersen P. Ovarian and extraovarian mucinous tumors with solid mural nodules. *Int J Gynecol Pathol* 1991; **10**: 145–55.

18 Puls LE, Powell DE, DePriest PD. Transition from benign to malignant epithelium in mucinous and serous cystadenocarcinoma, *Gynecol Oncol* 1992; **47**: 53–7.

19 Ulbright TM, Roth LM, Stehman FB. Secondary ovarian neoplasms: a clinicopathologic study of 35 cases. *Cancer* 1984; **53**: 1164–74.

20 Tavassoli FA. Serous tumor of low malignant potential with early stromal invasion (serous LMP with microinvasion). *Mod Pathol* 1988; **1**: 407–14.

21 Bell DA, Scully RE. Ovarian serous borderline tumors with stromal microinvasion: a report of 21 cases. *Hum Pathol* 1990; **21**: 397–403.

22 Fox H. The concept of borderline malignancy in ovarian tumours: a reappraisal. *Curr Top Pathol* 1989; **78**: 111–34.

23 Mostoufizadeh M, Scully RE. Malignant tumors arising in endometriosis. *Clin Obstet Gynecol* 1980; **23**: 951–63.

24 Czernobilsky B. Endometriosis. In: Fox H, ed. *Haines and Taylor: Obstetrical and gynaecological pathology.* Edinburgh: Churchill Livingstone, 1987; 763–77.

25 Nilbert OM, Pejovic T, Mandahl, N, Lofif S, Willen H, Mitelman F. Monoclonal origin of endometriotic cysts. *International Journal of Gynecological Cancer.* 1995; **5**: 61–3.

26 Czernobilsky B. Endometrioid neoplasia of the ovary: a reappraisal. *Int J Gynecol Pathol* 1982; **1**: 203–10.

27 van Niekerk CC, Boerman OC, Ramaekers FCS, Poels LG. Marker profile of different phases in the transition of normal human ovarian epithelium to ovarian carcinomas. *Am J Pathol* 1991; **138**: 455–63.

3 Biology of epithelial ovarian cancer

ALAN A ELBENDARY, ANDREW BERCHUCK,
ROBERT C BAST JR

Malignant cells are genetically unstable, proliferate uncontrollably, invade adjacent tissues, and metastasise to distant sites. In the past our ability to define these processes at a molecular level was limited both by available concepts and by techniques. During the last decade, however, a revolution in molecular biology has allowed us to characterise the phenotype and function of cancer cells to a degree not previously possible.

In normal cells, DNA replication and proliferation are tightly regulated. Cellular growth and division are triggered by both intracellular and extracellular signals. Once triggered proliferation is further controlled at several checkpoints that are critically positioned in the cell cycle and which must be traversed before the initiation of DNA synthesis (G_1/S) or of mitosis (G_2/M). These checkpoints provide a fail-safe mechanism that blocks cellular proliferation if conditions for continued cell growth are not fulfilled. For cells to traverse checkpoints, the timely activation of certain growth-stimulatory genes (proto-oncogenes) and the inactivation of growth-inhibitory genes (tumour suppressor genes) are required. Peptide growth factors also influence the ability of cells to traverse these checkpoints by activation of proto-oncogenes and inhibition of tumour suppressor function. Disruption of these pathways is common to the aetiology of many cancers, although tumours from different sites and from different patients have different profiles of molecular alterations.

Normal cells maintain a distinctive phenotype and function within the three-dimensional matrix of tissues and organs. Epithelial cells remain in contact with one another or with the extracellular matrix of the basement membrane. During malignant transformation, changes in cell adhesion and motility are linked to increases in proliferation or loss of growth inhibition, which permits local invasion and metastasis by a subset of neoplastic cells.

More than 90% of the ovarian cancers that develop in adults arise from a simple layer of epithelial cells that cover the normal ovary or that line cysts immediately beneath the ovarian surface. These cells are generally quiescent and remain in contact with a basement membrane. At the time of ovulation the surface epithelial cells covering a follicle express proteases

20

that facilitate follicular rupture. After ovulation epithelial cells proliferate to repair the defect created in the ovarian surface. These physiological changes are usually controlled by autocrine and paracrine mechanisms. During oncogenesis, alterations occur both in cell growth regulation and in differentiation of the transformed epithelial cells. Repeated ovulation may favour the development or expression of such changes. Epidemiological studies have indicated that an increased risk of ovarian cancer is associated with early menarche, late menopause, nulliparity, and the use of agents that induce ovulation. Conversely, a reduced risk of ovarian cancer has been linked to multiple pregnancies, prolonged lactation, and the use of oral contraceptive drugs that suppress ovulation. The use of oral contraceptives for as long as five years during the time that a woman is menstruating cuts the risk of epithelial ovarian cancer in later life by half.[1]

Most ovarian cancers, like other epithelial cancers, are derived from single cells that have undergone multiple genetic alterations. More than 90% of epithelial ovarian cancers are clonal, as judged by identical p53 mutation, X-chromosome inactivation, and loss of heterozygosity on different chromosomes in the primary tumour and in the metastases.[2,3] During malignant transformation the cumulative effect of multiple genetic changes may be more important than their sequence. Genetic abnormalities, activation of proto-oncogenes, loss of tumour suppressor genes, and aberrant regulation by peptide growth factors have all been implicated in the aetiology of ovarian cancer.

Genetic abnormalities

DNA ploidy

DNA content (ploidy) is often abnormal in ovarian cancers. In different reports from 52% to 83% of the tumours were aneuploid, whereas from 17% to 48% were diploid.[4] The prevalence of aneuploidy seemed to correlate with worsening grade of tumour. Considerable variation was reported among the different studies, however, and in some cases no significant differences were noted in the incidence of aneuploidy and grade of tumour.[5–9] Benign neoplasms and non-metastatic borderline tumours have been analysed and uniformly found to be diploid. This suggests that aneuploidy is a characteristic of malignancy. Furthermore, a remarkable correlation was noted between the ploidy of the primary tumour and the metastatic and recurrent foci, which suggests that DNA ploidy is a stable characteristic of the particular tumour.

Aneuploidy is associated with a subset of aggressive ovarian cancers. Patients with diploid ovarian cancers had a median survival of 48 months, whereas those with aneuploid cancers had a median survival of only 19 months.[4,7] Multivariate analysis indicated that aneuploidy is an independent predicator of poor outcome.[7] DNA ploidy may be of greatest clinical use in identifying borderline tumours with a poor prognosis which is related to their capacity to metastasise.

Chromosomal alterations

Karyotypic analyses of epithelial ovarian cancers have disclosed frequent abnormalities of chromosomes 1, 3, 6, 11, 17, and 19, with less frequent abnormalities of chromosomes 2, 4, 5, and 21.[8–17] Deletions and rearrangements have been found in chromosomes 1, 3, and 6, with breakpoints at 1p34p36, 1q21-23, 3p12-p21, and 6q15-q23. Other abnormalities in chromosomes 7p and 19q have also been identified. A specific translocation, t(6;14)(q21;24), was reported in a series of papillary serous adenocarcinomas[14] but was not detected in subsequent studies. The multiplicity and great complexity of chromosomal aberrations encountered in ovarian cancers suggest that many of these changes may result from an inherent genetic instability of the tumour cells and may not contribute to aetiological events in ovarian carcinogenesis.

Loss of heterozygosity

The development of polymorphic genetic markers has provided more precise probes to detect genetic injury. By identifying loss of heterozygosity at specific allelic loci, recurring sites of chromosomal deletions have been identified in various human cancers. Allelotyping of ovarian cancers has shown that loss of heterozygosity correlates with significant frequency (>30%) for loci on chromosomes 4p, 6p, 6q, 7p, 8q, 11p, 12q, 13q, 16p, 17p, 17q, 18q, 19p, and Xp.[18–23] The pattern of loss of heterozygosity seems to correlate with grade of tumour. For example, Zheng *et al*[21] reported that loss of heterozygosity on chromosomes 3 and 11 was associated with high grade tumours. More recently, loss of heterozygosity at 11p15.1-11p15.5 was noted in 47.5% of all ovarian cancers and strongly correlated with poor differentiation.[23] On the other hand, Dodson *et al*[22] found that loss of heterozygosity on chromosome arms 13q and 15q was associated with high-grade tumours, whereas loss of heterozygosity on 3p was common in low grade tumours. Furthermore, when serous cystadenocarcinomas were compared with non-serous tumours, a significantly higher incidence of loss of heterozygosity on chromosomes 6q, 13q, and 19q was noted in the former group, suggesting that loss of heterozygosity patterns may affect the histology. This clustering of genetic changes at specific foci suggests that genes at these sites have critical roles in maintaining normal phenotypes and proliferation. Indeed, two tumour suppressor genes have been mapped to these loci, p53 at 17p and *BRCA* at 17q.

BRCA-1 and familial ovarian cancer

Although most ovarian cancers are sporadic, familial predisposition to ovarian cancer has been well described, and several familial ovarian cancer syndromes have been defined. Perhaps the most common of these syndromes is the hereditary breast/ovarian cancer syndrome in which patients have an autosomally dominant genetic susceptibility to both ovarian and breast cancer. The molecular events surrounding this disorder have recently come under intense scrutiny. Using genetic linkage analysis several investigators have linked this disorder to allele loss on chromosome 17 and in particular to loci 17q12-23.[24–28] Recently, the *BRCA*-1 gene was isolated, and mutations of it have been found in the germ line of affected kindreds.[29] Interestingly, few mutations have been detected in sporadic breast and ovarian cancers.[30] Identification of the gene should prove useful in counselling most patients with familial ovarian cancers. Whether *BRCA*-1 will also prove useful for identifying women at risk of apparently sporadic ovarian cancers outside these well-defined kindreds remains to be seen. The molecule seems to have a zinc finger motif, consistent with a potential role as a transcriptional regulatory factor.

Oncogenes

Oncogenes were originally discovered in oncogenic retroviruses where they were linked to the viral transformation of host cells. Shortly after their discovery, significant homology was noted between these viral oncogenes and apparently normal host genes called *proto-oncogenes*. In their native state proto-oncogenes are not carcinogenic, but function to regulate normal cellular growth and differentiation. Through mutation, deletion, or translocation, proto-oncogenes are activated to oncogenes the protein products of which mediate malignant transformation. The transforming potential of proto-oncogenes may be unmasked by alteration in only a single allele, so activated oncogenes function in a dominant manner. To date, over 60 oncogenes have been identified and classified into families based on their structural and functional properties (table 3.1). The products of different oncogenes resemble growth factors, growth factor receptors, signal-transducing proteins, and DNA-binding proteins that regulate gene expression. Although oncogenes have been implicated in the aetiology of many human cancers, activation of a single oncogene is generally not sufficient for carcinogenesis; multiple genetic abnormalities are required.

Although the role of most oncogenes in ovarian carcinogenesis is still undefined, mutation or aberrant expression of some have been identified. In early studies, somatic mutation of Ki-*ras* was found in the tumour cells but not in the mesothelial cells of a single patient with epithelial ovarian cancer.[31] In subsequent reports, amplification or somatic mutation in Ha-*ras* or Ki-*ras* was shown in only 2%–12% of epithelial ovarian cancers,[31–33] contrasting with the substantially higher prevalence of *ras* abnormalities in

23

Table 3.1 Families of oncogenes

Peptide growth factors
  sis
  int
  hst
  FGF-5

Tyrosine kinases

Receptor	*src family*	*Non-receptor* *abl family*	*Miscellaneous*
EGF receptor (*erb*B)	*src*	*abl*	*fes*
HER-2/*neu* (*erb*B-2)	yes	*arg*	
CSF-1 receptor (*fms*)	*fgr*		
met	*lck*		
trk	*fyn*		
kit	*lyn*		
sea	*hck*		
ret			
ras			

Serine/threonine kinases

raf family	*Protein kinase C family*	*Miscellaneous*
c-*raf*-1	Protein kinase β-1	*mos*
A-*raf*	Protein kinase γ	*pim*-1
B-*raf*		

G proteins

	ras family	*Miscellaneous*
	N-*ras*	*gip*
	Ha-*ras*	*gsp*
	Ki-*ras*	

Nuclear oncogenes
AP-1 transcription factor:

Components	*myc family*	*Hormone receptors*	*Miscellaneous*
jun	c-*myc*	*erb*A	*myb*
jun-B	L-*myc*		*ets*-1
jun-D			*ets*-2
fos			*ski*
fra-1			*rel*

pancreatic or colonic cancers. Although no correlation between *ras* level of expression and histological state or grade was found, other authors have linked *ras* mutation to borderline and mucinous ovarian cancers.[34 35] Oncogenic activation of *myc* has been shown in up to a third of ovarian cancers,[36–40] and here again there was no relationship to the biological behaviour of the tumour or the impact on survival.

Tyrosine kinases represent the largest family of the oncogenes. These enzymes transfer a phosphate molecule from adenosine triphosphate to tyrosine residues on specific proteins. Phosphorylation of these substrates often alters their function within cells. The actions of the tyrosine kinases are balanced by those of the phosphatases that dephosphorylate proteins. In general, phosphorylation and dephosphorylation of proteins is an important mechanism for controlling protein activity. Tyrosine phosphorylation, in particular, is of singular importance in cellular signalling. Oncogenic

24

activation of these kinases leads to aberrant phosphorylation of proteins that cancels signal transduction and contributes to transformation of the affected cells. Two types of tyrosine kinases have been described. The first are membrane-spanning tyrosine kinases that act as receptors for exogenous growth factors. After binding of the ligand to the extracellular domain of the receptor, the cytoplasmic domain is activated by autophosphorylation. The activated cytoplasmic domain phosphorylates other cytoplasmic proteins that trigger the molecular cascade that is needed for signal transduction. The second type of tyrosine kinases, exemplified by the *src* family, are anchored to the inner surface of the cytoplasmic membrane. These tyrosine kinases also phosphorylate cellular proteins that are involved in signal transduction. Unlike the receptor tyrosine kinases that are regulated by peptide growth factors, less is known about the regulation of the non-receptor tyrosine kinases.

Among the receptor tyrosine kinases, the *fms* proto-oncogene encodes the receptor for macrophage colony-stimulating factor. Normal ovarian epithelium and 70% of ovarian cancers secrete biologically active macrophage colony-stimulating factor, which is a potent chemoattractant for macrophages. In normal ovarian epithelium, little if any *fms* is expressed, but more than half of epithelial ovarian cancers express appreciable amounts of this receptor. The coexpression of macrophage colony-stimulating factor and its receptors in ovarian cancers could establish a stimulatory autocrine loop. Furthermore, the overexpressed macrophage colony-stimulating factor may then act in a paracrine manner to recruit macrophages. The ovarian cancer cells can be further stimulated by macrophage-derived cytokines, including interleukin-1, interleukin-6, and tumour necrosis factor. It is consistent with this postulated mechanism that concentrations of tumour necrosis factor and interleukin-6 are considerably raised in the ascites and serum of patients with ovarian cancer.[41] Concomitant expression of macrophage colony-stimulating factor and *fms* in human ovarian cancers has been associated with advanced stage and poor differentiation,[42-44] and recently increased concentrations of macrophage colony-stimulating factor have been associated with shorter survival. Increased concentrations of this marker may be related to tumour burden, but could also reflect increased proliferative potential or increased invasiveness.

HER-2/*neu* encodes another receptor tyrosine kinase that may contribute to regulation of cell growth in ovarian cancer. Slamon *et al* reported the first large study of HER-2/*neu* overexpression in ovarian cancer,[45] in which overexpression of HER-2/*neu* was found in 30% of ovarian cancers, and was associated with very poor prognosis. Subsequent studies have reported an incidence of 15% to 40%[46-50] and several of these have suggested that overexpression of HER-2/*neu* is associated with short survival.[45 46] In studies at Duke University Medical Center,[46] overexpression of HER-2/*neu* was found in 32% of stages III and IV ovarian cancers. When analysed with respect to other prognostic factors, HER-2/*neu* overexpression was associated with a poor outcome. Patients whose cancers did not overexpress

HER-2/*neu* were five times more likely to undergo a negative second-look operation and had a median survival of 32 months compared with 16 months for patients whose cancers overexpressed HER-2/*neu*.

Two additional cytoplasmic kinases seem to be involved in signal transduction from the periphery of the cell to the nucleus. The serine and threonine kinases—including the *raf* family, the protein kinase C family, *mos*, and *pim*-1—phosphorylate serine and threonine residues of specific proteins. Although their role in ovarian carcinogenesis remains undefined, their oncogenic activation has been shown in retroviruses and in vitro. Furthermore, AKT2 serine/threonine kinase was overexpressed in two of eight ovarian cancer cell lines and in two of 15 primary epithelial ovarian cancers.[51]

Tumour supressor genes

Unlike proto-oncogenes that promote proliferation, tumour suppressor genes retard or inhibit proliferation. In contrast to the dominantly acting oncogenes, tumour suppressor genes function in a recessive manner, so alterations are required in both alleles to eliminate gene function. One allele is inactivated by somatic or germ line mutation, and the remaining allele is then lost through various possible mechanisms such as replacement by a reduplicated copy of the original mutation, mitotic recombination, or gene conversion. Several tumour suppressor genes have been isolated and implicated in the aetiology of various human cancers (table 3.2). We have examined several of these tumour suppressor genes, including RB, WT-1, APC, and VHL, but we could not detect any mutations in them in epithelial ovarian cancer (unpublished data). To date p53 remains the only tumour suppressor gene that has been shown to be inactivated in an appreciable proportion of ovarian cancers.

The p53 gene encodes a 53-kDa nuclear phosphoprotein. Wild type p53 protein binds directly to specific regions of the DNA and blocks progress through the cell cycle. Recent evidence has suggested that p53 acts as a "policeman", protecting the genome and ensuring fidelity of replication.[52 53] The p53 concentrations increase after damage to cellular DNA and inhibit the cell in the G1 phase of the cell cycle. This inhibition of proliferation allows the cellular machinery time to repair the incurred damage to the DNA; if the damage is irreparable, p53 can also trigger apoptosis (programmed cell death).

Mutations in p53 are usually point mutations and lead to an alteration in the structure of the protein. The mutant protein has a longer half life and is overexpressed in cells. Unlike wild type p53 the mutant form neither binds to DNA nor activates genes, so the inactivation of p53 leads to the loss of its negative regulatory function and leaves the cells more vulnerable to mutations. Mutant p53 can also cooperate with activated *ras* oncogenes to enhance transforming potential.[54] Interestingly, it has been observed in vitro that mutant p53 protein complexes with the wild type and can inhibit

26

Table 3.2 *Tumour suppressor genes*

Gene	Locus	Function	Outcome
Retinoblastoma	13q	Regulates transcription	Retinoblastoma
Wilms' tumour gene 1	11p	Transcription factor	Wilms' tumour
Von Hippel-Lindau	3p	Unknown	Renal/CNC tumours
Adenomatous polyposis coli	5q	Unknown	Colonic cancer
Deleted in colonic carcinoma	18q	Cell-membrane adhesion molecule	Colonic cancer
Mutated in colonic carcinoma	5q	Unknown	Colonic cancer
Neurofibromatosis type 1	17q	Guanosine triphosphatase activating protein	Neurofibroma
Neurofibromatosis type 2	22q	Cytoskeletal membrane link	Neurofibroma
p53	17p	Transcription	Li-Fraumeni syndrome and other cancers
Protein tyrosine phosphatase-γ	3p	Phosphatase	Mammary and renal cancers
BRCA-1	17q	Unknown	Mammary and ovarian cancers

Table 3.3 p53 Overexpression in epithelial ovarian tumours at Duke University Medical Center

Tumour type	No (%) that overexpressed p53
Invasive stage III/IV	46/92 (50)
Invasive stage IC/II	11/25 (44)
Invasive stage IA/IB	4/27 (15)
Borderline stage III	2/8 (25)
Borderline stage I/II	0/41
Benign	0/17

its function, which suggests that mutant p53 may act in a dominant negative fashion.

Clinically, mutation of the p53 gene is the single most common genetic alteration seen in human cancers. Mutations have been reported in 90% of pancreatic cancers, 60% of colonic cancers, 50% of bronchial cancers, and 25% of breast cancers. In ovarian cancers, p53 mutations have been observed in up to half of cases.[55-57] None of the individual studies is of sufficient size, however, to permit correlation between p53 status and other clinicopathological features of ovarian cancer. In the series from Duke University Medical Center, p53 overexpression and mutation were only rarely observed in benign or borderline lesions (table 3.3). Overexpression was found in 15% of stage IA lesions and in half of cancers diagnosed at more advanced stages,[57] so p53 mutations seem to be a late event in the pathogenesis of ovarian cancer.

The absence of p53 mutations in non-metastatic borderline tumours and the extremely low incidence of such mutations, even in advanced-stage borderline tumours, stands out in contrast to the high incidence of these mutations in frankly invasive epithelial ovarian cancers. Conversely, the incidence of Ki-*ras* mutations, which has been reported to be between 20% and 50%,[35 58] is several times higher than the incidence of these mutations in epithelial cancers. These observations, together with the fact that borderline tumours are uniformly diploid whereas epithelial cancers are aneuploid, suggest that borderline tumours may be a variant of ovarian cancer rather than an obligatory precursor of most invasive epithelial ovarian cancers.

Some time ago, Li and Fraumeni[59] described a syndrome in which affected patients were susceptible at an early age to multiple, separate, primary cancers, including ovarian cancer. Germ line mutations in p53 have been shown to be important in the pathogenesis of the Li-Fraumeni syndrome. More recently, Buller *et al*[60] analysed p53 status in four families with familial ovarian cancer. Germ line p53 mutations were noted in three of the four families and in six of 11 affected patients, so it seems that familial ovarian cancers are heterogeneous, and more than one mechanism may contribute to their pathogenesis.

Growth factors

Peptide growth factors are low-molecular-weight glycoproteins that stimulate proliferation or differentiation by binding to specific cell-membrane receptors. The binding of the growth factor to its receptor leads to the autophosphorylation of the intracellular domain, which may then phosphorylate specific tyrosine residues on protein substrates to initiate the complex mechanism of signal transduction. Unlike hormones that can act on distant targets, peptide growth factors act by only one of three mechanisms in their local environment. In an autocrine pathway the growth factor is secreted in the extracellular space to act on receptors in the same cells. This mechanism seems to be an important pathway by which cancers escape regulation and is exemplified by the previously discussed *fms* oncogene. In a paracrine model the released growth factor acts on a nearby cell of a different type. In a juxtacrine model the growth factor released from one cell regulates directly adjacent cells. Both autocrine and paracrine regulations have been documented in ovarian cancer.

Epidermal growth factor, its homologue transforming growth factor α, or both, are expressed by normal ovarian epithelial cells and by some ovarian cancer cells. Binding of these growth factors to the epidermal growth factor receptor stimulates proliferation. Treatment of these tumour cell lines with antibodies directed at transforming growth factor α can inhibit proliferation, and interrupt autocrine growth stimulation. Epidermal growth factor receptors have been detected in normal epithelial cells and in 77% of patients with advanced stage ovarian cancer. Continued expression of epidermal growth factor receptors has been associated with a poor prognosis.[61 62]

In contrast to the effects of epidermal growth factor and transforming growth factor α, transforming growth factor β inhibits proliferation of epithelial cells. Three closely related forms of transforming growth factor β have been identified in human tissues, transforming growth factors β 1, β 2 and β 3. Transforming growth factors β 1 and β 2 have been detected in normal human ovarian epithelial cells. The addition of transforming growth factor β to cultures of normal human ovarian surface epithelial cells inhibits growth. This inhibition is, however, lost in many ovarian cancer lines. Some fail to express the factor, others cannot activate the growth factor, and most are not inhibited by the addition of exogenous transforming growth factor β. In a cell line in which all of these functions remained intact, growth could be stimulated by incubation with an antibody that neutralised transforming growth factor β, consistent with the possibility that transforming growth factor β acts by an autocrine mechanism to inhibit growth.[63] When tumour cells were isolated directly from ascitic fluid in ovarian cancer, however, transforming growth factor β inhibited growth in more than 95%, whereas expression of transforming growth factor β was lost in 40%.[64] In contrast to tumour cell lines that have lost the potential for both autocrine and paracrine growth inhibition, many ascitic tumour cells in ovarian cancer seem to have lost autocrine, but not paracrine, growth inhibition during ovarian oncogenesis.

Invasion and metastasis

More than 80% of epithelial ovarian cancers have metastasised by the time the disease is diagnosed. Although lymphangitic and haematogenous metastases can occur, tumour cells are more likely to spread over the surface of the peritoneal cavity forming multiple nodules. Relatively little is known about the events that control this distinctive pattern of invasion and metastasis. Studies in other cancer models have shown that tumour dissemination is a multistep process that requires loss of adherence to adjacent cells, increased motility, proteolytic degradation of basement membranes, migration, adherence to vascular endothelium, proliferation, and tumour-induced angiogenesis. Once successfully completed, the process can be repeated in a cyclical manner, resulting in further dissemination of tumour cells.

The peritoneal cavity is the most common site for ovarian cancer metastases. Ovarian cancer spreads in one of two ways, either carried passively in the peritoneal fluid or migrating actively in response to specific host and tumour derived chemoattractants. Many clinical data support the role of peritoneal fluid as a medium for migration of ovarian cancer cells. Recent studies have identified several chemoattractants that stimulate migration of ovarian cancer cells, suggesting that directed migration may also have an important role in the metastasis of ovarian cancers. Fibronectin and laminin are associated with ovarian cancers and have been shown to have chemoattractant properties. In addition, insulin-like growth factor, which is expressed by some ovarian cancers, enhances the migration of ovarian cancer cell lines in vitro.

Migrating ovarian cancer cells must be able to adhere to mesothelial cells that line the peritoneal cavity. Cannistra *et al*[65] have recently shown that the binding of ovarian cancer cells to peritoneal mesothelial cells is mediated by CD44H, a surface-membrane-bound proteoglycan. Both normal and malignant ovarian epithelial cells express CD44H, so that expression of this molecule may be necessary, but not sufficient, for metastasis to occur. In addition, binding of the cancer cells to the mesothelium was not completely inhibited by treatment with anti-CD44 antibody, suggesting that other cell-surface-associated molecules may play a significant part in this process. Several adhesion molecules that bind to laminin, type IV collagen, and fibronectin can anchor tumour cells. Transcription of the laminin receptor is increased in some ovarian cancers compared with normal ovarian epithelium.[66] Similarly, fibronectin is overexpressed in some ovarian cancer cell lines.[67] The overexpression of both these molecules facilitates the binding of the tumour cells to the peritoneal surfaces. Non-covalent binding to laminin and fibronectin is reversible. Tissue transglutaminase is increased in some ovarian cancers[68] and can create more stable covalent bonds between molecules at the cell surface and in the extracellular matrix.

Degradation of the basement membrane is the next important step in tumour dissemination. Several enzymes, including type IV collagenase, plasminogen activator, and cathepsin, are necessary for the proteolytic digestion of the basement membrane. Raised cathepsin concentrations have been found in serum samples from women with advanced stage ovarian cancer, suggesting that this enzyme has a role in metastases of ovarian cancer. We have shown that urinary-type plasminogen activator receptor and its ligand are upregulated in ovarian cancer,[69] and while we have shown that the invasiveness of the ovarian cancer cell lines correlated significantly with urinary-type plasminogen activator production, others have shown that this invasiveness can be inhibited by treating the cells with antibodies directed against urinary-type plasminogen activator. Similarly, treatment with anticathepsin antibodies reduced the invasive potential of these cells. Ovarian cancers express increased concentrations of matrix metalloproteinases 2 and 9.[70–73] Treatment of cancer cells that overexpress the HER-2/*neu* receptor with the ligand heregulin increases their matrix metalloproteinases expression and their ability to invade Matrigel membranes.[74] Under these conditions, heregulin paradoxically decreases the potential for anchorage-independent growth. Consequently, the poor prognosis observed in ovarian and breast cancers that overexpress HER-2/*neu* may be related to increased invasiveness rather than to increased proliferative potential.

Another important event in tumour metastasis is the formation of new blood vessels. Angiogenesis occurs in many normal physiological conditions such as wound healing. In this setting, tumour necrosis factors α and β and fibroblast growth factor are angiogenic. Because these factors are expressed by some ovarian cancers, it is possible that they are involved in tumour-induced angiogenesis.

Conclusion

In clinical practice, ovarian cancer is heterogeneous, which is reflected in tumours of different stage, grade, histological type, and susceptibility to surgical cytoreduction. Even when these factors are taken into account, there are pronounced differences among patients in response to treatment and in prognosis. Recent molecular studies have begun to provide a basis for understanding the heterogeneity of ovarian cancer. Transformation of individual ovarian epithelial cells can occur through different combinations of genetic alterations. Although our understanding of the pathogenesis of ovarian cancer at a molecular level is still fragmented, it seems that different fractions of ovarian cancers exhibit inappropriate maintenance of autocrine growth stimulation by transforming growth factor α, loss of paracrine growth inhibition by transforming growth factor β, novel or increased expression of tyrosine kinase growth factor receptors, and loss of p53 tumour suppressor function. As further knowledge of such changes is acquired, it should be possible not only to improve prognostication but also to identify specific molecular targets for prevention, diagnosis, and treatment. In the future, detection of specific molecular lesions should prompt the prescription of exactly those agents that inhibit the activated oncogenes or replace the tumour suppressor functions that were dysregulated in a particular patient's cancer. Recent studies in which transfer of single tumour suppressor genes have reversed the malignant behaviour of tumour cells argue for the feasibility of this approach.

References

1 Stanford JL. Oral contraceptives and neoplasia of the ovary. *Contraception* 1991; **43**: 543–56.

2 Jacobs IJ, Kholer MF, Wiseman R, *et al*. The clonal origin of ovarian cancer: analysis by loss of heterozygosity, p53 mutation and X chromosome inactivation. *J Natl Cancer Inst* 1992; **84**: 1793–8.

3 Mok C-H, Tsao W-W, Knapp RC, Fishbaugh PM, Lau CC. Unifocal origin of advanced human epithelial ovarian cancer. *Cancer Res* 1992; **52**: 5119–22.

4 Brescia RJ, Barakat RA, Beller U, *et al*. The prognostic significance of nuclear DNA content in malignant epithelial tumors of the ovary. *Cancer* 1990; **65**: 141–7.

5 Iversen OE. Skaarland E. Ploidy assessment of benign and malignant ovarian tumors by flow cytometry: a clinicopathological study. *Cancer* 1987; **60**: 82–7.

6 Rodenburg CJ, Cornelisse CJ, Hermans J, Fleuren GJ. DNA flow cytometry and morphometry as prognostic indicators in advanced ovarian cancer: a step forward in predicting the clinical outcome. *Gynecol Oncol* 1988; **29**: 176–87.

7 Blumenfeld D, Braly PS, Ben-Ezra J, Klevecz RR. Tumor DNA content as a prognostic feature in advanced epithelial ovarian carcinoma. *Gynecol Oncol* 1987; **27**: 389–402.

8 Berchuck A, Boente MP, Kerns BJ, *et al*. Ploidy analysis of epithelial ovarian cancers using image cytometry. *Gynecol Oncol* 1992; **44**: 61–5.

9 Kallioniemi OP, Punnonen R, Mattila J, Lehtinen M, Koivula T. Prognostic significance of DNA index, multiploidy, and S-phase fraction in ovarian cancer. *Cancer* 1988; **61**: 334–9.

10 Gallion HH, Powell DE, Smith LW, *et al*. Chromosome abnormalities in human epithelial ovarian malignancies. *Gynecol Oncol* 1990; **38**: 473–7.

11 Bello MJ, Rey JA. Chromosome aberrations in metastatic ovarian cancer: relationship with abnormalities in primary tumors. *Int J Cancer* 1990; **45**: 50–4.

12 Tanaka K, Boice CR, Testa JR. Chromosome aberrations in nine patients with ovarian cancer. *Cancer Genet Cytogenet* 1989; **43**: 1–14.

13 Whang-Peng J, Knutsen T, Douglass EC, Chu E, Ozols RF, Hogan WM, *et al*. Cytogenetic studies in ovarian cancer. *Cancer Genet Cytogenet* 1984; **11**: 91–106.

14 Trent JM, Salmon SE. Karyotypic analysis of human ovarian carcinoma cells cloned in short term agar culture. *Cancer Genet Cytogenet* 1981; **3**: 279–91.

15 Pejovic T, Heim S, Mandahl N, *et al*. Chromosome aberrations in 35 primary ovarian carcinomas. *Genes Chromosom Cancer* 1992; **4**: 58–68.

16 Huber H, Knogler W, Karlic H, Akrad M, Soregi G, Schweizer D. Structural chromosomal abnormalities in gynecologic malignancies. *Cancer Genet Cytogenet* 1990; **50**: 189–97.

17 Roberts CG, Tattersall MH. Cytogenetic study of solid ovarian tumors. *Cancer Genet Cytogenet* 1990; **48**: 243–53.

18 Lee JH, Kavanagh JJ, Wildrick DM, Wharton JT, Blick M. Frequent loss of heterozygosity on chromosomes 6q, 11, and 17 in human ovarian carcinomas. *Cancer Res* 1990; **50**: 2724–8.

19 Ehlen T, Dubeau L. Loss of heterozygosity on chromosomal segments 3p, 6q and 11p in human ovarian carcinomas. *Oncogene* 1990; **5**: 219–23.

20 Foulkes WD, Trowsdale J. Isolating tumour suppressor genes relevant to ovarian carcinoma – the loss of heterozygosity. In: Sharp F, Mason P, Blackett T, Berek J, eds. *Ovarian cancer*. London: Chapman & Hall, 1995; **3**: 23–38.

21 Zheng JP, Robinson WR, Ehlen T, Yu MC, Dubeau L. Distinction of low grade from high grade human ovarian carcinomas on the basis of losses of heterozygosity on chromosomes 3, 6, and 11 and HER-2/neu gene amplification. *Cancer Res* 1991; **51**: 4045–51.

22 Dodson MK, Hartmann LC, Cliby WA, *et al.* Comparison of loss of heterozygosity patterns in invasive low-grade and high-grade epithelial ovarian carcinomas. *Cancer Res* 1993; **53**: 4456–60.

23 Kiechle-Schwarz M, Bauknecht T, Wienker T, Walz L, Pfleiderer A. Loss of constitutional heterozygosity on chromosome 11p in human ovarian cancer: positive correlation with grade of differentiation. *Cancer* 1993; **72**: 2423–32.

24 Margaritte P, Bonaiti-Pellie C, King MC, Clerget-Darpoux F. Linkage of familial breast cancer to chromosome 17q21 may not be restricted to early-onset disease. *Am J Hum Genet* 1992; **50**: 1231–4.

25 Narod SA, Feunteun J, Lynch HT, *et al.* Familial breast-ovarian cancer locus on chromosome 17q12-q23. *Lancet* 1991; **338**: 82–3.

26 Hall JM, Lee MK, Newman B, *et al.* Linkage of early-onset familial breast cancer to chromosome 17q21. *Science* 1990; **250**: 1684–9.

27 Newman B, Austin MA, Lee M, King MC. Inheritance of human breast cancer: evidence for autosomal dominant transmission in high-risk families. *Proc Natl Acad Sci USA* 1988; **85**: 3044–8.

28 Saito H, Inazawa J, Saito S, *et al.* Detailed deletion mapping of chromosome 17q in ovarian and breast cancers: 2-cM region on 17q21.3 often and commonly deleted in tumors. *Cancer Res* 1993; **53**: 3382–5.

29 Miki Y, Swensen J, Shattuck-Eidens D, *et al.* A strong candidate for the breast and ovarian cancer susceptibility gene BRCA1. *Science* 1994; **266**: 66–71.

30 Futreal PA, Liu Q, Shattuck-Eidens D, *et al.* BRCA1 mutations in primary breast and ovarian carcinomas. *Science* 1994; **266**: 120–2.

31 Feig LA, Bast RC Jr, Knapp RC, Cooper GM. Somatic activation of the rasK gene in a human ovarian carcinoma. *Science* 1984; **223**: 698–701.

32 Enomoto T, Inoue M, Perantoni AO, Terakawa N, Tanizawa O, Rice JM. K-ras activation in neoplasms of the human female reproductive tract. *Cancer Res* 1990; **50**: 6139–45.

33 Haas M, Isakov J, Howell SB. Evidence against ras activation in human ovarian carcinomas. *Molecular Biology and Medicine* 1987; **4**: 265–75.

34 Teneriello MG, Ebina M, Linnoila RI, *et al.* p53 and Ki-ras gene mutations in epithelial ovarian neoplasms. *Cancer Res* 1993; **53**: 3103–8.

35 Mok SC, Bell DA, Knapp RC, *et al.* Mutation of K-ras protooncogene in human ovarian epithelial tumors of borderline malignancy. *Cancer Res* 1993; **53**: 1489–92.

36 Baker VV, Borst MP, Dixon D, Hatch KD, Shingleton HM, Miller D. c-myc amplification in ovarian cancer. *Gynecol Oncol* 1990; **38**: 340–2.

37 Zhou DJ, Gonzalez-Cadavid N, Ahuja H, Battifora H, Moore GE, Cline MJ. A unique pattern of proto-oncogene abnormalities in ovarian adenocarcinomas. *Cancer* 1988; **62**: 1573–6.

38 Serova OM. Amplification of c-myc proto-oncogene in primary tumors, metastases and blood leukocytes of patients with ovarian cancer. *Eksperimentalnaia Onkologiia* 1987; **9**: 25–7.

39 Sasano H, Garrett CT, Wilkinson DS, Silverberg S, Comerford J, Hyde J. Protooncogene amplification and tumor ploidy in human ovarian neoplasms. *Hum Pathol* 1990; **4**: 382–91.

40 Berns EM, Klijn JG, Henzen-Logmans SC, Rodenburg CJ, van der Burg ME, Foekens JA. Receptors for hormones and growth factors and (onco)-gene amplification in human ovarian cancer. *Int J Cancer* 1992; **52**: 218–24.

41 Moradi MM, Carson LF, Weinberg B, Haney AF, Twiggs LB, Ramakrishnan S. Serum and ascitic fluid levels of interleukin-1, interleukin-6, and tumor necrosis factor-alpha in patients with ovarian epithelial cancer. *Cancer* 1993; **72**: 2433–40.

42 Ramakrishnan S, Xu FJ, Brandt SJ, Niedel JE, Bast RC Jr, Brown EL. Constitutive production of macrophage colony-stimulating factor by human ovarian and breast cancer cell lines. *J Clin Invest* 1989; **83**: 921–6.

43 Kacinski BM, Stanley ER, Carter D, *et al.* Circulating levels of CSF-1 (M-CSF) a lymphohematopoietic cytokine may be a useful marker of disease status in patients with malignant ovarian neoplasms. *Int J Radiat Oncol Biol Phys* 1989; **17**: 159–64.

44 Kacinski BM, Carter D, Mittal K, *et al.* Ovarian adenocarcinomas express fms-complementary transcripts and fms antigen, often with coexpression of CSF-1. *Am J Pathol* 1990; **137**: 135–47.

45 Slamon DJ, Godolphin W, Jones LA, *et al.* Studies of the HER-2/neu proto-oncogene in human breast and ovarian cancer. *Science* 1989; **244**: 707–12.

46 Berchuck A, Kamel A, Whitaker R, *et al.* Overexpression of HER-2/neu is associated with poor survival in advanced epithelial ovarian cancer. *Cancer Res* 1990; **50**: 4087–91.

47 Zhang X, Silva E, Gershenson D, Hung MC. Amplification and rearrangement of c-erb B proto-oncogenes in cancer of human female genital tract. *Oncogene* 1989; **4**: 985–9.

48 Zheng JP, Robinson WR, Ehlen T, Yu MC, Dubeau L. Distinction of low grade from high grade human ovarian carcinomas on the basis of losses of heterozygosity on chromosomes 3, 6, and 11 and HER-2/neu gene amplification. *Cancer Res* 1991; **51**: 4045–51.

49 Rubin SC, Finstad CL, Wong GY, Almadrones L, Plante M, Lloyd KO. Prognostic significance of HER-2/neu expression in advanced epithelial ovarian cancer: a multivariate analysis. *Am J Obstet Gynecol* 1993; **168**: 162–9.

50 Kacinski BM, Mayer AG, King BL, Carter D, Chambers SK. NEU protein overexpression in benign, borderline, and malignant ovarian neoplasms. *Gynecol Oncol* 1992; **44**: 245–53.

51 Cheng JQ, Godwin AK, Bellacosa A, *et al.* AKT2, a putative oncogene encoding a member of a subfamily of protein-serine/threonine kinases, is amplified in human ovarian carcinomas. *Proc Natl Acad Sci USA* 1992; **89**: 9267–71.

52 Lane DP. Cancer. p53, guardian of the genome. *Nature* 1992; **358**: 15–16.

53 Zambetti GP, Levine AJ. A comparison of the biological activities of wild-type and mutant p53. *FASEB J* 1993; **7**: 855–65.

54 Hinds P, Finlay C, Levine AJ. Mutation is required to activate the p53 gene for cooperation with the ras oncogene and transformation. *J Virol* 1989; **63**: 739–46.

55 Marks JR, Davidoff AM, Kerns BJ, *et al.* Overexpression and mutation of p53 in epithelial ovarian cancer. *Cancer Res* 1991; **51**: 2979–84.

56 Eccles DM, Brett L, Lessells A, *et al.* Overexpression of the p53 protein and allele loss at 17p13 in ovarian carcinoma. *Br J Cancer* 1992; **65**: 40–4.

57 Kohler MF, Kerns BJ, Humphrey PA, Marks JR, Bast RC Jr, Berchuck A. Mutation and overexpression of p53 in early-stage epithelial ovarian cancer. *Obstet Gynecol* 1993; **81**: 643–50.

58 Teneriello MG, Ebina M, Linnoila RI, *et al.* p53 and Ki-ras gene mutations in epithelial ovarian neoplasms. *Cancer Res* 1993; **53**: 3103–8.

59 Li FP, Fraumeni JF Jr. Soft-tissue sarcomas, breast cancer, and other neoplasms: A familial syndrome? *Ann Intern Med* 1969; **71**: 747.

60 Buller RE, Anderson B, Connor JP, Robinson R. Familial ovarian cancer. *Gynecol Oncol* 1993; **51**: 160–6.

61 Berchuck A, Rodriguez GC, Kamel A, *et al.* Epidermal growth factor receptor expression in normal ovarian epithelium and ovarian cancer. Correlation of receptor expression with prognostic factors in patients with ovarian cancer. *Am J Obstet Gynecol* 1991; **164**: 669–74.

62 Kohler M, Janz I, Wintzer HO, Wagner E, Bauknecht T. The expression of EGF receptors, EGF-like factors and c-myc in ovarian and cervical carcinomas and their potential clinical significance. *Anticancer Res* 1989; **9**: 1537–47.

63 Berchuck A, Rodriguez G, Olt GJ, *et al.* Regulation of growth of normal ovarian epithelial cells and ovarian cancer cell lines by transforming growth factor-β. *Am J Obstet Gynecol* 1992; **166**: 676–84.

64 Hurteau J, Rodriguez GC, Whitaker RS, *et al.* Transforming growth factor-beta inhibits proliferation of human ovarian cancer cells obtained from ascites. *Cancer* 1994; **74**: 93–9.

65 Cannistra SA, Kansas GS, Niloff J, DeFranzo B, Kim Y, Ottensmeier C. Binding of ovarian cancer cells to peritoneal mesothelium in vitro is partly mediated by CD44H. *Cancer Res* 1993; **53**: 3830–8.

66 van den Brule FA, Berchuck A, Bast RC, *et al.* Differential expression of the 67 kD laminin receptor and 31 kD human laminin binding protein in human ovarian carcinomas. *Br J Cancer* (in press).

67 Olt G, Berchuck A, Soisson AP, Boyer CM, Bast RC Jr. Fibronectin is an immunosuppressive substance associated with epithelial ovarian cancer. *Cancer* 1992; **70**: 2137–42.

68 Greenberg CS, Hettasch JM, Hurteau J, Bast RC Jr. The tissue transglutaminase gene is expressed in ovarian carcinoma cells *in vitro* and *in vivo*. *Proceedings of the American Association for Cancer Research* 1994; **35**: A202.

69 Rodriguez G, Berchuck A, Whitaker R, *et al*. The urinary-type plasminogen activator receptor and its ligand are up-regulated in epithelial ovarian cancer. *Proceedings of the American Association for Cancer Research* 1993; **34**: A486.
70 Brown PD. Clinical trials of a low molecular weight matrix metalloproteinase inhibitor in cancer. *Ann NY Acad Sci* 1994; **732**: 217–21.
71 Davies B, Brown PD, East N, Crimmin MJ, Balkwill FR. A synthetic matrix metalloproteinase inhibitor decreases tumor burden and prolongs survival of mice bearing human ovarian carcinoma xenografts. *Cancer Res* 1993; **53**: 2087–91.
72 Naylor MS, Stamp GW, Davies BD, Balkwill FR. Expression and activity of MMPS and their regulators in ovarian cancer. *Int J Cancer* 1994; **58**: 50–6.
73 Moser TL, Young TN, Rodriguez GC, Pizzo SV, Bast RC Jr, Stack MS. Secretion of extracellular matrix-degrading proteinases is increased in epithelial ovarian carcinoma. *Int J Cancer* 1994; **56**: 552–9.
74 Xu FJ, Rodriguez G, Bae DS, *et al*. Heregulin and anti-p185$^{c\text{-}erbB\text{-}2}$ antibodies inhibit proliferation, increase invasiveness and enhance tyrosine autophosphorylation of breast cancer cells that overexpress p185$^{c\text{-}erbB\text{-}2}$. *Proceedings of the Annual Meetings of the American Association for Cancer Research* 1994; **35**: A225.

4 Screening: potential benefits and adverse consequences

SUSAN E MACKEY, WILLIAM T CREASMAN

Any physician who is involved in the care of patients with ovarian cancer can attest to the devastating nature of the disease. Patients usually present with advanced disease, and the available treatment regimens are unfortunately unsuccessful for most patients. The newer chemotherapy regimens have increased the rate of response but have had little impact on the final results. With these poor results, it is easy to see why detection of this disease at an early, highly curable stage is a desirable goal.

Epidemiology

Despite its grim prognosis, the overall incidence of ovarian cancer is comparatively low. Estimates for 1994 in the United States note ovarian cancer tied with urinary tract cancer for sixth place in terms of incidence in women, each accounting for 4% of new cancers diagnosed (fig. 4.1). In comparison, breast cancers will account for 32% of new cancers, followed by colorectal, lung, uterus, and leukaemia and lymphoma. Estimated mortality rates for women in 1994 indicate that deaths from ovarian cancer will account for 5% of all deaths from cancer (fig. 4.2). Based on 1994 census population projections, 24 000 new cases of ovarian cancer will be diagnosed in the United States and 13 600 patients will die. By comparison, there will be 182 000 new cases and 46 000 deaths from breast cancer, 72 000 new cases and 59 000 deaths from lung cancer, and 74 000 new cases and 28 200 deaths from colorectal cancer in women in 1994.[1]

The five-year survival rate for all stages of ovarian cancer based on patients treated from 1982–1986 is 41%, but when the figure is evaluated by stage at presentation there is a great deal of disparity. The five-year survival rate for stage I disease is 79%, for stage II 57%, for stage III 23%, and for stage IV 8%.[2] Only about 23% of patients present with localised disease,[1] most having advanced cancers that are not easily amenable to cure.

36

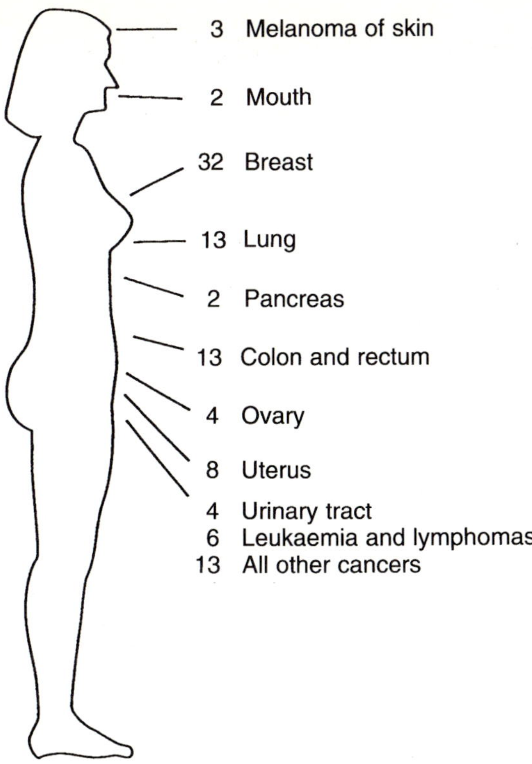

Fig. 4.1 Estimated percentage incidence of cancer in women.

Background

Because of the extreme differences in survival rates between early and advanced ovarian cancers, much attention has been focused on the search for a screening test or tests that can detect precancerous or early disease. This search has increased in the last decade because of the focus on women's health issues. It has become politically fashionable to concentrate on prevention, early detection, and treatment of those diseases that are found predominantly in women, or those cancers in which there have been significant increases in incidence in women. Compared with similar research into men's health issues this new consideration is long overdue and will benefit women's health.

Detection of early ovarian neoplastic change is difficult, and most cases of localised disease are identified by chance. Ovaries are susceptible to benign changes, and differentiating between the two processes can be problematic. Pelvic examination routinely includes palpation for adnexal masses to try and identify abnormalities, but this has proved to be inadequate.[3] Early attempts to screen for precancerous and early stage disease included cervical and vaginal cytological examination and "cul-de-sac" aspiration of mesothelial cells.[4] Neither of these methods proved

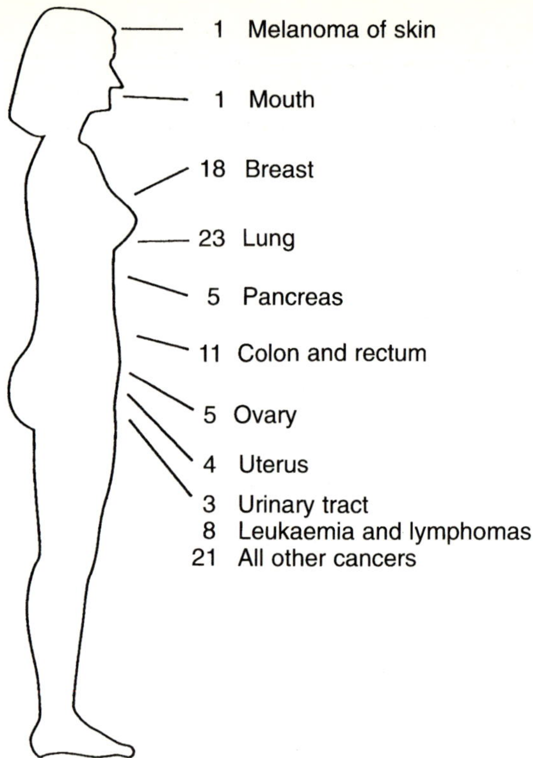

Fig. 4.2 Estimated percentage deaths of cancer in women.

useful, and they have therefore been abandoned. Methods being explored today as potential screening tools include tumour markers, particularly CA125, and ultrasonography including colour Doppler imaging. These methods have not been fully evaluated in randomised controlled trials, and care must be taken to avoid instituting them routinely before they have undergone full scientific investigation.

Natural history

Screening for cancer implies finding the disease in a precancerous stage before it has undergone malignant transformation, or at an early, curable stage. Unfortunately little is known about the natural history of ovarian cancer. It is generally assumed to progress in an orderly fashion from stage I to stage IV, though the length of time spent in each stage is unknown. Individual cases may not progress this way, and multifocal disease may indicate multifocal origins.[5] There have been case reports of women who have developed peritoneal carcinomatosis after oophorectomy,[67] which suggests that the peritoneum itself (which shares a common embryological origin with the ovarian epithelium) can undergo malignant transformation

38

either simultaneously with the ovary, or as the primary cancer. There is also debate over the malignant potential of benign ovarian cysts. Benign mucinous cystadenocarcinomas have been associated with borderline tumours, but direct transformation to an adenocarcinoma is difficult to document. It is therefore not known if a premalignant stage exists for ovarian cancer.[8]

Risk factors

Risk factors for the development of ovarian cancer have been studied to understand the disease better and to help identify a population for screening. The highest known risk for the development of the disease is in women who belong to families that are known to be prone to hereditary cancers.[9] Interest in this phenomenon was first reported in the 1950s with the publication of a case report of ovarian cancer in a mother and five daughters.[10] As extensive pedigrees from cancer-prone families were developed, an autosomal dominant pattern of inheritance of multiple cancer types was seen.[11-13] Family cancer patterns can be divided into three subsets: site-specific familial cancer, which is associated with only epithelial ovarian cancer; breast and ovarian familial cancer syndrome, which is associated with both breast and epithelial ovarian cancer; and Lynch II syndrome, which is associated with familial clustering of breast, ovarian, endometrial, gastrointestinal, and genitourinary cancers. In all three types men are thought to be capable of carrying the affected gene and the degree of risk to an individual patient is dependent upon the frequency of the occurrence of cancer in first and second degree relatives.[14] There is also a significantly higher incidence of ovarian cancer among family members of patients known to have the disease, but who are not part of these hereditary cancer-prone families.[15-17] The age at diagnosis of patients with hereditary types of ovarian cancer is significantly earlier than that of the general population,[17 18] and this is important in terms of appropriate age at screening and adequate counselling of patients.

Reproductive and environmental factors have also been investigated for their contributions to the aetiology and pathophysiology of ovarian cancer. Many studies have shown an association between pregnancy history and risk of developing the disease. It seems that the greater the number of ovulatory cycles a woman experiences, the greater her risk of developing ovarian cancer. This has been substantiated by studies that have shown a significant increase in the incidence of disease in nulliparous patients or those who have had fewer pregnancies.[19] Breast feeding,[20 21] which is associated with prolonged postpartum amenorrhoea, and the use of oral contraceptives[22-24] have been related to a decreased risk of ovarian cancer.

Two theories have been proposed to explain these associations. The first, proposed by Fathalla[25] asserts that persistent, uninterrupted ovulation results in repetitive trauma to the ovarian epithelium which is followed by cellular proliferation. This can lead to entrapment of epithelial inclusion cysts which may undergo malignant transformation. Ovulation also exposes the ovarian epithelium to steroid-rich follicular fluid and pituitary gonadotropins which may induce a neoplasm. A second theory contends that nulliparity is a sign of an underlying reproductive disorder which is also associated with the development of ovarian cancer.[25] Several case-controlled studies have shown an increased prevalence of infertility among patients with ovarian cancer compared with controls.[27-30] None of this information is concrete, and it is doubtful how useful it could be in the development of a screening programme.

Screening principles

The general principles and goals of cancer screening have been well defined (table 4.1).[31] The objective of a screening programme is to provide a simple, inexpensive test to a population that is free of symptoms which will classify a large number of people as likely or unlikely to have a particular cancer. The ultimate goal is to reduce the morbidity and mortality of the disease.

Table 4.1 Principles of screening

Characteristics of diseases suitable for screening
- Serious consequences
- Treatment of presymptomatic disease more effective than treatment after symptoms have developed
- Detectable preclinical phase
- High prevalence of preclinical phase

Characteristics of tests suitable for screening
- Simple to do
- Inexpensive
- Acceptable to patients and providers
- Safe
- Relatively painless
- Valid

To be suitable for a screening programme a disease should meet the following criteria. It should have serious consequences such as appreciable complications or a high mortality rate. Ovarian cancer undoubtedly meets this criterion with an overall five-year survival rate of only 41%, and multiple associated complications. Treatment must be available that is more effective when applied to early or presymptomatic disease than when applied to symptomatic patients. The disease should have a detectable preclinical phase that is long enough to allow manageable screening intervals. These latter two points are controversial because they imply a consistent progression from stage I to stage IV, while little is understood about the natural history of the disease. There is no question that preclinical disease exists, and this is highlighted by the fact that most stage I cases are asymptomatic and discovered only by chance. Nothing can be said about the length of this phase, however, and it has never been proved that all stage III and IV cancers pass through stages I and II. It cannot be stated therefore that identification and treatment of preclinical disease will affect the diagnosis of later stages by interrupting a natural progression of the cancer, as implied by these criteria. The preclinical phase should have a high prevalence among the people screened to justify the expense of a screening programme. Little is known or understood about the preclinical phase of ovarian cancer, and therefore little can be said about its prevalence. Most patients present with regional or distant disease as opposed to localised disease,[1] which probably reflects the fact that these patients have symptoms and are therefore presenting for evaluation. Some of these patients could have cancers with multifocal origins that were never truly localised initially.

Obviously more research into the natural history of ovarian cancer is needed. Some workers have concentrated their screening efforts on patients with a family history of ovarian cancer to increase the prevalence of the disease in their study population. This type of information is difficult to obtain reliably, and applies only to a small percentage of women who will ultimately develop the disease. Its overall impact on the mortality and complication rates of ovarian cancer must therefore be questioned.

Disease

Present — Absent

	Present	Absent
Positive	a True Positive	b False Positive
Negative	c False Negative	d True Negative

Test

Sensitivity = a/a+c
Specificity = d/b+d
Positive predictive value = a/a+b
Negative predictive value = d/c+d

Fig. 4.3 Statistical principles.

Several criteria have been identified as important characteristics of suitable screening tests.[31] A screening test should be simple to perform, inexpensive, acceptable to the patient and to the people doing the procedure, safe, relatively painless, and valid. The validity of a test is measured by its sensitivity, specificity, and predictive values (fig. 4.3). The sensitivity of a test indicates the probability that the test will be positive when the disease is present, while specificity indicates the probability that the test will be negative when the disease is not present. The positive predictive value of a test is the probability that the disease is not present when the test is negative. The positive predictive value is an important measure of a screening test, because it indicates the number of diagnostic procedures done for patients without the disease for every procedure done for a patient with the disease. For example, if a test had a 10% positive predictive value, 10 diagnostic procedures would have to be done for patients with a positive screening test for every patient actually found to have the disease. For ovarian cancer, this would mean nine unnecessary operations for every case of cancer found.

The positive predictive value of a screening test is dependent on the specificity of the test, which in turn is dependent on the sensitivity of the test (fig. 4.3). Ideally a test should have both a high sensitivity and a high specificity but this can be difficult, as an increase in one measure often leads to a reduction in the other. The positive predictive value is also dependent on the prevalence of the disease in the study population—the more prevalent the disease, the higher the probability that the disease is actually present when the test is positive. The level of specificity required to maintain a positive predictive value of 10% could be lowered by increasing the prevalence of the disease in the screened population. For example, if only patients with a positive family history of ovarian cancer were screened, a much less specific test could be used to keep the positive predictive value at 10%. Unfortunately most women destined to develop ovarian cancer have no family history and would not benefit from the screening programme. It has been calculated that a specificity of 99.6% is required of any ovarian cancer screening test to make a true impact on the disease.[32] All potential screening procedures should be evaluated for their sensitivities, specificities, and positive predictive values in randomised, controlled trials before being used in the general population.

Serum tumour markers

Many serum tumour markers have been evaluated to assess their role in screening for ovarian cancer in women without symptoms. The marker most extensively studied has been CA125, a 200 kDa glycoprotein detected by the murine monoclonal antibody OC125 produced by somatic hybridisation of spleen cells from mice immunised with ovarian cancer cell lines. CA125 was first described by Bast in 1981,[33] and was the 125th attempt to produce a monoclonal antibody to an epithelial ovarian cancer.[34] In 1983, Bast *et al* published a study in which they evaluated 888 apparently healthy blood donors and 101 patients with surgically diagnosed ovarian cancer with the CA125 assay.[35] They found that only 1% of healthy donors had CA125 concentrations of more than 35 U/ml, while 82% of patients with ovarian cancer did. They also found that 29% of patients with other cancers and 6% of patients with benign disease had raised serum concentrations of CA125 (table 4.2). Small amounts of CA125 are found in many normal adult tissues, including pleuropericardium, peritoneum, fallopian tube epithelium, endometrium, and endocervical tissue, all of which develop from coelomic epithelium or the Mullerian duct system.[36] A study of 2544 healthy women showed that menopausal status, previous hysterectomy (in premenopausal women), and age (in postmenopausal

Table 4.2 Serum CA125 concentrations in various diseases

Group	Total No Tested	No (%) with antigen concentration	
		>35 U/ml	>65 U/ml
Apparently healthy controls	888	9 (1.0)	2 (0.2)
Men	537	4 (0.7)	2 (0.4)
Women	351	5 (1.4)	0
Patients with benign diseases	143	9 (6.3)	3 (2.1)
Patients with other cancers	200	57 (28.5)	44 (22.0)
Pancreatic	29	17 (58.6)	13 (44.8)
Lung	25	8 (32.0)	6 (24.0)
Breast	25	3 (12.0)	2 (8.0)
Colorectal	71	16 (22.5)	12 (16.9)
Other gastrointestinal	30	8 (26.7)	6 (20.0)
Other not gastrointestinal	20	5 (25.0)	5 (25.0)
Patients with ovarian carcinoma	101	83 (82.2)	75 (74.3)

women) significantly influenced the serum CA125 concentration, which suggests that the reference limit of 35 U/ml should be adjusted for these conditions.[37] Benign states associated with raised CA125 concentrations include hepatic cirrhosis with ascites, granulomatous peritonitis, acute pancreatitis, endometriosis, adenomyosis, uterine leiomyomas, benign ovarian cysts, pelvic inflammatory disease, menstruation, the first trimester of pregnancy, and recent laparotomy.[38]

The CA125 assay was then developed and evaluated for its usefulness in monitoring the treatment of ovarian cancer, detecting a relapse of the disease, and distinguishing benign from malignant conditions. In the study by Bast *et al* they found that rising or falling concentrations of CA125 correlated with progression or regression of disease 93% of the time, making it a reliable, non-invasive way to monitor the response to treatment.[35] In 1987 Schwartz *et al* found an 83% correlation between an increased CA125 concentration and recurrence of disease, though a concentration within the reference range did not always reflect absence of disease at second-look laparotomy.[39]

Many studies have assessed the value of CA125 concentrations in the preoperative evaluation of pelvic masses. Vasilev *et al* found that only 78% of patients with cancer had values of more than 35 U/ml, whereas 22% of benign pelvic masses were associated with CA125 concentrations of more than 35 U/ml.[40] Malkasian *et al* raised the cutoff value to 65 U/ml,[41] and found that threequarters of the cancers were associated with raised CA125 concentrations, while only 8% of benign lesions were.[42–45] Investigators have attempted to increase the diagnostic accuracy of the CA125 assay by including ultrasonography and other tumour markers in the preoperative evaluation. All the studies agree that the CA125 concentration is most helpful in the evaluation of postmenopausal women. Vasilev *et al* found that 80% of women over the age of 50 with pelvic masses and CA125 concentrations of more than 35 U/ml had cancer compared with only 15% of patients under the age of 50.[40] This information is important when deciding whether a patient should be managed by a gynaecological oncologist or a general gynaecologist.

The value of the CA125 assay in detecting early ovarian cancer was first assessed in 1985 by Bast *et al* in a case report.[46] A patient with hypogammaglobulinaemia had regularly stored serum samples dating back to five years before her diagnosis of ovarian cancer. When these serum samples were evaluated it was noted that about one year before the clinical presentation of her disease, her CA125 concentrations had begun to rise. This rise was linear until she underwent cytoreductive surgery, after which the concentration fell abruptly. It remained low during her course of chemotherapy, and stayed within the reference range despite a second-look laparotomy that indicated recurrent disease. This serendipitous finding suggested that the CA125 assay could be useful in detecting early ovarian cancers.

Three years later, Zurawski *et al* published a study in which they measured serum CA125 concentrations in 105 women who had already developed ovarian cancer.[47] The blood was obtained from the JANUS serum bank, a collection of specimens acquired from more than 100 000 people in Norway.[47] Half the specimens collected within 18 months of the diagnosis of ovarian cancer had CA125 concentrations of more than 30 U/ml, and about a quarter of those collected 60 months or more before the diagnosis showed a rise. In comparison, only 7% of controls had raised CA125 concentrations. In the same year Zurawski with another group measured preoperative CA125 concentrations in 36 patients with stage I and stage II ovarian carcinomas.[48] They noted that the mean CA125 concentrations for stage I and stage II cases were 133 and 382 U/ml, respectively. More importantly, though, while 83% of stage II cases had CA125 concentrations of 35 U/ml or more, only half the stage I cases had a similar rise. These percentages included data from mucinous tumours which tended to be associated with lower CA125 values. In a study by Mann *et al* preoperative serum concentrations were raised in only 23% of patients with stage I disease.[49] In a review by Jacobs and Bast, CA125 concentrations were raised in about 50% of cases of stage I disease.[50]

An interesting hypothesis was put forward by Jacobs *et al* as to why serum CA125 concentrations are routinely raised in disseminated ovarian cancer but not in localised disease.[51] In stage I cancer there is usually a high tissue expression of CA125, despite a serum concentration within the reference range. This also occurs in many benign circumstances. Serum CA125 concentrations seem to rise when normal tissue barriers are disrupted, as in malignant invasion, so an increase in serum CA125 may be related not only to the production of the CA125 antigen but also to its release into the circulation. If this hypothesis is correct, then screening for early ovarian cancer with the CA125 assay must take place after enough tissue disruption has occurred to increase serum concentrations of the antigen, but before the cancer has metastasised widely. Unfortunately we have a poor understanding of the natural history of the disease.

One of the earliest attempts to use the CA125 assay as a screening tool was by Jacobs *et al* in 1988 when they evaluated 1010 apparently healthy postmenopausal women by pelvic examination and serum CA125 measurement.[52] Participants with abnormal findings underwent abdominal ultrasonography, and those with abnormal ultrasound scans were referred to a gynaecologist for evaluation. Thirty-one women had serum CA125 concentrations of more than 30 U/ml, one of whom was diagnosed as having stage Ia clear-cell carcinoma of the ovary. None of the women with a CA125 concentration within the reference range developed ovarian cancer within a year of follow-up. The specificity of the CA125 assay was calculated to be 97%, which was below the estimated specificity of 99.6% necessary to keep the positive predictive value greater than or equal to 10%. The addition of ultrasonography as a follow-up evaluation increased the specificity to 99.8%. The sensitivity of the assay was not calculated in this study.

In 1992 Einhorn *et al* evaluated the serum CA125 assay as a screen for ovarian cancer in 5550 apparently healthy women.[53] Women whose CA125 concentrations were raised and an equal number of age-matched controls in whom they were within the reference range were followed by means of pelvic examinations, transabdominal ultrasonography, and serial CA125 measurements. One hundred and seventy-five women had raised CA125 concentrations, six of whom had ovarian cancer. Of the women with CA125 concentrations within the reference range, three were subsequently diagnosed as having ovarian cancer. Using thresholds of 30 and 35 U/ml, the specificity of the CA125 assay was 97% and 98.5%, respectively, for women aged 50 or older, and 91% and 94.5%, respectively, for women aged 49 or younger.

Helzlsouer *et al* evaluated the sensitivity and specificity of serum CA125 measurements for the detection of ovarian cancer by designing a case-controlled study nested within a cohort of women who donated blood to a community-based serum bank[54]; 20 305 specimens were collected from county residents. Thirty-seven cases of ovarian cancer were identified through the county cancer registry over a 15-year follow-up period, and matched with 73 controls. CA125 concentrations were evaluated with respect to time from donation to diagnosis. The sensitivity and specificity of the CA125 assay were 24% and 96%, respectively, for the entire follow-up period and improved with shorter follow-up intervals. Within the first three years of follow-up the sensitivity of the assay was 57.1% and the specificity 100%. Despite a high specificity, the authors concluded that the CA125 assay alone did not have sufficiently high sensitivity to be recommended for routine screening for ovarian cancer.

Attempts have been made to increase the specificity of the CA125 assay by combining it with other tumour markers. In 1987, Einhorn *et al* evaluated

CA125 concentrations together with those of CA15-3 (a tumour-associated antigen commonly raised in breast and ovarian cancer[34]) and TAG-72 (a tumour-associated antigen raised in patients with ovarian cancer, but rarely in benign disease[34]) in 219 patients undergoing diagnostic laparotomy for pelvic masses.[44] They found that combining the three tests improved the specificity for detecting ovarian cancer, but sacrificed the higher sensitivity of the CA125 assay. This was confirmed in 1990 by Soper *et al.*[45] Preoperative serum samples were assayed for CA125, TAG-72, and CA15-3 concentrations in 100 women with pelvic masses. CA125 concentrations alone yielded a sensitivity of 88% and a specificity of 83% for epithelial ovarian carcinomas. Coordinate rises of CA125 (above 65 U/ml) and TAG-72 or CA15-3, distinguished ovarian epithelial carcinomas from benign masses with increased specificity (98%), but reduced sensitivity (73%).

In 1991 Bast *et al* assayed multiple markers in a panel of serum samples from 47 patients with ovarian cancer and 50 patients with benign disease whose CA125 concentrations exceeded 35 U/ml.[55] These additional markers included TAG-72, CA15-3, PLAP (placental alkaline phosphatase), HMFG 1 and 2 (human milk fat globule protein), and NB/70K (a nonbound 70 kD transmembrane glycoprotein[34]). Only TAG-72 and CA15-3 were consistently raised in the serum of patients with cancer and within the reference ranges in the serum of patients with benign disease. In 1992, Jacobs *et al* measured the concentrations of CA125, CA15-3, and TAG-72 in the serum of 217 of their 1010 original study participants to increase the specificity of the CA125 assay.[56] Specificity was increased from 97.0% to 98.9% using a cut-off of 30 U/ml, and from 99.5% to 99.9% using a cut-off of 50 U/ml when a positive test was defined as an increased serum CA125 concentration combined with either a raised CA15-3 or TAG-72 concentration. Sensitivity was not studied, but will always fall to the level of the least sensitive marker.

Serial measurements of the CA125 antigen have also been evaluated for their use in increasing the specificity of the assay for detecting ovarian cancer. In 1990 Zurawski *et al* measured CA125 concentrations in 1082 apparently healthy women.[57] Thirty-six women with concentrations of 35 U/ml or more and their age-matched controls underwent serial CA125 measurements every three months, and a pelvic examination at least every six months. Ultrasound scans were done when the clinical examination warranted. Of the 36 subjects with initially raised serum CA125 concentrations, only two showed a doubling of the value and in only one of these two was this increase sustained. A diagnosis of stage III ovarian cancer was eventually made in this participant, whereas no cancers were identified in the remaining participants. This gave a specificity of 99.9% for the doubling of the assay and 100% for a sustained increase.

In the study by Jacobs *et al* in 1992 serial CA125 concentrations were measured in 30 of the participants with initially raised values, and in 30 participants in whom they were not.[56] Serum CA125 values in women with initial CA125 concentrations of less than 30 U/ml remained less than

46

30 U/ml in serial samples. During follow-up of the 30 women with initially raised serum CA125 values, serial concentrations fell towards the reference range. A CA125 concentration of more than 30 U/ml at three months after an initial value above 30, 50, or 70 U/ml increased the specificity of the assay to 98.4%, 99.6%, or 99.8%, respectively.

In the study by Jacobs *et al* in 1988 the specificity of the CA125 assay (97%) was greatly improved by the addition of a pelvic examination (100%) or a follow-up ultrasound examination (99.8%).[52] Despite the high specificity of the CA125 assay combined with the pelvic examination, the value of this approach was thought by the authors to be limited by the poor sensitivity of the pelvic examination. The combination of the CA125 assay and ultrasonography was thought to offer the most hope of a specific and sensitive method for the early detection of ovarian cancer.

A follow-up to this study was published in 1993 in which 22 000 apparently healthy postmenopausal volunteers underwent serum CA125 measurements, and those with values of 30 U/ml or more were recalled for abdominal ultrasonography.[58] Forty-one women showed evidence of abnormality and were investigated surgically. Eleven of these volunteers had ovarian cancer, while 30 had benign disease or no abnormality. Of the remaining 21 959 volunteers who had no abnormality detected, eight were subsequently diagnosed as having ovarian cancer during follow-up. This screening protocol achieved a specificity of 99.9% and a positive predictive value of 26.8%, but a sensitivity of only 78.6% at one year and 57.9% at two years of follow-up.

Ultrasound

A drawback to the use of serum tumour markers in screening for ovarian cancer has been their lack of specificity for ovarian disease. Many researchers have preferred to concentrate their efforts on a direct evaluation of the pelvic organs with ultrasonography. Though ultrasound technology is well established in obstetrics, its clinical application in gynaecology has lagged behind, probably because of the poor quality images obtained with earlier instruments. With today's refined equipment clear pictures of pelvic organs and diseases can easily be evaluated. Abnormal ovarian size, morphology, and vascularisation—possible early signs of ovarian cancer—can be evaluated.

Much of the initial work in the sonographic assessment of pelvic masses dates back to the early 1970s. The images obtained with the primitive equipment available at the time were capable of characterising masses into only three major groups: cystic, complex, or solid.[59] As the equipment became more refined, so did the evaluations. In 1978, Fleischer *et al* published a study in which he correlated the sonographic appearance of surgically proved pelvic masses with their morphological characteristics.[60] Using grey scale transabdominal ultrasonography they were able to identify 91% of the lesions accurately based on size, location, internal consistency, and definition of borders. A study from Meire *et al* presented a retrospective analysis of ultrasound examinations obtained from patients referred with pelvic masses.[61] They identified eight ultrasound characteristics and noted how often each occurred in both benign and malignant lesions. Applied prospectively, they predicted these characteristics would have properly identified 91% of the masses as benign or malignant.

Ultrasonography is now often used to evaluate abnormal findings on gynaecological examination, and has been used to challenge the concept that all postmenopausal women with minimally enlarged ovaries should undergo laparotomy. Several studies concluded that small anechoic lesions are rarely malignant, even in the postmenopausal population,[62–66] though not all agree that a negative sonogram is a completely reliable finding.[67] Ultrasonography has also been compared with pelvic examination, and is a much more reliable way of evaluating the ovaries.[62 68] In a study by Rulin and Preston in 1987 10% of pelvic masses less than 10 cm in diameter were missed on physical examination, and of those palpated the examination predicted the size of the mass to within 2 cm in only 68% of cases.[62] Ultrasonographic prediction of size was accurate in 87% of the cases scanned.

Transabdominal ultrasonography

The use of ultrasonography as a potential screening method for ovarian cancer was first proposed by Campbell *et al* in 1982.[69] Initially, 11 climacteric women with benign disease underwent transabdominal ultrasonographic evaluation of their ovaries the evening before operation. At the laparotomy the diameter and morphology of the ovaries were calculated and compared with the ultrasound findings. The correlation coefficient between ovarian volumes measured by ultrasonography and those obtained by direct measurement at operation was 0.97. This led to a further study in which 31 postmenopausal women underwent an ultrasound evaluation. Of the ovaries studied, 84% were identified in this group, the exceptions being three women who were obese and two women with distorted anatomy after hysterectomy. The ovaries of the remaining 25 women were normal except for one cystic ovary, which had not been identified by pelvic examination.

Based on this information the authors thought that pelvic ultrasonography warranted further study as a screening method for the early detection of ovarian cancer. They established an ovarian scanning clinic to evaluate the use of abdominal ultrasonography to detect ovarian cancer in a large group of self-referred women over 45 years old.[70] A total of 1084 women were scanned, and the ovaries were identified 99.4% of the time. The authors were able to identify the morphology of a normal ovary as being uniformly hypoechoic, ovoid in shape, and having a smooth echogenic outline. Normal postmenopausal ovarian volume, calculated from the formula for ovoid volume—that is, the product of the largest diameter and the two diameters perpendicular to it [$(d1) \times (d2) \times (d3) \times 0.523$], was identified as roughly

3.7 cm^3. They described morphological abnormalities as entirely cystic, cystic with loculations, cystic with solid areas (nodules or papillary growths), or solid with an irregular outline. Every patient who had an abnormality detected on ultrasound examination had some type of ovarian disease at operation, including one clear cell carcinoma. Encouraged by this information, further studies of the ultrasonographic appearance of ovaries were undertaken. A large study of postmenopausal women conducted by Goswamy *et al* identified factors that influenced gonadal size.[71] A retrospective analysis by Granberg *et al* classified ovarian tumours according to their macroscopic appearances on ultrasound examination and compared this information with the corresponding histological diagnoses.[65]

The largest study of transabdominal ultrasonography as a screening method for early ovarian cancer was published by Campbell *et al* in 1989.[72] Over 5000 women who were symptom free were enrolled in the study, and 4201 (77%) completed the entire course. Each women was scheduled to undergo three annual screenings to detect grossly abnormal ovaries or non-regressing masses. An ovary was considered abnormal if it had hyperechogenic or hypoechogenic morphology, an irregular outline, or a volume of more than 20 ml. Abnormal scans were repeated after three to eight weeks to exclude transient changes. If an abnormality persisted the patient was referred for surgical investigation. Of 338 abnormal screens, 326 women underwent surgical evaluation, and 379 ovarian masses were found. This included only six primary ovarian cancers and six metastatic ovarian cancers, giving an overall specificity of 97.7%, a false positive rate of 2.3%, and an odds ratio of 1:67 that a positive result was indicative of a primary ovarian cancer. Furthermore, they were unable to identify any ultrasound characteristics that could be used to differentiate consistently between benign and malignant ovarian tumours.

In a follow-up study they re-evaluated the data from the earlier study to improve the false positive rate and odds ratio.[73] By changing the criteria for an abnormal scan to abnormal ovarian morphology or volume plus a defined volume change at the second scan they reduced the false positive rate to 1.6%, but the odds ratio was still 1:50, meaning that 51 laparotomies would be required to find one case of primary ovarian cancer.

Transvaginal ultrasonography

To try and overcome the disadvantages of transabdominal sonography, many investigators began to explore the technique of transvaginal scanning. One of the early descriptions of its use was published in 1988 by Timor-Tritsch *et al*.[74] They listed many advantages of the approach, including: there is no need for the patient to have a full bladder, which can be both uncomfortable and time-consuming; obese patients can easily be examined; and the transducer probe is placed close to the pelvic organs allowing the use of a higher ultrasound frequency to produce better resolution of the picture and clearer images. Those clearer images provided an opportunity to define ovarian morphology more accurately and to help discriminate benign from malignant lesions.[66][75–77]

The first investigation of transvaginal ultrasonography as a screening method for ovarian cancer was published in 1989 by Higgins *et al*.[78] In their study, 506 symptom free women aged 40 years or older underwent transvaginal ultrasonography. Ovarian volume was calculated, and ovarian morphology was classified as uniformly hypoechogenic, cystic, solid, or complex. Measurements of ovarian dimensions were possible in 95% of the premenopausal patients, and 86.5% of the postmenopausal patients. Twelve patients (2.4%) had abnormal findings, and 10 agreed to surgical investigation. Four serous cystadenomas, three endometriomas, two cystic teratomas, and one metastatic adenocarcinoma from the colon were found. A follow-up study was published the following year in which they reported the results of the first 1000 patients screened.[79] Using the criteria noted above, a total of 31 patients (3.1%) had abnormal scans, and 24 underwent exploratory laparotomy. In addition to their previous findings they discovered another four serous cystadenomas, three endometriomas, one cystic teratoma, two follicular cysts, three hydrosalpinxes, and one cervical myoma. A third study in their series published in 1991 evaluated only postmenopausal women,[80] which avoided the confusion of normal variation in ovarian size noted during different phases of the menstrual cycle. Of the 1300 symptom free postmenopausal women who underwent transvaginal ultrasonography, 33 (2.5%) had ovarian abnormalities, and 27 consented to operation. Ovarian enlargement was apparent on clinical examination in only 10. At the time of operation three adenocarcinomas, 14 serous cystadenomas, three epithelial cysts, three leiomyomas, two hydrosalpinxes, one endometrioma, and one thecoma were found. Of the three adenocarcinomas, one was the metastatic colonic cancer previously noted, but the other two were primary stage I ovarian carcinomas, neither of which were palpable on examination or associated with increased CA125 concentrations. Though these studies proved that the technique of transvaginal ultrasonography vastly improved the ability to detect ovarian abnormalities, they also highlighted the difficulties associated with differentiating benign from malignant diseases.

Many attempts have been made to improve the specificity of ovarian cancer detection by transvaginal sonography. Bourne *et al* evaluated a cohort of women with a family history of ovarian cancer to assess the prevalence of the disease in the study population.[81] Using transvaginal ultrasonography they screened 776 women who had at least one first or second degree relative with ovarian cancer. Forty-three women (5.5%) were referred for surgical investigation, and 39 underwent laparotomy. They found three primary stage I ovarian cancers, consistent with a false positive rate of 5.2%, a positive predictive value of 7.7%, and an odds ratio of 1:13. As an extension of the University of Kentucky Ovarian Screening Project a study was published in 1993 in which a morphology index was applied to each ovary seen on ultrasonography (table 4.3).[82] Of 3220

50

Table 4.3 Sonographic morphology index for ovarian tumours

	Category				
	0	1	2	3	4
Volume (cm³)	<10	10–50	>50–200	>200–500	>500
Structure of cyst wall	Smooth <3 mm thickness	Smooth >3 mm thickness	Papillary projection <3 mm	Papillary projection ≥3 mm	Predominantly solid
Structure of septa	No septa	Thin septal >3 mm	Thick septal 3 mm–1 cm	Solid area ≥1 cm	Predominantly solid

postmenopausal women without symptoms who were screened, 44 (1.4%) had persistent ovarian abnormalities, and three of these had primary ovarian cancer at the time of laparotomy. Nine of the 44 patients who were operated on had a morphology index of more than five, including the three with ovarian cancer. None of the patients with cancer had a raised serum CA125 concentration. Though both these studies seemed to offer improvements on the specificity of transvaginal sonography, they both risk a significant loss of sensitivity.

Transvaginal colour Doppler imaging

The most successful means of improving the diagnostic accuracy of transvaginal ultrasonography is the addition of colour Doppler flow studies. This technique involves the detection of blood vessels by measuring the change in frequency of reflected ultrasound waves from moving red blood cells. This information is colour-coded according to the direction of flow toward or away from the transducer, so that areas of vascularisation can be identified as areas of colour. Flow velocity waveforms may be displayed on the screen to illustrate the distribution and frequency of the shifted Doppler frequencies over time. An analysis of these waveforms can be used to measure the impedance to blood flow in the imaged vessels, expressed as either the resistance index or the pulsatility index (fig. 4.4). A reduction in either index reflects a reduction in the impedance to blood flow distal to the point of sampling[83]; the development of an adequate vascular supply is critical for the early development and unrestricted growth of a cancer.[84] These tumour vessels have a decreased amount of smooth muscle in their walls compared with normal blood vessels and therefore provide less resistance to blood flow through arterioles.[85] Using colour Doppler imaging this low impedance can be detected and expressed as a low resistance or pulsatility index.

In 1989 three studies were published of colour Doppler evaluations of tumour vascularity in gynaecological disorders. Hata *et al* did transabdominal ultrasound examinations with colour Doppler imaging on eight normal volunteers and 97 patients with various gynaecological

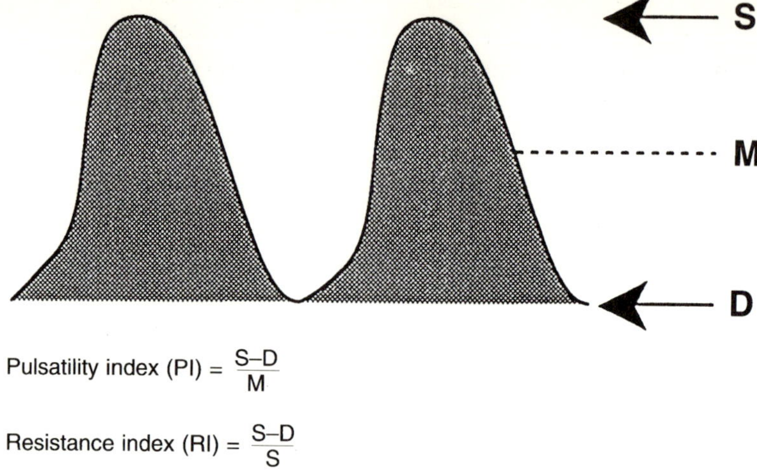

Pulsatility index (PI) = $\dfrac{S-D}{M}$

Resistance index (RI) = $\dfrac{S-D}{S}$

Fig. 4.4 Calculation of pulsatility and resistance indexes. S=peak Doppler shifted frequency at systole; D=minimum Doppler shifted frequency at diastole; and M=mean maximum shifted frequency during the cardiac cycle.

problems.[86] They found that all cases of ovarian carcinoma, endometrial carcinoma, and trophoblastic disease were associated with abnormal resistance indexes. Eight of 44 patients with benign tumours had abnormal flow, but all were associated with leiomyomas or adenomyosis. Kurjak *et al* used transvaginal colour Doppler to evaluate the pelvic circulation of 41 patients with pelvic masses and 15 patients with infertility before laparoscopy or laparotomy.[87] Among the infertility patients, Doppler studies showed low resistance indexes in three cases, all of which were associated with corpus luteum cysts. Among the patients with pelvic masses there were seven cancers, and all were associated with low resistance indexes. All but one of the 34 benign masses were associated with high resistance indexes, the exception being a benign granulosa cell tumour.

The third study (by Bourne *et al*) was the first attempt to evaluate transvaginal colour flow imaging as a possible screening technique for ovarian cancer.[88] They studied three groups of volunteers who had had previous transvaginal ultrasound examinations: 10 premenopausal women with normal ovaries on days 1–8 of the menstrual cycle; 20 postmenopausal women with normal ovaries; and 20 patients with abnormal findings on ultrasound examination who subsequently underwent laparotomy. The pulsatility index was calculated to reflect intraovarian resistance to blood flow. None of the women with healthy ovaries by ultrasound had an abnormal pulsatility index. Of the 20 women with abnormal ultrasound scans, eight had primary ovarian cancer diagnosed at laparotomy. Seven of these patients had a low pulsatility index and the remaining subject had an early intraepithelial or borderline serous cystadenocarcinoma associated with a normal pulsatility index. The remaining 12 patients who underwent

52

laparotomy for abnormalities noted on ultrasonography had benign findings, and 11 of these cases were associated with a normal pulsatility index. The exception occurred in a patient with bilateral dermoid cysts. The authors concluded that the absence of intratumoural neovascularisation and a high pulsatility index could be used to exclude invasive primary ovarian cancer and reduce the false positive rate associated with conventional screening techniques without sacrificing sensitivity.

These findings led to an explosion of studies evaluating the efficacy of colour Doppler imaging as a screening technique for ovarian cancer. The largest was by Kurjak *et al* in which they presented data from 14 317 asymptomatic or minimally symptomatic women.[89] Each patient underwent an ultrasound evaluation including a transvaginal colour Doppler examination and those diagnosed as having adnexal masses were surgically investigated. A total of 624 benign adnexal tumours were discovered each of which had normal resistance indexes; there was one exception, a patient with chronic appendicitis and hydrosalpinx. There were 56 ovarian cancers, 54 of which had low resistance indexes. These findings were consistent with a sensitivity of 96.4% and a specificity of 99.8%.

Weiner *et al* studied 53 patients with ovarian masses who underwent transvaginal colour Doppler imaging and serum CA125 measurement before laparotomy.[90] There were 36 benign and 17 malignant tumours. Intratumoural blood vessels in 16 of the cancers showed low impedance to flow as reflected by a low pulsatility index. Normal impedance to flow was found in 35 of the benign masses, 11 of which had suspicious sonographic findings and 14 of which had raised CA125 concentrations. The sensitivity and specificity of preoperative pulsatility index in detecting ovarian cancers were 94% and 97%, respectively. By comparison, the sensitivity and specificity of abnormal preoperative ultrasound findings were 94% and 69%, respectively, and those of raised preoperative serum CA125 concentrations were 82% and 61%, respectively.

Kawai *et al* evaluated transvaginal colour Doppler imaging in 24 patients with ovarian tumours.[91] They were able to predict whether a mass was benign or malignant on an abnormal pulsatility index in 95.8% of the cases. Bourne *et al* evaluated 1601 self-referred women with a family history of ovarian cancer.[92] All the patients were screened with transvaginal ultrasound, and 61 women had persistent abnormal adnexal masses. Patients with abnormal findings were also examined by colour Doppler imaging for signs of vascular changes and reduced resistance to blood flow. A morphology score was assigned to each mass based on the abnormal ultrasound findings. Sixty-one patients were referred for operation and six primary ovarian cancers were found. They calculated an odds ratio of 1:9 using only abnormal ultrasound findings. Use of a high morphology score or a low pulsatility index would have decreased these odds to 2:5. It is of interest that three of the six ovarian cancers were tumours of low malignant potential and one of the invasive lesions was stage III at time of diagnosis. The significant odds ratio is therefore probably much lower.

Summary

From the information available it seems clear that transvaginal sonography possibly with colour Doppler imaging offers the most hope as a screening tool for early ovarian cancer. It is easy to do, acceptable to patients, and seems to be highly sensitive as well as specific. Unfortunately, early detection of a disease is only significant if it decreases mortality, and no evidence has yet been presented to suggest that the use of any of the methods currently available would reduce the mortality from ovarian cancer. What is needed is a large-scale study over several decades that compares the mortality rates of a screened population compared with an unscreened population. It has been calculated that the detection of a reduction in mortality of a third with 80% power at the 5% level of significance would require almost 100 000 women for the screened population with an additional 100 000 women in the control group.[93] Until such a study is undertaken, no conclusions about the efficacy of any screening method can be drawn.

Because there is a low prevalence of the disease in the general population, and the lack of information regarding any possible screening success, routine screening for ovarian concern cannot be recommended. It would be more efficient to concentrate on preventing ovarian cancer through the use of oral contraceptives, tubal ligations, and removing ovaries at the time of hysterectomy. The use of screening techniques may be appropriate in those patients with a documented family history of ovarian cancer. It must be remembered that such a history accounts for less than 1% of all patients with ovarian cancers.

References

1 Boring C, Squires T, Tong T, Montgomery S. Cancer statistics 1994. *CA Cancer Journal for Clinicians* 1994; **44**: 7–26.

2 Petterson F, ed. Annual report on the results of treatment in gynecological cancer. *Int J Gynaecol Obstet* 1990; **21**: 245.

3 McGowan L, Stein D, Miller W. Cul-de-sac aspiration for diagnostic cytologic study. *Am J Obstet Gynecol* 1966; **96**: 413–17.

4 Garrett W. On the reduction of cancer mortality: Graham Crawford's experiment, 1949–1969. *Med J Aust* 1970; **1**: 1239–41.

5 Bourne T, Hampson J, Reynolds K, Collins W, Campbell S. Screening for early ovarian cancer. *Br J Hosp Med* 1992; **48**: 454–9.

6 Chen K, Schooley J, Flam M. Peritoneal carcinomatosis after prophylactic oophorectomy in familial cancer syndrome. *Obstet Gynecol* 1985; **66**: 93–4s.

7 Tobacman J, Tucker M, Kase R, Greene M, Costa J, Fraumeni J. Intra-abdominal carcinomatosis after prophylactic oophorectomy in ovarian-cancer-prone families. *Lancet* 1982; **2**: 795–7.

8 Anderson M. Malignant potential of benign ovarian cysts: The case "for." In: Sharp F, Mason W, Leake R, eds. *Ovarian cancer*. London: Chapman and Hall Medical, 1989: 187–96.

9 Lynch H, Fitzsimmons R, Conway T, Bewtra C, Lynch J. Hereditary carcinoma of the ovary and associated cancers: A study of two families. *Gynecol Oncol* 1990; **36**: 48–55.

10 Liber A. Ovarian cancer in mother and five daughters. *Arch Pathol* 1950; **49**: 280–90.

11 Lynch H, Guirgis H, Albert S, Brennan M, Lynch J, Kraft C, *et al.* Familial association of carcinoma of the breast and ovary. *Surg Gynecol Obstet* 1974; **138**: 717–24.

12 Lynch H, Harris R, Guirgis H, Maloney K, Carmody L, Lynch J. Familial association of breast/ovarian carcinoma. *Cancer* 1978; **41**: 1543–9.

13 Lynch H, Lynch P. Tumor variation on the cancer family syndrome: ovarian cancer. *Am J Surg* 1979; **138**: 439–42.

14 American College of Obstetricians and Gynecologists. *Genetic risk and screening techniques for epithelial ovarian cancer.* Committee Opinion, no. 117. Washington DC: American College of Obstetricians and Gynecologists, 1992.

15 Schildkraut J, Thompson W. Familial ovarian cancer: A population-based case-control study. *Am J Epidemiol* 1988; **128**: 456–66.

16 Koch M, Gaedke H, Jenkins H. Family history of ovarian cancer patients: A case-control study. *Int J Epidemiol* 1989; **18**: 782–5.

17 Amos C, Shaw G, Tucker M, Hartge P. Age at onset for familial epithelial ovarian cancer. *JAMA* 1992; **268**: 1896–9.

18 Lynch H, Watson P, Bewtra C, Conway T, Hippee C, Kaur P, *et al.* Hereditary ovarian cancer: heterogeneity in age at diagnosis. *Cancer* 1991; **67**: 1460–6.

19 Koch M, Jenkins H, Gaedke H. Risk factors of ovarian cancer of epithelial origin: A case-control study. *Cancer Detect Prev* 1988; **13**: 131–6.

20 Cramer D, Hutchinson G, Welch W, Scully R, Knapp R. Factors affecting the association of oral contraceptives and ovarian cancer. *N Engl J Med* 1987; **317**: 1047–51.

21 Whittemore A, Harris R, Itnyre J. Characteristics relating to ovarian cancer risk: Collaborative analysis of 12 US case-control studies. II. Invasive epithelial ovarian cancers in white women. *Am J Epidemiol* 1992; **136**: 1184–1203.

22 Schneider A. Risk factor for ovarian cancer. *N Engl J Med* 1987; **317**: 508–9.

23 Prentice R, Thomas D. On the epidemiology of oral contraceptives and disease. *Adv Cancer Res* 1987; **49**: 285–401.

24 Centers for Disease Control. The reduction in risk of ovarian cancer associated with oral-contraceptive use. *N Engl J Med* 1987; **316**: 650–5.

25 Fathalla M. Factors in the causation and incidence of ovarian cancer. *Obstet Gynecol Surv* 1972; **27**: 751–768.

26 Daly M. Ovarian Cancer. The epidemiology of ovarian cancer. *Hematol Oncol Clin North Am* 1992; **6**: 729–38.

27 Joly D, Lilienfeld A, Diamond E, Bross I. An epidemiologic study of the relationship of reproductive experience to cancer of the ovary. *Am J Epidemiol* 1974; **99**: 190–209.

28 McGowan L, Parent L, Lednar W, Norris H. The woman at risk for developing ovarian cancer. *Gynecol Oncol* 1979; 7: 325–44.

29 Nasca P, Greenwald P, Chorost S, Richard R, Caputo T. An epidemiologic case-control study of ovarian cancer and reproductive factors. *Am J Epidemiol* 1984; **119**: 705–13.

30 Whittemore A, Wu M, Paffenbarger R, Sarles D, Kampert J, Grosser S, *et al.* Epithelial ovarian cancer and the ability to conceive. *Cancer Res* 1989; **49**: 4047–52.

31 Hulka B. Cancer Screening: Degrees of proof and practical application. *Cancer* 1988; **62**: 1776–80.

32 Jacobs I, Oram D. Potential screening tests for ovarian cancer. In: Sharp F, Mason W, Leake R, eds. *Ovarian cancer.* London: Chapman and Hall Medical, 1989: 197–205.

33 Bast C, Feeney M, Lazarus H, Madler L, Colvin R, Knapp R. Reactivity of a monoclonal antibody with human ovarian carcinoma. *J Clin Invest* 1981; **68**: 1331–7.

34 Schwartz P. The role of tumor markers in the preoperative diagnosis of ovarian cysts. *Clin Obstet Gynecol* 1993; **36**: 384–94.

35 Bast R, Klug T, St John E, Jenison C, Niloff J, Lazarus H, *et al.* A radioimmunoassay using a monoclonal antibody to monitor the course of epithelial ovarian cancer. *N Engl J Med* 1983; **309**: 883–7.

36 Einhorn N. Ovarian Cancer. Early diagnosis and screening. *Hematol Oncol Clin North Am* 1992; **6**: 843–50.

37 Grover S, Quinn M, Weidman P, Koh H. Factors influencing serum CA125 levels in normal women. *Obstet Gynecol* 1992; **79**: 511–14.

38 Bast R, Boyer C, Olt G, Berchuck A, Soper J, Clarke-Pearson D, *et al.* Identification of markers for early detection of epithelial ovarian cancer. In: Sharp F, Mason W, Leake R, eds. *Ovarian cancer.* London: Chapman and Hall Medical, 1989; 265–75.

39 Schwartz P, Chambers S, Chambers J, Gutmann J, Katopodis N, Foemmel R. Circulating tumor markers in the monitoring of gynecologic malignancies. *Cancer* 1987; **60**: 353–361.

40 Vasilev S, Schlaerth J, Campeau J, Morrow C. Serum CA125 levels in preoperative evaluation of pelvic masses. *Obstet Gynecol* 1988; **71**: 751–6.

41 Malkasian G, Knapp R, Lavin P, Zurawski V, Podratz K, Stanhope R, *et al*. Preoperative evaluation of serum CA125 levels in premenopausal and postmenopausal patients with pelvic masses: Discrimination of benign from malignant disease. *Am J Obstet Gynecol* 1988; **159**: 341–6.

42 Finkler N, Benacerraf B, Lavin P, Wojciechowski C, Knapp R. Comparison of serum CA125, clinical impression, and ultrasound in the preoperative evaluation of ovarian masses. *Obstet Gynecol* 1988; **72**: 659–64.

43 Cole L, Nam J, Chambers J, Schwartz P. Urinary gonadotropin fragment, a new tumor marker. II. Differentiating a benign from a malignant pelvic mass. *Gynecol Oncol* 1990; **36**: 391–4.

44 Einhorn N, Knapp R, Zurawski V. CA125 assay used in conjunction with CA15-3 and TAG-72 assays for discrimination between malignant and non-malignant diseases of the ovary. *Acta Oncol* 1989; **28**: 655–7.

45 Soper J, Hunter V, Daly L, Tanner M, Creasman W, Bast R. Preoperative serum tumor-associated antigen level in women with pelvic masses. *Obstet Gynecol* 1990; **75**: 249–54.

46 Bast R, Seigal F, Runowicz C, Klug T, Zurawski R, Schonholz D, *et al*. Elevation of serum CA125 prior to diagnosis of an epithelial ovarian carcinoma. *Gynecol Oncol* 1985; **22**: 115–20.

47 Zurawski R, Orjaseter H, Andersen A, Jellum E. Elevated serum CA125 levels prior to diagnosis of ovarian neoplasia: relevance for early detection of ovarian cancer. *Int J Cancer* 1988; **42**: 677–80.

48 Zurawski V, Knapp R, Einhorn N, Kenemans P, Mortel R, Ohmi K, *et al*. An initial analysis of preoperative serum CA125 levels in patients with early stage ovarian carcinoma. *Gynecol Oncol* 1988; **30**: 7–14.

49 Mann W, Patsner B, Cohen H, Loesch M. Preoperative serum CA125 levels in patients with surgical stage I invasive ovarian adenocarcinoma. *J Natl Cancer Inst* 1988; **80**: 208–9.

50 Jacobs I, Bast R. The CA125 tumour associated antigen; a review of the literature. *Hum Reprod* 1989; **4**: 1–12.

51 Jacobs I, Prys Davies A, Oram D. Role of CA125 in screening for ovarian cancer. In: Sharp F, Mason W, Creasman W, eds. *Ovarian cancer 2*. London: Chapman and Hall Medical, 1992: 265–75.

52 Jacobs I, Bridges J, Reynolds C, Stabile I, Kemsley P, Grudzinskas J, *et al*. Multimodal approach to screening for ovarian cancer. *Lancet* 1988; **1**: 268–71.

53 Einhorn N, Sjovall K, Knapp R, Hall P, Scully R, Bast R, *et al*. Prospective evaluation of serum CA125 levels for early detection of ovarian cancer. *Obstet Gynecol* 1992; **80**: 14–18.

54 Helzlsouer K, Bush T, Alberg A, Bass K, Zacur H, Comstock G. Prospective study of serum CA 125 levels as markers of ovarian cancer. *JAMA* 1993; **269**: 1123–6.

55 Bast R, Knouf S, Epenetos M, Dhokia B, Daly L, Tanner M, *et al*. Coordinate elevation of serum markers in ovarian cancer but not in benign disease. *Cancer* 1991; **68**: 1758–63.

56 Jacobs I, Oram D, Bast R. Strategies for improving the specificity of screening for ovarian cancer with tumor-associated antigens CA125, CA15-3, and TAG 72.3. *Obstet Gynecol* 1992; **80**: 396–9.

57 Zurawski V, Sjovall K, Schoenfeld D, Broderick S, Hall P, Bast R, *et al*. Prospective evaluation of serum CA125 levels in a normal population, phase I: The specificities for single and serial determinations in testing for ovarian cancer. *Gynecol Oncol* 1990; **36**: 299–305.

58 Jacobs I, Prys Davies A, Bridges J, Stabile I, Fay T, Lower A, *et al*. Prevalence screening for ovarian cancer in postmenopausal women by CA125 measurement and ultrasonography. *BMJ* 1993; **306**: 1030–4.

59 Morley P, Barnett E. The use of ultrasound in the diagnosis of pelvic masses. *Br J Radiol* 1970; **43**: 602–16.

60 Fleischer A, James E, Mills J, Julian C. Differential diagnosis of pelvic masses by gray scale sonography. *AJR* 1978; **131**: 469–76.

61 Meire H, Farrant P, Guha T. Distinction of benign from malignant ovarian cysts by ultrasound. *Br J Obstet Gynaecol* 1978; **85**: 893–9.

62 Rulin M, Preston A. Adnexal masses in postmenopausal women. *Obstet Gynecol* 1987; **70**: 578–81.

63 Goldstein S, Bala S, Snyder J, Beller U, Raghavendra N, Beckman M. The postmenopausal cystic adnexal mass: the potential role of ultrasound in conservative management. *Obstet Gynecol* 1989; **73**: 8–10.

64 Andolf E, Jorgensen C. Cystic lesions in elderly women, diagnosed by ultrasound. *Br J Obstet Gynaecol* 1989; **96**: 1076–9.

65 Granberg S, Wikland M, Jansson I. Macroscopic characterization of ovarian tumors and the relation to the histologic diagnosis: criteria to be used for ultrasound evaluation. *Gynecol Oncol* 1989; **35**: 139–44.

66 Granberg S, Norstrom A, Wikland M. Tumors in the lower pelvis as imaged by vaginal sonography. *Gynecol Oncol* 1990; **37**: 224–9.

67 Luxman D, Bergman A, Sagi J, David M. The postmenopausal adnexal mass: correlation between ultrasonic and pathologic findings. *Obstet Gynecol* 1991; **77**: 726–8.

68 Granberg S, Wikland M. A comparison between ultrasound and gynecologic examination for detection of enlarged ovaries in a group of women at risk for ovarian carcinoma. *J Ultrasound Med* 1988; **7**: 59–64.

69 Campbell S, Goessens J, Goswamy R, Whitehead M. Real-time ultrasonography for determination of ovarian morphology and volume. *Lancet* 1982; **1**: 425–6.

70 Goswamy R, Campbell S, Whitehead M. Screening for ovarian cancer. *Clin Obstet Gynecol* 1983; **10**: 621–43.

71 Goswamy R, Campbell S, Royston J, Bhan V, Battersby R, Hall J, *et al.* Ovarian size in postmenopausal women. *Br J Obstet Gynaecol* 1988; **95**: 795–801.

72 Campbell S, Bhan V, Royston P, Whitehead M, Collins W. Transabdominal ultrasound screening for early ovarian cancer. *BMJ* 1989; **299**: 1363–7.

73 Campbell S, Royston P, Bhan V, Whitehead M, Collins W. Novel screening strategies for early ovarian cancer by transabdominal ultrasonography. *Br J Obstet Gynaecol* 1990; **97**: 304–11.

74 Timor-Tritsch I, Bar-Yam Y, Elgali S, Rottem S. The technique of transvaginal sonography with the use of a 6.5 MHz probe. *Am J Obstet Gynecol* 1988; **158**: 1019–24.

75 Sasson A, Timor-Tritsch I, Artner A, Westoff C, Warren W. Transvaginal sonographic characterization of ovarian disease: Evaluation of a new scoring system to predict ovarian malignancy. *Obstet Gynecol* 1991; **78**: 70–6.

76 Rodriguez M, Platt L, Medearis A, Lacarra M, Lobo R. The use of transvaginal sonography for evaluation of postmenopausal ovarian size and morphology. *Am J Obstet Gynecol* 1988; **159**: 810–14.

77 Rottem S, Levit N, Thaler I, Yoffe N, Bronshtein M, Manor D, *et al.* Classification of ovarian lesions by high frequency transvaginal sonography. *Journal of Clinical Ultrasound* 1990; **18**: 359–63.

78 Higgins R, van Nagell J, Donaldson E, Gallion H, Pavlik E, Endicott B, *et al.* Transvaginal sonography as a screening method for ovarian cancer. *Gynecol Oncol* 1989; **34**: 402–6.

79 van Nagell J, Higgins R, Donaldson E, Gallion H, Powell D, Pavlik E, *et al.* Transvaginal sonography as a screening method for ovarian cancer. *Cancer* 1990; **65**: 573–7.

80 van Nagell J, DePriest P, Puls L, Donaldson E, Gallion H, Pavlik E, *et al.* Ovarian cancer screening in asymptomatic postmenopausal women by transvaginal sonography. *Cancer* 1991; **68**: 458–62.

81 Bourne T, Whitehead M, Campbell S, Royston P, Bhan V, Collins W. Ultrasound screening for familial ovarian cancer. *Gynecol Oncol* 1991; **43**: 92–7.

82 DePriest P, van Nagell J, Gallion H, Shenson D, Hunter J, Andrews S, *et al.* Ovarian cancer screening in asymptomatic postmenopausal women. *Gynecol Oncol* 1993; **51**: 205–9.

83 Campbell S, Bourne T, Reynolds K, Hampson J, Royston P, Whitehead M, *et al.* Role of colour Doppler in an ultrasound-based screening programme. In: Sharp F, Mason W, Creasman W, eds, *Ovarian cancer 2*. London: Chapman and Hall Medical, 1992; 237–47.

84 Folkman J, Watson K, Ingber D, Hanahan D. Induction of angiogenesis during the transition from hyperplasia to neoplasia. *Nature* 1989; **339**: 58–61.

85 Kurjak A, Zalud I. Transvaginal colour Doppler in the differentiation between benign and malignant ovarian masses. In: Sharp F, Mason W, Creasman W, eds. *Ovarian cancer 2*. London: Chapman and Hall Medical, 1992; 249–64.

86 Hata T, Hata K, Senoh D, Makihara K, Aoki S, Takamiya O, *et al.* Doppler ultrasound assessment of tumor vascularity in gynecologic disorders. *J Ultrasound Med* 1989; **8**: 309–14.

87 Kurjak A, Zalud I, Jurkovic D, Alfirevic Z, Miljan M. Transvaginal color Doppler for the assessment of pelvic circulation. *Acta Obstet Gynecol Scand* 1989; **68**: 131–5.

88 Bourne T, Campbell S, Steer C, Whitehead M, Collins W. Transvaginal colour flow imaging: a possible new screening technique for ovarian cancer. *BMJ* 1989; **299**: 1367–70.

89 Kurjak A, Zalud I, Alfirevic Z. Evaluation of adnexal masses with transvaginal color ultrasound. *J Ultrasound Med* 1991; **10**: 295–7.
90 Weiner Z, Thaler I, Beck D, Rottem S, Deutsch M, Brandes J. Differentiating malignant from benign ovarian tumors with transvaginal color flow imaging. *Obstet Gynecol* 1992; **72**: 159–62.
91 Kawai M, Kano T, Kikkawa F, Maeda O, Oguchi H, Tomoda Y. Transvaginal Doppler ultrasound with color flow imaging in the diagnosis of ovarian cancer. *Obstet Gynecol* 1992; **79**: 163–7.
92 Bourne T, Campbell S, Reynolds K, Whitehead M, Hampson J, Royston P, *et al*. Screening for early familial ovarian cancer with transvaginal ultrasonography and colour blood flow imaging. *BMJ* 1993; **306**: 1025–9.
93 Andolf E. Ultrasound screening in women at risk for ovarian cancer. *Clin Obstet Gynecol* 1993; **36**: 423–32.

5 Screening for ovarian cancer: epidemiological aspects, study design, and target population

JOHN K GOHAGAN, PHILIP C PROROK,
BARNETT S KRAMER, JOYCE E CORNETT

Epidemiology

The incidence of ovarian cancer varies throughout the world. Data from five continents show three distinct patterns: low incidence in Asia, high incidence in western Europe and North America, and intermediate incidence in Central and South America (Table 5.1).[1 2]

Mortality rates also vary. Because most incident cases present clinically at a late stage with incurable disease, mortality rates closely parallel incidence rates. Northern Europe and the United Kingdom are high risk regions. In these populations age standardised death rates range from 7.5 to 9.9/100 000 women based on the world standard population.[3] Ovarian cancer is the fifth most common cause of cancer deaths among American women and is the leading cause of death from gynaecological malignancies. Mortality rates in America have remained fairly constant for decades, ranging from 8.4/100 000 in 1973 to 7.8/100 000 in 1990, adjusted to the 1970 United States standard population.[4]

Ozols' recent review reflected the essential elements that affect mortality outcomes.[5] Tumours that develop in the surface epithelium of the ovary readily penetrate the ovarian capsule and exfoliate malignant cells into the

Table 5.1 Incidence of ovarian cancer[1]

Regional grouping	Age adjusted incidence*
Western Europe and North America	11–14/100 000
Central and South America	5–8/100 000
Asia (China/Japan)	4–5/100 000

* Adjusted to the world population

peritoneal cavity where they colonise abdominal organs, particularly the right hemidiaphragm, omentum, bladder, and colon. Dissemination occurs also through the lymphatic and vascular systems.

Ovarian cancer spreads insidiously without symptoms until the function of major organs is affected. Common early symptoms of disseminated disease are vague and include nausea, constipation, and dyspepsia. Early detection of asymptomatic ovarian cancer is usually as a result of incidental findings of palpable adnexal masses or perhaps during hysterectomy. Investigations of adnexal masses may include abdominal or transvaginal ultrasonography, serum CA125 assay, and ultimately exploratory laparotomy or laparoscopy.

Screening for ovarian cancer

It is often assumed that earlier diagnosis of cancer automatically confers benefit and that any diagnostic test that identifies early stage disease must therefore be useful for screening. This is not necessarily the case. For screening to be effective, the screening test must detect cancers at a stage in their development when treatment can improve outcome more than treatment of symptoms. Demonstration of an improvement in the clinical or histological stage of the disease is essential but is not sufficient evidence of the benefit of screening.

Currently available tests such as CA125 assay and transvaginal ultrasound scanning have been shown to detect at least some ovarian cancers at a presymptomatic, clinically localised stage. Survival if cancers are detected early is better than if they are detected late, but survival that correlates with stage (early stage at detection implies longer survival after the date of detection) does not necessarily imply a benefit of early detection. The benefit of early detection can be conferred only in terms of extension of life—that is, in terms of a reduction in mortality.

At present there is no evidence that early detection by screening or by chance is effective in reducing mortality from ovarian cancer, as pointed out so eloquently by Granai.[6] This observation was strongly affirmed in the National Cancer Institute Consensus Development Conference on Ovarian Cancer, 5–7 April, 1994.

A randomised trial of ovarian screening

The National Cancer Institute is conducting a randomised controlled trial to measure the effects of screening on the reduction in mortality from ovarian cancer. While measuring reduction in mortality (extension of life) is the primary objective of the trial, secondary objectives include calculating the true sensitivity, specificity, and positive predictive value of each type of screening, and assessing the incidence, stage, and survival of patients with cancer.

Ovarian cancer is one of four cancers that are being screened for in the Prostate, Lung, Colorectal, and Ovarian Cancer Screening Trial. The design allows for 37 000 women and 37 000 men aged 60–74 at entry in each of the two arms of the trial. The regimens are given below. Men will be screened for prostate, lung, and colorectal cancer, and women for ovarian, lung, and colorectal cancer. Patients in the control arm will receive usual medical care. Men in the screened arm will undergo digital rectal examination; assay of the serum concentration of prostate-specific antigen, flexible sigmoidoscopy, and chest radiography. Women in the screened arm will undergo pelvic examination, assay of the serum concentration of CA125, transvaginal ultrasound, flexible sigmoidoscopy, and chest radiography.

The frequency and duration of screening will be the initial examination then annually for three years except for sigmoidoscopy when the initial examination will be repeated after three years.

The timing of major events in the trial is: years 1–2—development of the protocol and pilot studies; years 2–3—pilot review and if needed adjustments of the protocol; years 3–5—recruitment and initial screening of subjects in the main phase of the study; years 6–8—follow-up and completion of screening; years 9–15—further follow-up; and year 16—final follow-up and analysis of data.

Screening methods

The CA125 assay is made on a blood sample drawn in a few minutes by a technician. Transvaginal ultrasound images the ovaries from within the vagina with little discomfort in about 5–10 minutes and is also done by a technician.

In small preoperative studies of women with ovarian masses (50 to 150 patients), serum CA125 titres have been raised (typically above 35 U/ml) in from 68% to 100% of cases over all stages and in 40% to 50% of patients with stage I disease. The serum CA125 titre may be raised during pregnancy, endometriosis, menstruation, benign ovarian tumours, and with breast, colon, pancreatic, lung, gastric, and liver cancers. Jacobs and Bast reported that in screening 1010 postmenopausal women with both pelvic examination and CA125 assay, the only cancer diagnosed was detected by CA125 assay.[7] In a screening study in Sweden of 5550 women over 40 years old, nine cancers were detected, six (in women over 50 years of age) of the nine by CA125 radioimmunoassay.[8] The reported specificities in these two trials were 97% and 98.5%, respectively, for women aged 50 years or older with cut off points of 30 U/ml and 35 U/ml. These preoperative and prospective studies together suggest that CA125 may have the potential for some early detection. No studies have been conducted to measure sensitivity and specificity in a large screened population, and no randomised trials have been done to assess the impact of screening with CA125 titres on mortality from ovarian cancer.

Experience with transvaginal ultrasound, a newer higher resolution technique than transabdominal ultrasound, is more limited. In a series of 1017 cases 0.3% of unilocular ovarian tumours or cystic masses seen on ultrasound were malignant on histological examination compared with 8% of multilocular, and 39% of solid tumours.[9] van Nagell has been using transvaginal ultrasound for screening women over the age of 40 since 1987. He reported that using 8 cm^3 as the upper limit of normal ovarian volume, 31 abnormal ultrasound scans were obtained from 1000 women, 24 of whom underwent laparotomy (adenocarcinoma n = 1, serous cystadenoma n = 8, endometrioma n = 6, cystic teratoma n = 2, hydrosalpinx n = 3, follicle cyst n = 2, and pedunculated myoma n = 1; one case unspecified).[10] He later reported that of three cancers detected in 1550 screens, transvaginal ultrasound identified all three (Ovarian Cancer Working Conference 13 September 1990).

To some extent CA125 assay and transvaginal ultrasound may be complementary, but no prospective screening trials to investigate this question have been done. Pelvic examination has never been evaluated as a method of screening.

Organisational structure

Overall management and scientific direction of the Prostate, Lung, Colorectal, and Ovarian Cancer Screening Trial is by the National Cancer Institute, and screening is done at 10 competitively selected clinical centres under contract to the National Cancer Institute.

Screening centres follow all abnormal screening results to find out the incidence of prostate, lung, colorectal, and ovarian cancers. An annual survey of general health is the primary method of diagnosing cancers in the control group, interval cancers in the intervention group, cancers among participants in the intervention group who have withdrawn from screening, and of following the general health of all participants.

The coordinating centre developed and implemented a data entry system, oversees data management and quality control, and provides central coordination. A single laboratory conducts and reports all CA125 assays.

A biorepository has been established for the serial, prospective collection of whole blood, buffy coat, and serum samples from the screened subjects in the trial. This is to make possible studies to evaluate new early detection markers. It makes possible molecular epidemiological and aetiological risk assessment studies.

The steering committee of principal investigators from the screening centres, the coordinating centre, the laboratory, and the biorepository provides scientific input for the major operational aspects of the study, including recruitment, screening, data reporting, and publication.

The monitoring and advisory panel (a data safety and monitoring board) comprises extramural experts not associated with the trial who oversee all aspects. The panel will participate in the evaluation of the pilot phase of the trial, undertake general scientific overseeing and data monitoring as the trial progresses, and review any suggested changes in the protocol.

Pilot phase

Pilot studies are being conducted during the first two years of the trial. On the basis of these, decisions on the long term commitment to the trial are to be made. Key areas to be evaluated are: recruitment, randomisation, protocol for screening and following participants, screening logistics, quality assurance, contamination, compliance, biorepository processes, and data collection and management.

Main phase

On successful completion of the pilot phase, full scale recruitment will begin and continue to the end of the fifth year of the study. During the main phase assessments of cancer incidence and deaths, quality assurance, recruitment, follow-up, contamination, compliance, and information systems will continue. In addition, sensitivity, specificity, predictive value, prevalence, incidence, stage, histology, survival, interval compared with screening detection rates, lead-time, morbidity, and mortality will be investigated.

Calculations of sample size

The sample size for this trial was calculated by the method of Taylor and Fontana[11] modified to allow for arbitrary magnitude of screening impact, fixed sample size ratio between screened and control groups, and estimated degrees of compliance in the screened and control groups. Let N_c be the number of subjects randomised to the control group, and N_s the number randomised to the screened group, with $N_s = fN_c$. For $0 \leq r \leq 1$, assume that the study is designed to detect a $(1-r) \times 100\%$ reduction in the cumulative disease-specific death rate over the duration of the trial. Let P_c be the proportion of subjects in the control group who comply with the control group intervention (in this case usual medical care) and P_s be the proportion of subjects in the screened group who comply with the screened group intervention. The total number of disease-specific deaths needed for a one-sided α-level significance test with power $1-\beta$ is then:

$$D = \{ (Q_1 + f Q_2) Z_{1-a} - \sqrt{Q_1 Q_2} \ (1+f) Z_b \}^2 / f (Q_1 - Q_2)^2,$$

where $Q_1 = r + (1-r)P_c$

and

$Q_2 = 1 - (1-r)P_s.$

Table 5.2 Cancer mortality rates/person year ($\times 10^{-5}$), estimated from 1983–1987 data

Age (years)	Ovary	
	White women	Black women
50–54	14.2	10.4
55–59	20.3	15.3
60–64	27.5	23.3
65–69	35.3	27.4
70–74	41.5	33.8
75–79	45.2	34.5
80–84	49.8	41.1
85+	44.7	35.0

The number of participants required in the control group is $N_c = D/(Q_1 + fQ_2) R_c Y$, where Y is the duration of the trial from entry to end of follow-up in years, and R_c is the average annual disease-specific death rate in the control group expressed in deaths/person/year.

Because those recruited for screening trials are expected to be healthier than the general population (the healthy volunteer effect), we anticipate that the usual cancer mortality rate obtained from national or registry data will overestimate the mortality rate of the participants, at least during the early part of the trial. The sample size for the Prostate, Lung, Colorectal, and Ovarian Cancer Screening Trial is based on the rates of prostate cancer. For a 10 year screening trial for prostate cancer with men entered between the ages of 60 and 74, it is assumed that for the first two years the control group mortality rate will be a quarter the usual rate, for the next three years it will be half the usual rate, and for the last five years it will equal the usual rate. The usual mortality rate was estimated by the unweighted average mortality rate for prostate cancer for men aged 65 to 79. This age range was used to adjust for aging over the 10 years of the trial. The usual mortality rates were obtained from national data.[12]

The calculations were made assuming a 10 year trial using a one-sided 0.05 level test, $P_c = P_s = 1$, a target reduction in mortality of 20%, and a screened group equal in size to the control group ($f = 1$). This resulted in a sample size of 37 000 men. Equal numbers of men and women are included for diseases that affect both sexes—37 000 screened and 37 000 controls of each sex. These sample sizes are based on mortality rates among white people.

One can do similar calculations for ovarian cancer. Based on the death rates from ovarian cancer shown in table 5.2, one can arrive at the estimate of sample size given in table 5.3 for various combinations of reductions in mortality and power. For a sample size of 37 000 women in each arm, the power to detect various magnitudes of reduction in mortality for ovarian cancer is given in table 5.4.

If the reduction in mortality from screening for ovarian cancer was 35%, there would be almost 90% power with this design. There is qualitative

Table 5.3 Number of participants aged 60–74 at entry that would be needed in each arm of the trial in ovarian cancer

Power	Percentage reduction in mortality		
	20	30	35
0.9	134 697	56 069	39 733
0.8	97 365	40 606	28 817

Table 5.4 Power by percentage reduction in mortality for the 37 000 women in each arm of the trial of screening for carcinoma of the ovary

	Percentage reduction in mortality			
	20	25	30	35
Power	0.45	0.62	0.77	0.88

Table 5.5 Expected contamination and compliance ranges by screening method

	Compliance	Contamination
CA125	>90%	<10%
Ovarian palpation	>85%	<10%
Transvaginal ultrasound	>85%	<10%

evidence that the potential reduction in mortality from ovarian screening could be large, and British workers have estimated that a reduction of a third could be expected (Cuckle H. Unpublished observation). Alternatively, if the mortality effect was only a quarter, 84 000 screened women and an equal number of controls would be required to achieve 90% power. If it became apparent during the study that it was necessary and appropriate to do so, the female population base of this trial could be increased.

We recognise that compliance will not be perfect in either randomised group. The control group may become contaminated if people decide to be screened of their own volition ($P_c < 1$) and some of those randomised to screening may withdraw their consent during the trial ($P_s < 1$). Non-compliance includes those who refuse to participate when first contacted.

Inquiries about potential compliance and contamination at the four sites indicated that the ranges of probable values would be as shown in table 5.5. In addition to direct contact with health maintenance organisations and existing screening centres, published data from the 1987 National Health Interview Survey were used to gauge these effects.[13 14] These numbers are necessarily subjective. Accurate estimates are being obtained during the pilot phase of the trial. In the context of these levels of contamination and compliance, the estimated true effects on mortality required for the 88% power are no higher than 47%.

Conclusions

The Prostate, Lung, Colorectal, and Ovarian Cancer Screening Trial is designed to answer the pressing question of whether screening for ovarian cancer using CA125 assay and transvaginal ultrasound followed by appropriate treatment reduces mortality from ovarian cancer. It is the only study currently under way which can do so.

It may be necessary to supplement the population size for the ovarian arm of the Prostate, Lung, Colorectal, and Ovarian Cancer Screening Trial, but a decision will not be made until a thorough evaluation of the pilot results is completed. Any increase in the study population could be either direct or through collaboration with other countries.

Ovarian cancer is an ominous disease; the fact that it is asymptomatic until the late incurable stages is frightening. Practically nothing is known about the potential to reduce mortality by early detection and treatment. To quote Granai, "While embracing hope and searching for treatment, we must also, when appropriate, tell the emperor—whether the emperor is a patient, a colleague, a health care executive, a demanding institution, a government agency, a university, society, or ourselves—that at least for now, there are no clothes."[6] The ovarian arm of the Prostate, Lung, Colorectal, and Ovarian Cancer Screening Trial is essential at this juncture, for through it we will learn more about early detection by CA125 assay and transvaginal ultrasound than by any other effort in the forseeable future.

References

1 Muir C, Waterhouse J, Mack T, Powell J, Welan S, eds. *Cancer incidence in five continents.* Vol 5, IARC Pub. No. 88. Lyon, France: World Health Organization, International Agency for Research on Cancer and International Association of Cancer Registries, 1987.

2 Petitti DB, Porterfield D. Worldwide variations in the lifetime probability of reproductive cancer in women: implications of best-case, worse-case, and likely-case assumptions about the effect of oral contraceptive use. *Contraception* 1992; **45**: 93–104.

3 La Vecchia C, Levi F, Lucchini F, Negri E, Franceschi S. Descriptive epidemiology of ovarian cancer in Europe. *Gynecol Oncol* 1992; **46**: 208–15.

4 Miller BA, Ries LA, Hankey BF, Kosary CL, Harras A, Devesa SS, *et al*, eds. *SEER Cancer Statistics Review 1973–1990*; NHI Pub. No. 93; 2789. Bethesda: National Cancer Institute, 1993.

5 Ozols RF. Ovarian cancer, part II: treatment. Ozols RF, ed. *Current problems in cancer.* Philadelphia: Mosby Year Book, March/April 1992; 69–77.

6 Granai CP. Ovarian cancer—unrealistic expectations. *Sounding Board: N Eng J Med* 1992; **327**: 197–99.

7 Jacobs I, Bast RC Jr. The CA125 tumor associated antigen: a review of the literature. *Hum Reprod* 1989; **4**: 1–12.

8 Einhorn N, Sjovall K, Knapp R, Hall P, Scully R, Bast R, *et al*. Prospective evaluation of serum CA125 levels for early detection of ovarian cancer. *Obstet Gynecol* 1992; **80**: 14–8.

9 Granberg S, Wikland M, Janssen I. Macroscopic characterization of tumors and the relation to the histological diagnosis: Criteria to be used for ultrasound evaluation. *Gynecol Oncol* 1989; **35**: 139–44.

10 van Nagell J, Jr, Higgins R, Donaldson E, Gallion H, Powell D, Pavlik E, *et al*. Transvaginal sonography as a screening method for ovarian cancer. *Cancer* 1990; **65**: 573–7.

11 Taylor WF, Fontana RS. Biometric design of the Mayo lung project for early detection and localization of bronchogenic carcinoma. *Cancer* 1972; **30**: 1344–7.
12 1987 Annual Cancer Statistics Review, including cancer trends: 1950–1985. NIH Publication No 88; 2789.
13 Brown ML, Potosky, AL, Thompson GB, Kessler LG. The knowledge and use of screening tests for colorectal and prostate cancer: Data from the 1987 National Health Interview survey. *Preventive Medicine* 1990; **19**: 562–74.
14 Polednak AP. Knowledge of colorectal cancer and use of screening tests in persons 40–74 years of age. *Preventive Medicine* 1990; **19**: 213–26.

6 Prevention of ovarian cancer

HUGH R K BARBER

At present early diagnosis of ovarian cancer is a matter of chance rather than scientific achievement. There is no effective prophylaxis against it except removal of the entire infundibulopelvic ligament and the ovaries. If the ligament is not removed completely, supernumerary or accessory ovaries or ovarian remnants may be left behind. In addition, ovaries are covered with mesothelial tissue (as is the whole peritoneal cavity) and it is possible that whatever is stimulating the ovary to undergo malignant change may also be stimulating other parts of the peritoneal cavity. Woodruff stated that because the surface of the ovary is a mesothelial structure and the rest of the peritoneal cavity also consists of mesothelial tissue the presence of stage III ovarian cancer may indicate multiple primary tumours rather than metastatic spread.[1] It may be possible to identify the origin of such lesions with flow cytometry, DNA probes, and chromosomal studies. If the lesions are multiple primary tumours there is little chance that early diagnosis and prophylaxis will have an important role until such time as it is possible to draw up predictable regimens of treatment.

Familial ovarian cancer

For many years we have been aware that ovarian cancer occurs in families more often than can be accounted for by chance. From 1929–1970 there were about five reports of familial ovarian cancer,[2,3] and the number has increased considerably since then. The Familial Ovarian Cancer Registry was established in 1981 and continues to register families in which there are two or more first degree blood relatives with ovarian cancer.

Li *et al* reported on familial ovarian cancer in *JAMA* in 1970,[4] and since then there have been many papers. Recently Piver *et al* at the Roswell Park Memorial Institute in Buffalo, New York, established the Gilda Radner Ovarian Cancer Registry.[5] They reported that the age of diagnosis of the familial ovarian cancer syndrome is usually earlier than that of sporadic cases (mean age 47 compared with 53 years); the histological diagnosis is usually poorly differentiated serous adenocarcinoma; there is an autosomal dominant pattern of inheritance with variable penetrance; transmission is

68

in a vertical fashion from either parent; and the prognosis is uniformly poor with less than 10% of the women surviving for more than five years.

Lynch *et al* have also contributed a great deal to research on genetics and patients with cancer.[6] They described three different genetic and hereditary syndromes in patients with ovarian cancer: women at risk of developing only ovarian cancer; women at risk of developing breast and ovarian cancer; and the "cancer family syndrome" in which men are at risk of developing adenocarcinoma of the colon and women are at risk of developing carcinoma of the colon, ovary, and uterus. Because of the autosomal dominant nature of the inheritance of the familial syndrome, the risk for female first degree relatives of affected patients can be as high as 50%. A first degree relative is a mother, a sister, or a daughter; aunts and grandmothers are second degree blood relatives. When only one first degree blood relative has been treated for common epithelial cancer the risk factor is in the range of 5%–10%. Familial ovarian cancer comprises only about 5%–8% of all ovarian cancers.

The 50% risk relates to two first degree blood relatives having cancer, but if any of these patients has a cancer in the late thirties or early forties the recurrence of cancer in the first degree blood relatives in the premenopausal years increases the risk; at present, however, it is not known whether this risk is as high as 50%.

Genetic counselling about prophylactic oophorectomy should be given at a time when an affected woman has completed her family, but no later than the age of 35 years. This stage is critical to all women with familial ovarian cancer because the disease is more common in sister to sister and mother to daughter pedigrees. Sisters and daughters in families with a history of familial ovarian cancer have a 50% chance of developing the disease. This rate compares with a 1.4% chance in women with no such history. In view of these figures, genetic counselling should start when the patient is in her early twenties, and physical surveillance should begin when she is in her early thirties. This consists of pelvic and abdominal examinations, and assays of the tumour markers carcinoembryonic antigen, CA125, NB/70K, and LASA-P, every six months, together with pelvic or vaginal colour Doppler ultrasound scan, or both, every year.

The Ovarian Cancer Registry recommends that patients with a family history of ovarian cancer who are under 35 years old should have a pelvic examination, vaginal colour Doppler and ultrasound scans, and assays of tumour markers every six months. Patients 35 years old or more who have completed their families are advised to undergo prophylactic hysterectomy and oophorectomy.

The importance of careful histological examination of the specimen was illustrated by a case report.[7] A woman who had undergone prophylactic oophorectomy three years earlier presented with intra-abdominal carcinomatosis compatible with a primary ovarian cancer. On retrospective serial sectioning of the ovaries a small focus of primary ovarian cancer was found. It is important, therefore, that at the time of laparotomy the entire abdomen is thoroughly examined including the diaphragm, omentum, and retroperitoneal nodes. Washings of the abdomen and pelvis should be examined cytologically, and both ovaries should be examined histopathologically rather than just one or more sections of each ovary. It is also important that the infundibulopelvic ligament is clamped high up; ideally this would be at the level of the renal vessels on the left and at the level at which the renal vessels arise from the aorta and empty into the vena cava on the right. This will ensure removal of any supernumerary or embryonic remnants.

Although the purpose of prophylactic oophorectomy is to prevent the development of ovarian cancer in most cases, both patients and their physicians must be warned about the possibility of extraovarian papillary carcinoma. It has been reported that multiple metastatic lesions (involving the omentum and liver after prophylactic oophorectomy with a histopathological diagnosis of metastatic papillary serous adenocarcinoma) from a primary ovarian lesion have occurred. It is likely that this syndrome is rare, but it is important in the counselling of the patient about prophylactic oophorectomy to explain that total prevention of any cancer is difficult to predict.

Hormonal changes and ovarian cancer

Several studies have suggested that disordered endocrine function may have a role in the aetiology of ovarian cancer. Women with a history of few or no pregnancies seem to have a higher incidence; epidemiological studies have indicated that pregnancy protects against ovarian cancer by interrupting incessant ovulation. Analogous findings occur in animals and ovarian cancer is particularly uncommon in animals with oestrus cycles. Egg-laying hens (which ovulate frequently) have high rates of ovarian cancer whereas broiler hens (which ovulate less frequently) have considerably lower rates. "Ovulatory age", the total number of years that a woman has ovulated, is a powerful predictor of the risk of ovarian cancer. The mechanism of this carcinogenesis is not clear, but it has been suggested that epithelial inclusions may arise during the reparative phase after ovulation. At the time of ovulation the mesothelial structure is literally torn as the ovum extrudes. During the reparative phase there is mitotic proliferation at the site of ovulation and it is possible that any carcinogen, or viral particle, or other substance, may cause the proliferation to smoulder and continue.

Since 1977, when the first epidemiological study was published that reported that oral contraceptives had a protective effect against ovarian cancer, many studies have confirmed it. After oral contraceptives have been used for three or more years the reduction in risk is about 40%. Longer use is associated with even greater protection—that is, about a 50% reduction after four years and 60%–80% reduction after seven or more years. The benefit derived from using oral contraceptives is greatest for the women at higher risk (those with lower parity) and persists for many years after cessation of oral contraceptives. The cancer and steroid hormone study done by the Centers for Disease Control and the National Institute of Child Health and Human Development reported on the reduction of risk of ovarian cancer associated with oral contraceptives.[8] The paper included a summary of studies about oral contraceptives and ovarian cancer from 1977–87, all of which showed a decreased risk of ovarian cancer when oral contraceptives were used. Young women who have a family history of ovarian cancer might benefit from taking oral contraceptive agents for at least five years, but there is no supporting evidence that this will confer the same benefit that it does to those who do not have such a history. It is, however, reasonable to offer these women oral contraceptive treatment for at least five years in the hope that it will produce a similar protective effect as that seen in women with no family history.

Postmenopausal palpable ovary syndrome

The contrast in size between premenopausal and postmenopausal ovaries is striking. Whereas premenopausal ovaries are about $3.5 \times 2 \times 1.1$ cm in diameter, postmenopausal ovaries shrink and atrophy to about $1.5 \times 0.75 \times 0.5$ cm by three to five years after the cessation of menstruation. In addition, menopausal ovaries often lie against the side wall of the pelvis, so the combination of size and position make palpation of a normal menopausal ovary difficult.

Postmenopausal palpable ovary syndrome is a clearly established entity which was first described over two decades ago. What would be interpreted as a normal sized premenopausal ovary indicates an ovarian tumour in a woman who is three or more years postmenopausal and the nature of the tumour should be investigated.[9]

The advent of more sensitive ultrasound techniques has increased the reporting of small, clinically undetectable ovarian cysts. This causes consternation among those who expect a postmenopausal ovary to be a fibrous tissue structure. Although this is generally so, there are instances in which there is a persistence of small cysts that seem to be held in "suspended animation". The ultrasonographer might report such cysts as an unusual finding in women of this age group, which would require the gynaecologist to make a decision about management. It has been suggested that if the scan does not look suspicious and pelvic examination is normal the patient should be reviewed in two months or so with a further pelvic examination and scan. Evacuation of the bowel with an enema may aid pelvic examination. If the findings are unchanged (as they almost certainly will be) the patient does not require laparotomy; she should be seen again in another two or three months for a further pelvic examination and scan. If the findings are still unchanged further examinations are required at only six month intervals.

Prophylaxis at the time of hysterectomy

There are three papers that provide a baseline for evaluating patients with ovarian cancer who have had an exploratory laparotomy or hysterectomy in the past. In the first a total of 624 cases of ovarian cancer were studied.[10] Fifty four had arisen in patients who had previously had laparotomies for benign lesions but in whom the ovaries had been preserved; if oophorectomy had been done at that time, therefore, cancer would have been prevented in 9% of patients seen during the period September 1947–December 1960. In 35% of the patients pelvic laparotomy had been done within five years of the diagnosis of ovarian cancer. The five year cure among patients with cancer in retained ovaries was 18%. Among the 34 patients in this group who had undergone pelvic laparotomy when they were over 40 years old, in whom one or both of the ovaries had been left in situ, and in whom carcinoma had eventually developed, only two were subsequently cured. Among 21 patients who had undergone laparotomy with conservation of the ovaries when they were less than 40 years old, three survived five years after their operation for ovarian cancer. At the time that the paper was published the authors stated that the evidence might be interpreted as justification for doing prophylactic bilateral oophorectomies for all patients aged 40 years or over who were having pelvic laparotomies for benign lesions. They concluded that there was little justification for preserving any ovary at any age if there were atrophic changes, if cytological studies indicated appreciable oestrogen deficiency, or if there was evidence of arterial sclerosis when the abdomen was opened. They included the observation that ovaries should be excised at the time of laparotomy for carcinoma of the colon, particularly the descending colon.

The second article was a 20 year series from the Butterworth Hospital in Grand Rapids, Michigan, of 236 cases of ovarian cancer seen between 1949 and 1960.[11] Of these 86% were in patients 41 years old or more; 61% were stages III or IV and the overall five year survival was 16%. A total of 126 patients had had previous operations and 28 had had hysterectomies. The author suggested that a radical approach to pelvic operations in women over 35 years old could prevent 20% of all ovarian cancers.

The third study reported 755 women who were evaluated and treated between 1977 and 1990.[12] Ninety five of them (13%) had previously undergone hysterectomy with preservation of one or both ovaries, and 60 of the 95 were over the age of 40. The authors concluded that prophylactic oophorectomy in women undergoing hysterectomy at the age of 40 or over would have prevented 138/2632 cases of ovarian cancer (5.2%) in the combined published series. Applied nationally such an approach would be expected to prevent more than 1000 cases of ovarian cancer annually. The authors recommended routine prophylactic oophorectomy in all women undergoing hysterectomy over the age of 40. Such a policy would have prevented 60 cases of ovarian cancer treated at the University of Miami during the 14 year period of the study.

It is interesting that the first and third papers reported about the same salvage rate whereas the second, in which the age group at the time of operation was younger, found that 20% of all ovarian cancers would have been prevented if the ovaries had been removed. In conclusion the first and third papers recommended 40 as the age after which the ovaries should be removed during hysterectomy, and the other paper recommended 35. The point to emphasise is that in women undergoing hysterectomy at this age or over, removal of the ovaries with an adequate amount of the infundibulopelvic ligament could prevent an appreciable number of ovarian cancers. As about 800 000 hysterectomies are done each year in the United States, and most of these are for women over the age of 40, the impact of such a policy could be substantial.

Screening for ovarian cancer: potential effect on mortality

Data from the United States on the incidence of ovarian cancer and survival were used in one study to estimate the potential benefit in terms of mortality from screening tests with various sensitivities.[13] A test with 80% sensitivity could reduce mortality by 50% if all cases detected by screening experienced current stage I survival rates, and the benefit would be greatest among women aged 45 or over. For each cancer detected, however, there would be at least 50 false positive screening tests unless the specificity of the test was more than 98%. If the most optimistic assumptions are made about screening, and if they could be met, then universal periodic screening of women aged 45–75 would result in about 5000 additional patients surviving five years or longer each year. Uptake of existing screening tests is far less than universal, so it is expected that the impact of any screening programme would fall short of these projections.

Young women have better survival, both because they tend to be diagnosed at earlier stages and because they have better stage-specific survival rates than older women. Nearly all the benefit of screening would be among women screened and diagnosed over the age of 45. Carcinoma of the ovary is a relatively rare disease and, as survival is quite good below the age of 45, more than 40 000 screening tests would be required to yield a single additional survivor in this age group. From the age of 45–74 there are 27 million women with at least one ovary who would be candidates for screening and this number will increase as the first wave of "baby boomers" come into the postmenopausal years.

Among women aged 45–74, however, about 6000 tests would be required to gain an additional survivor if the test was 80% sensitive. If it were 70% sensitive then 6850 tests would be required. If the sensitivity of this hypothetical test increased to 90%, then about 5300 tests would be required. It should be noted that to screen all women aged 45–74 would cost about US$14 billion.

Currently it is inappropriate to offer tests such as CA125 assay and ultrasonography for screening because no benefits have been shown. Meanwhile, it seems reasonable to advise younger patients who are worried about ovarian cancer that they can reduce the risk of the disease by about 50% if they use oral contraceptives for at least five years. This route of prevention is available to nearly all young women and could make as great a contribution to the prevention of deaths from ovarian cancer as the best possible screening programme.

References

1 Woodruff JD. The pathogenesis of ovarian neoplasia. *Johns Hopkins Med J* 1979; **144**: 117–20.

2 Piver MS, Mettlin CJ, Tsukada Y, Nasca P, Greenwald P, McPhee ME. Familial ovarian cancer registry. *Obstet Gynecol* 1984; **64**: 195–9.

3 Piver MS. Newsletter: Familial Ovarian Cancer Registry. Buffalo: Roswell Park Memorial Institute, New York State Department of Health. 1989: 1–2.

4 Li FP, Rapoport AH, Fraumeni JR Jr, Jensen RD. Familial ovarian carcinoma. *JAMA* 1970; **214**: 1559–61.

5 Piver MS, Mettlin CJ, Tsukada Y, Nasca P, Greenwald P, McPhee ME. Familial ovarian cancer registry. *Obstet Gynecol* 1984; **64**: 195–9.

6 Lynch HT, Albano WA, Lynch JR, Lynch PM, Campbell A. Surveillance and management of patients at high genetic risk for ovarian carcinoma. *Obstet Gynecol* 1982; **59**: 589–96.

7 Chen KTK, Schooley JL, Flam MS. Peritoneal carcinomatosis after oophorectomy in familial ovarian cancer syndrome. *Obstet Gynecol* 1985; **66**: 93S–4S.

8 Centers for Disease Control. The reduction in risk of ovarian cancer associated with oral contraceptive use. *N Engl J Med* 1987; **316**: 650–5.

9 Barber HRK, Graber EA. The PMPO syndrome (postmenopausal palpable ovary syndrome). *Obstet Gynecol* 1971; **38**: 921–3.

10 Terz JJ, Barber HRK, Brunschwig A. Incidence of carcinoma in the retained ovary. *Am J Surg* 1967; **113**: 511–5.

11 Gibbs EK. Suggested prophylaxis for ovarian cancer. *Am J Obstet Gynecol* 1971; **111**: 756–65.

12 Sightler SE, Boike GM, Estape RE, Averette HE. Ovarian cancer in women with prior hysterectomy: a 14-year experience at the University of Miami. *Obstet Gynecol* 1991; **78**: 681–4.

13 Westhoff C, Randall MC. Ovarian Cancer Screening: Potential effect on mortality. *Am J Obstet Gynecol* 1991; **65**: 502–5.

7 Management of familial and hereditary ovarian cancer

THOMAS H BOURNE, HENRY T LYNCH

Genealogy is fashionable. This interest in our ancestry has led to concern about how this heritage may influence the health of subsequent generations. Furthermore, the stimulus provided by prodigious scientific advances in clinical and molecular genetics has aroused interest in the family history of cancer by physicians as well as the public. This interest will probably increase since the identification of the BRCA1 gene in the hereditary breast and ovarian cancer syndrome[1] and its ultimate cloning.[2]

The identification of familial susceptibility to cancer requires a family history of cancer at all anatomical sites.[3] Knowledge of the cardinal phenotypic features of cancer provides useful clues to the diagnosis of a hereditary cancer syndrome. These features include early age at onset, a higher than expected incidence of bilateral disease in paired organs, a higher than expected incidence of multiple primary cancers, with specific combinations such as breast and ovarian carcinoma,[1] or colorectal cancer, endometrial carcinoma, ovarian cancer, and several others in the Lynch syndrome II.[4 5] In addition, premonitory physical signs such as multiple adenomatous polyps of the colon[6] or multiple atypical naevi[7] may be present in certain families, all of which accord with Mendelian patterns of inheritance. Using these clinical clues together with meticulous attention to the history of cancer in first degree relatives of the proband and selected second degree relatives (both sets of grandparents, aunts, and uncles) will in most cases enable the diagnosis of hereditary cancer syndrome to be made should it exist in the family.

A genetic diagnosis will become more secure, particularly in the absence of premonitory signs of genetic susceptibility to cancer, after linkage analysis. This may be seen in early onset hereditary breast cancer and the hereditary breast and ovarian cancer syndrome in the search for evidence of BRCA1 at chromosome 17q.[1] Linkage studies have also proved to be useful in familial adenomatous polyposis where the APC gene is located at chromosome 5q,[6] and in the Lynch syndromes, where gene mutations have been located at 2p (MSH2), (PMS2),[8 9 10] and 3p (MLH1), (PMS1).[11 12 13]

When the family history has been gathered, and whenever possible, the diagnoses of cancers have been documented by medical and pathology records, its potential for cancer control should be made available to all members of the extended family. Unfortunately, experience indicates that clinicians may not compile a family history of cancer in sufficient detail to diagnose a hereditary cancer syndrome even if it exists in a particular family.[14] Furthermore they rarely act on knowledge of the heriditary cancer syndrome for the extended family.[15]

Advances in molecular biology have been so rapid that we are still learning how to use the scientific knowledge we have gained for practical programmes of cancer control that might benefit patients. The cloning of genes should help to clarify the natural history and genetics of these cancers and facilitate their control. At the same time that this newfound ability to identify asymptomatic carriers has evolved, however, these new advances have also raised complex issues of ethical, legal, and medical management for which there is no precedent. Our discussions about genetic counselling and management issues in ovarian cancer combined with molecular genetic knowledge are therefore preliminary.

Clearly progress in molecular biology will continue. It is therefore likely that the genes for most forms of hereditary cancer will be identified within the next decade or two. Many conditions now thought of as familial aggregations or geographical clustering will be shown to have an hereditary component. Some conditions thought to be hereditary may be seen to be chance aggregations or related to environmental exposure. When considering these concerns, it is imperative to recognise that heterogeneity will pervade in most forms of hereditary cancer. The ovary is unlikely to be an exception. Collectively, these concerns will pose a diagnostic challenge to the clinical geneticist, molecular biologist, and the managing physician.

Definition of terms

The term "sporadic cancer" refers to those cases, ovary included, in which there is no family history of carcinoma through two generations involving siblings, offspring, parents, aunts and uncles, and both sets of grandparents. "Familial cancer" includes those cases with a family history including one or more first or second degree relatives with cancer of that organ that does not fit the definition of hereditary cancer. Chance must always be considered and common environmental factors may also explain familial clustering. A "hereditary cancer" is one in which there is a positive family history of cancer of that organ and, sometimes, syndrome-related cancers, with a high incidence and a distribution in the pedigree that is consistent with a Mendelian inheritance of a highly penetrating cancer susceptibility factor. Other factors that support the classification of hereditary cancer include early age at onset, and a higher than expected incidence of synchronous and metachronous tumours.

These are operational definitions and are useful for descriptive purposes providing that one understands their limitations. For example, an apparently sporadic ovarian cancer could possibly represent a new germline mutation (and so be hereditary). Other potential sources of misclassification include: incorrect family history, incomplete gene penetration, incorrect paternity, or the recessive inheritance from two parents who are heterozygous carriers of a deleterious gene when the cancer phenotype is only expressed in the homozygous state. As more deleterious genes are mapped and cloned, these diagnostic problems will be minimised. Nevertheless, careful clinical observation will always be important, particularly with regard to the often-neglected family history.

In addition, when applying these operational definitions to ovarian cancer it is essential to consider once again the fact that heterogeneity of cancer will be extant in hereditary forms of ovarian cancer. Specifically, in the hereditary breast and ovarian cancer syndrome, breast cancer predominates but ovarian cancer is an integral component. Hence, an excess of breast cancer in certain families with the hereditary breast and ovarian cancer syndrome may be the initial clue to the risk of ovarian cancer. In the same way, colonic and endometrial cancers will be the most common malignant neoplastic lesions in the Lynch syndrome II variant of hereditary non-polyposis colorectal cancer, yet the risk of ovarian cancer will be found to be excessive in that hereditary setting. Clearly, therefore, these definitions will have serious limitations when dealing with hereditary forms of ovarian cancer.

Finally, even the molecular genetics of hereditary ovarian cancer may in certain circumstances be confusing. This is particularly true for BRCA1. For example, it is estimated that BRCA1 accounts for about 45% of early onset, hereditary, site specific breast cancer, but it is an aetiological factor in virtually all cases of the hereditary breast and ovarian cancer syndrome. However, the more recently identified BRCA2 gene (located at chromosome 13q12-13) may account for most of the remaining families with early onset hereditary breast cancer, but it may not be important for ovarian cancer.[16]

Miki *et al* have provided strong evidence that the 17q-linked BRCA1 gene influences susceptibility to breast and ovarian cancer in the hereditary breast and ovarian cancer syndrome.[2] This was identified by positional cloning. Predisposing mutations were detected in five of eight kindreds which were presumed to segregate BRCA1 susceptibility alleles. It was noted that "... the mutations include an 11-base pair deletion, a 1-base pair insertion, a stop codon, a missense substitution, and an inferred regulatory mutation. The BRCA1 gene is expressed in numerous tissues, including breast and ovary, and encodes a predicted protein of 1863 amino acids. This protein contains a zinc finger domain in its amino-terminal region, but is otherwise unrelated to previously described proteins."

The authors are right to say that this research will facilitate early diagnosis of susceptibility to breast and ovarian cancer and should provide further elucidation of the biology of breast and ovarian cancer. We hope that a blood test will be available within a year or two, but, it will be essential for genetic counselling to be available for those who undergo this DNA testing.

Wooster *et al* did a genomic linkage search on 15 high risk families prone to breast cancer that were unlinked to the BRCA1 locus.[16] They localised a second breast cancer susceptibility locus which they termed BRCA2, ". . . to a 6-centimorgan interval on chromosome 13q12-13." Their evidence suggests that BRCA2, while conferring a high risk for breast cancer, differs from its BRCA1 counterpart in that it does not seem to increase the risk of ovarian cancer appreciably. In addition, BRCA2 seems to predispose to male breast cancer in the subject kindreds.

Management concerns

For those dealing directly with patients the management of women with a family history of ovarian carcinoma is not easy. When we are asked direct questions by women about their risks of developing the disease, or about the efficacy of any interventions, all we can do is be honest and say that we do not really know the answers. This provides little reassurance for the mother of two daughters who has watched her sister suffer and nursed her mother through the terminal stages of the disease. As doctors we have a tendency to want to "do" something; unfortunately there are few hard data to support "doing" anything at the moment. We can only adopt what might be thought of as a reasonable management strategy until we have better information available on which to base our decisions. The management of these women will therefore largely relate to screening for presumed early disease, and prophylactic oophorectomy.

The issue of screening for ovarian cancer, in common with most other types of health screening, is a controversial subject. Such an attitude is entirely appropriate, as the uncritical acceptance of any untested change to conventional approaches to the management of a disease would be irresponsible. Nevertheless, if we examine the impact our current treatment strategies have made on the disease it is apparent that a change in our clinical approach is required. At present the outcome for women with ovarian cancer depends largely on the surgical stage at the time of presentation. In most cases there are few symptoms associated with the early steps of the disease, so to detect the cancer in women without symptoms some sort of screening test is required.

For women with a family history of the disease it seems reasonable to offer screening even though there is no proved benefit. If one has informed many of the families about their increased risk, it is then difficult in practice to offer them nothing even though there is not really a valid justification. The question is, however, do we have the techniques available to detect this cancer early? To date the most useful are tumour markers and ultrasonography.

Tumour markers

The first study about screening women with a family history of ovarian cancer by measuring serum concentrations of CA125 was published by

Table 7.1 Effect of threshold value for Serum CA125 on the outcome of screening variables for familial ovarian cancer using transvaginal ultrasonography

Serum CA125 threshold value $\geq$ (U/ml)	No (%) of women referred for ultrasonography		Detection rate		False positive rate		Positive predicted volume (%)
			No	%	No	%	
2	1503	(100)	7	100	55	4	12.7
10	1083	(72)	6	86	43	3	14.0
15	652	(43)	5	71	31	2	16.1
20	380	(25)	5	71	16	1	31.3
25	243	(16)	4	57	13	0	30.8
30	130	(9)	3	43	10	1	30.0
35	84	(6)	3	43	7	0.5	42.9

Bourne *et al.*[17] This study allowed a direct comparison between the use of ultrasound and tumour markers because samples of venous blood were taken from 1503 women at the first stage of the screening procedure for the retrospective analysis of tumour associated antigens. Measurements of serum CA125 as the first stage screening procedure (using the conventional upper limit of 30 U/ml) would have missed four of the seven cancers detected (57%). Furthermore, for stage I disease the overall detection rate would have been just 40% (2/5) (Table 7.1). We can assume, therefore, that measurement of serum CA125 is not a viable screening test for familial ovarian cancer in view of its limited sensitivity for early disease; the detection rate of 40% is simply not good enough. The use of multiple tumour markers, however, may improve the detection rate for early disease to about 80%.[18]

Ultrasound

The alternative approach to screening is to use transvaginal ultrasonography. In purely practical terms there are potential advantages that can be gained by placing an ultrasound transducer in the vagina (closer to the area of interest) rather than on the abdomen. Higher frequency ultrasound can be used, and images of greater resolution obtained. In the context of screening for ovarian cancer, transvaginal ultrasonography has the practical advantage that it does not require that the bladder shall be full. In addition, in our practice most postmenopausal women who have had both transvaginal and transabdominal ultrasonography prefer the vaginal route.

In a recent study of 442 women, 13 were operated on, and one stage I serous cystadenocarcinoma was found (Grade M, Grade K. Personal communication). This impressive odds ratio of 13 operations being done for women with positive test results to find one cancer is to some extent a prevalence effect because about 30% of this group of patients had a positive family history of ovarian cancer.

Data from our clinic dedicated to screening high risk families for ovarian cancer were published in 1991.[19] All 776 women were free of symptoms and had at least one first or second degree relative who developed the disease (677, 87%, and 98, 13%, respectively). The mean age of the study population was 51 (range 24–78 years), 52% were premenopausal, 36% postmenopausal, and 12% had undergone hysterectomy. Forty three women were referred for operation (6%) and 39 of these underwent laparotomy; 48% (19/39) had bilateral disease, and 15% of the abnormal ovaries showed more than one pathological condition. Fifty five masses were detected of which 23 were tumours and 32 tumour-like conditions (greater than the number of women operated on because of the presence of bilateral disease in some cases). Of the three primary ovarian cancers detected, all were International Federation of Gynecology and Obstetrics (FIGO) stage Ia, so the disease prevalence within the study population was 0.39%. Forty women were operated on for benign disease, and so the false positive rate was 40/773, or 5%. The predictive value of a positive screen result was 8%.

The principle findings from this prospective study were:

- The prevalence of primary ovarian cancer was 3.9%, which is significantly higher than in the general population (0.4%).
- Within the limitations of the study design the detection rate of the screening procedure was 100%; this was based on the knowledge that no cases of ovarian carcinoma were reported in the study population within a year of a negative screening result.
- The overall false positive test rate was 5%.
- The prevalence of bilateral masses and mixed diseases within individual gonads were significantly higher than in the general population. This had been predicted by other workers,[20] but never shown before in a prospective screening study.
- The odds of a woman with a positive screening result having a primary ovarian cancer are considerably increased because of the disease prevalence. The odds of 12:1 against finding a primary ovarian cancer at operation is a considerable improvement in overall performance compared with the odds of over 50:1 previously reported in a study of the general population.[21] Some of the important outcomes of the study are shown in table 7.2.

The results of the study showed a tenfold increased risk of developing ovarian cancer in women who had a close relative that developed the disease. This is a highly significant increased risk and is consistent with a lifetime risk of developing the disease of 1:7. Furthermore the novel finding that there is also an increase in the prevalence of benign disease in these women suggests that some benign tumours have malignant potential. In terms of the development of a screening test, the apparently high detection rate for early cancer with ultrasound was encouraging. It might be argued, however, that ways of reducing the false positive rate further are needed,

Table 7.2 Screening for early ovarian cancer: comparison of results from two populations. Results are expressed as percentages, unless otherwise stated

Variable	Population	
	Community[21] (n = 5479)	Cancer families[19] (n = 776)
Outcome of screening		
Negative	94.0	82.0
Initially positive, finally negative	2.5	8.2
Positive throughout	3.6	5.5
Ovarian abnormalities:		
Bilateral masses	11.0	48.0
Mixed lesions	3.0	15.0
Tumour-like conditions	0.6	2.2
Benign epithelial tumours	0.7	2.6
Primary cancer (1000)	0.4	3.9*
Mean age at which cancer detected (years)		
Range	56 (46–61)	45 (30–63)

* 3.8 after 1601 screenings

as are ways of more accurately calculating the risk of ovarian cancer in more clearly defined types of family pedigree.

This begs the question as to which lesions need to be removed and which can be safely left *in situ*?

Some of the data about high resolution transvaginal ultrasonography have suggested that a degree of tissue characterisation may be possible with this technique. Though there must be exceptions, the consensus is that an ovarian mass less than 5.0 cm in diameter which is classified as a simple unilocular cyst on vaginal ultrasonography is highly unlikely to be malignant.[22]

In an attempt to make the assessment of such lesions less subjective, scoring systems have been evaluated. These give low values to simple unilocular cysts, and solid or cystic lesions with irregular margins are given higher scores depending on their degree of complexity.

The growth and progression of tumours is dependent on the process of neovascularisation and angiogenesis. This is a biological event common to many aspects of reproductive biology, and the understanding of its role in normal physiological events may well lead to a greater understanding of the biology of early cancer. Changes in tissue vascularity that are mediated by angiogenic factors are associated with the early stages of ovarian oncogenesis. Recent studies in transgenic mice have shown that for at least one type of cancer angiogenesis occurs during the transition from hyperplasia to neoplasia.[23]

It has been shown that ovarian carcinoma, even stage Ia, may be associated with areas of presumed angiogenesis that are detectable with colour Doppler.[24] In this study angiogenesis was shown in five late stages and two stage I ovarian carcinomas.

81

Table 7.3 Estimated lifetime risks of death from cancer in first degree female relatives of index patients with ovarian cancer.

Pedigree type	Cancer	Age of index patient (years)		All
		<55	55 and over	
No clear inheritance pattern	Ovarian		About 1 in 36	
Sure specific ovarian cancer	Ovarian		About 1 in 2	
Multiple site cancer family syndrome	Ovarian	1 in 5	1 in 16	1 in 10
	Mammary	1 in 5	1 in 10	1 in 7
	Uterine	1 in 144	1 in 195	1 in 173
	Colorectal	—	1 in 11	1 in 16
	Gastric	1 in 17	1 in 20	1 in 19
All	Mammary	1 in 9	—	—
	Ovarian	1 in 8	1 in 16	1 in 13

Assessing risk of cancer from the family history

The false positive rate can be reduced by improving the overall performance of a test. If the test is directed towards women at high risk of developing the disease in question, the positive predictive value of a positive screening result will be increased. The chances of finding a cancer among those women sent for operation will be significantly increased without the need for a second stage test.

Analysis of 391 pedigrees of women with ovarian cancer showed that the risks of a first degree female relative developing ovarian, stomach, breast, or endometrial cancer are 4.5, 1.4, 1.3 and 1.1, respectively.[25] The risks of developing cancer were lower if the index patient was less than 55 years old. Eighty two pedigrees were compatible with a diagnosis of mixed site cancer family syndrome and the relative risks in these families were 6.1, 2.8, 3.7, and 2.7 for ovarian, mammary, gastric, and colorectal cancer, respectively. Nineteen families showed evidence of being site specific ovarian cancer families for whom the relative risk of developing ovarian cancer was 39 and the lifetime risk 1:2 (table 7.3).

Where there was no obvious inheritance pattern the increased risk of ovarian cancer was minimal, whereas for site specific ovarian cancer pedigrees the lifetime risk was 1:2 irrespective of the age of the index patient. Those women in mixed site cancer families had a significantly increased risk of all cancers, which was greatest when the index case was less than 55 years old. At worst women in such families had a 1:5 lifetime risk of developing both breast and ovarian cancer, and a 1:11 risk of colorectal cancer. The overall lifetime risk of developing ovarian cancer irrespective of age or pedigree was 1:13, which is remarkably similar to the figure of 1:10 that we found in our prevalence screening study.

Until the widespread availability of genetic tests to calculate the risk of developing ovarian cancer, the information in this study is useful as it can be used to counsel women attending screening clinics and so plan appropriate

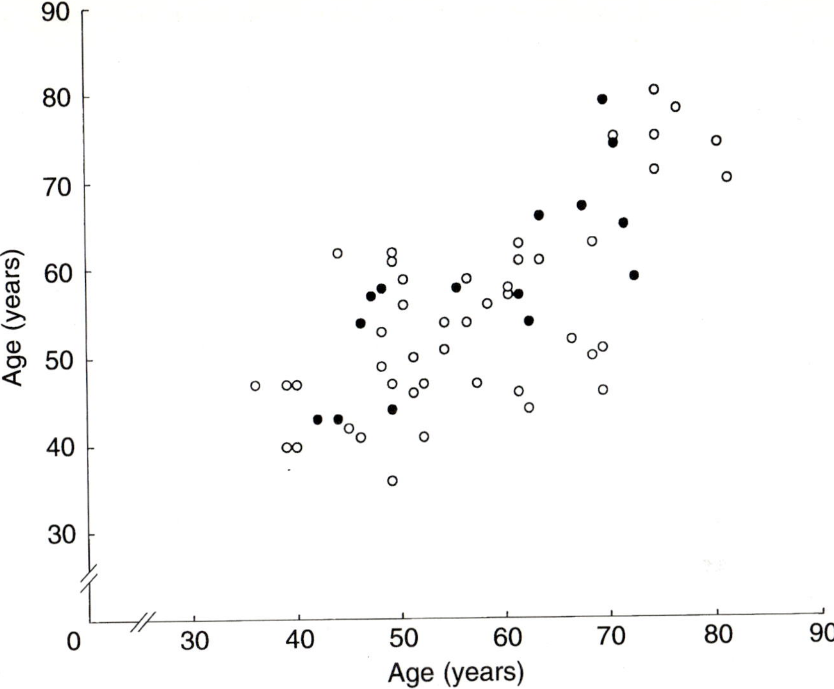

Fig. 7.1 The correlation between sisters' age of death from ovarian cancer.

management. It is important that where there was no clear inheritance pattern the relative risk of developing ovarian cancer was not significantly increased. It therefore seems inappropriate for women from such families to be offered either screening or—in particular—prophylactic oophorectomy. Within the site specific ovarian cancer families we were able to derive other useful information. Figure 7.1 shows the correlation between sisters' age of death from ovarian cancer.[26] The strong relationship between the dates of death in these high risk families is striking.

There is a good argument that in addition to screening for ovarian cancer in their earlier years, sisters of index cases in site specific ovarian cancer families should be offered prophylactic oophorectomy before the development of cancer in their sisters. An arbitrary working time interval of five years may provide a reasonably safe margin of error. It was also apparent from all the data that the age at which the index patients of a family developed cancer has a strong influence on the likelihood of first degree relatives developing cancer.[26] Figure 7.2 shows the relationship between the age of first degree relatives, the age of index patients at the time of diagnosis, and the risk of developing ovarian cancer. The younger the age of diagnosis, the greater the chance of a first degree relative developing ovarian cancer. Interestingly the age of first degree relatives is also relevant, as they seem to "outgrow" their risk. The risk does, however, remain relatively static in the age range 25 to 55 which are the years when

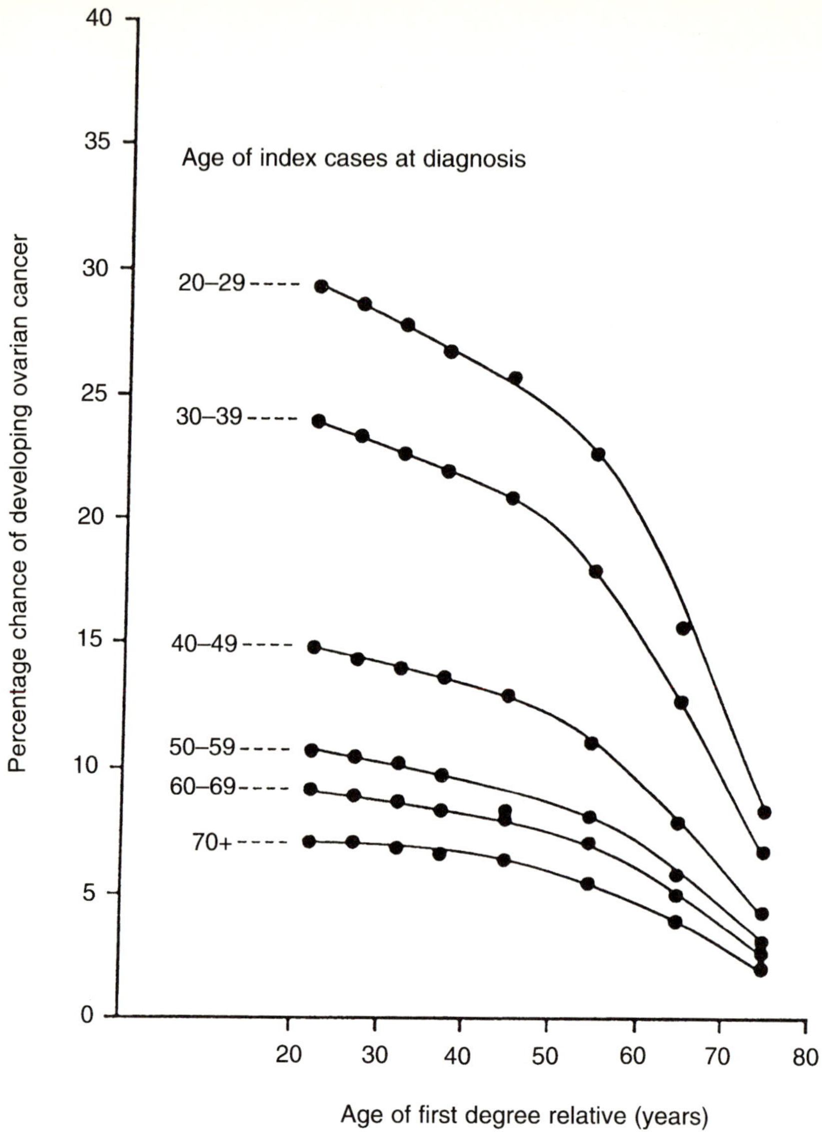

Fig. 7.2 The relation between the chances of developing ovarian cancer and both the age at diagnosis of index patients and the age of the first degree relatives at risk of the disease.

most years of life will be lost in the event of an ovarian cancer developing.

These data indicate that it may be reasonable to cease screening for ovarian cancer at the age of 70. The risk of cancer seems to have diminished significantly by that age, and the number of years of life saved by finding an early cancer will be small. Mixed site cancer may need a comprehensive screening programme involving mammography, colonoscopy, and ovarian ultrasonography. Offering prophylactic oophorectomy in such cases may be reasonable, but in younger women it is complicated by concerns over the use of hormone replacement therapy in women with a genetic predisposition to breast cancer. The prevalence of cancer is so much higher in site specific ovarian cancer and mixed site cancer families, that if a woman from such a family has a persistent ovarian mass, the chances of it being malignant will be significantly greater. Tests designed to discriminate between benign and malignant lesions in such cases may be unnecessary.

Genetic counselling

It will now be possible to use this information for highly definitive genetic risk assessment, particularly when a simple blood test for these germ line mutations becomes available. We can then tell high risk patients whether or not they have inherited a deleterious cancer-prone susceptibility gene.

We will then need to develop genetic counselling and management guidelines so that we use the results of blood testing (DNA) of these mutations accurately. This will enable the development of research projects which are likely to have a substantial impact on genetic counselling. For example, families who have these mutations can then be compared with those who do not. Questions about whether environmental exposure makes a difference in cancer expression might then be answered with greater precision. Genetic counsellors may ultimately be able to provide information about the avoidance of certain environmental factors that could give rise to cancer at specific anatomic sites.

Genetic counselling for any hereditary disorder must be based on highly accurate information about genetic risk. Patients must be fully informed about their genetic risks, but care must be taken to avoid alarming them unduly. It is imperative that patients be thoroughly educated about the natural history of the cancer phenotype so that they can appreciate why the recommended surveillance measures are so rigorous. Without this knowledge their compliance may be compromised. The success of genetic counselling also implies that the patients' general practitioners will be equally knowledgeable, compassionate, and understanding about the cancer risk and natural history of these disorders. Unfortunately, we have counselled many patients at risk of hereditary non-polyposis colorectal cancer and hereditary breast and ovarian cancer only to have their general practitioners tell them that they were too young to enter into a colonoscopy or ovarian transvaginal ultrasound screening programme and, furthermore, that cancer is predominantly an environmental disease, therefore "not to worry."

Education and genetic counselling

Education of patients from families at risk can be provided individually or in groups of as many as 15 or 20 family members. We have conducted group educational family counselling sessions followed by individual one-on-one genetic counselling for patients with hereditary breast and ovarian cancer.[28] We described the natural history of these disorders in detail and provided an accurate genetic risk for these multiple family members. We encouraged questions to evaluate the degree of the patients' understanding to reinforce the impact of the educational messages. This process often allows individual family members to become more relaxed, to discuss matters among themselves in the group session, and to resolve inhibitions about asking pertinent questions. Family members often find that many of the questions asked by their relatives reflect their own concerns. In turn, given the limitations of the physician's or genetic counsellor's time, coupled with the economy involved in seeing many family members in the same setting, it seems prudent to pay more attention to education and counselling in a family group. With respect to DNA-directed genetic counselling, however, the educational session can be provided to a group, but the counselling itself is better given privately face to face with the patient.

Our educational brochures include drawings of hypothetical pedigrees so that patients can appreciate the way in which the cancer is transmitted through an hereditary non-polyposis colorectal cancer or hereditary breast and ovarian cancer family. We also provide them with an accurate pedigree of their own family showing their position in the pedigree and indicating the significance of the risk. This "take-home" material can be referred to from time to time by the patients and is also made available to their general practitioners whom we encourage to refer to it during the patients' visits. In addition, we provide the doctor with a detailed letter explaining the patient's genetic risk of cancer, the type of information that we provided to the patient about the natural history of cancer in the family, and our recommendations for surveillance and management. We believe that this is essential so there will be no surprises when the patient visits the doctor and discusses openly the concerns about aspects of the natural history, genetics, and management of hereditary breast and ovarian cancer or hereditary non-polyposis colorectal cancer.

Individual counselling

Individual counselling must include the counsellor listening to the patient. The anxieties and concerns of the patient must be listened to, interpreted, and answered. There must be ample time allowed for the patient to ask questions and not feel the pressure of time. One or more additional sessions may be required before the patient fully understands and accepts the implications of the risk of cancer and the need to enter into the appropriate surveillance programme.

Results of counselling

We found that those women who were told they were linkage positive were more likely to want surveillance and prophylactic oophorectomy. With respect to prophylactic oophorectomy they were advised that even with ovaries that were normal at operation, based on careful histopathological sectioning, they still had about 5% lifetime risk for the development of peritoneal cystadenocarcinoma of ovarian origin. Most of the women who were told they were linkage negative indicated that they would not proceed with prophylactic surgery, but would continue careful surveillance. To date, there has been no evidence of serious emotional disturbance resulting from this disclosure. We think that this experience can be used by cancer geneticists and physicians in developing protocols for genetic counselling in cancer-associated hereditary disorders of all types. We concluded that physicians must understand current developments in cancer genetics and molecular genetics so that they can apply this knowledge effectively to genetic counselling as well as to the management of high risk patients.

Example of a case

Figure 7.3 contains haplotype information on "family 1816", which is a kindred prone to the hereditary breast-ovarian cancer syndrome and strongly linked (LOD score 4.2) to the BRCA1 gene. The vertical rectangle indicates the presence of the BRCA1 gene in their patient. Note, for example, patient IV-1 who has breast cancer and is positive for the BRCA1 gene. Her risk of ovarian cancer is extremely high (about 60%), and she would therefore be a candidate for prophylactic oophorectomy. We advised her, and all women in hereditary breast and ovarian cancer families who carry BRCA1, to consider having their children early and then consider prophylactic oophorectomy at the age of about 35. They are also advised of the option for the surveillance strategies that we have discussed, and are told of their limitations. Finally, these women who carry BRCA1 are told of the possibility (perhaps about 5%) of their developing peritoneal cystadenocarcinoma of ovarian origin even though their ovaries at the time of prophylactic oophorectomy may have been histologically normal.

Patient IV-469, a man with no symptoms of cancer and a first cousin of patient IV-1, also has the deleterious BRCA1 gene. He requires counselling so that he is aware of the genetic risk that could be conferred on any of his future children.

Patient IV-15, the sister of patient IV-1, has not inherited BRCA1. She can therefore be told that she is not at risk of the hereditary breast-ovarian cancer syndrome. However, because she is a woman, she must be told that her risk of breast cancer is in accord with population estimates which would be in the range of 11%–12% to the age of 85. We would therefore start her on a surveillance programme in accordance with the American Cancer Society recommendations, which would include a baseline management mammogram at the age of 40 which would be repeated every other year until the age of 50 and then annually thereafter. The patient should also be aware of her need for self examination of the breasts and she should see a physician annually for a breast examination.

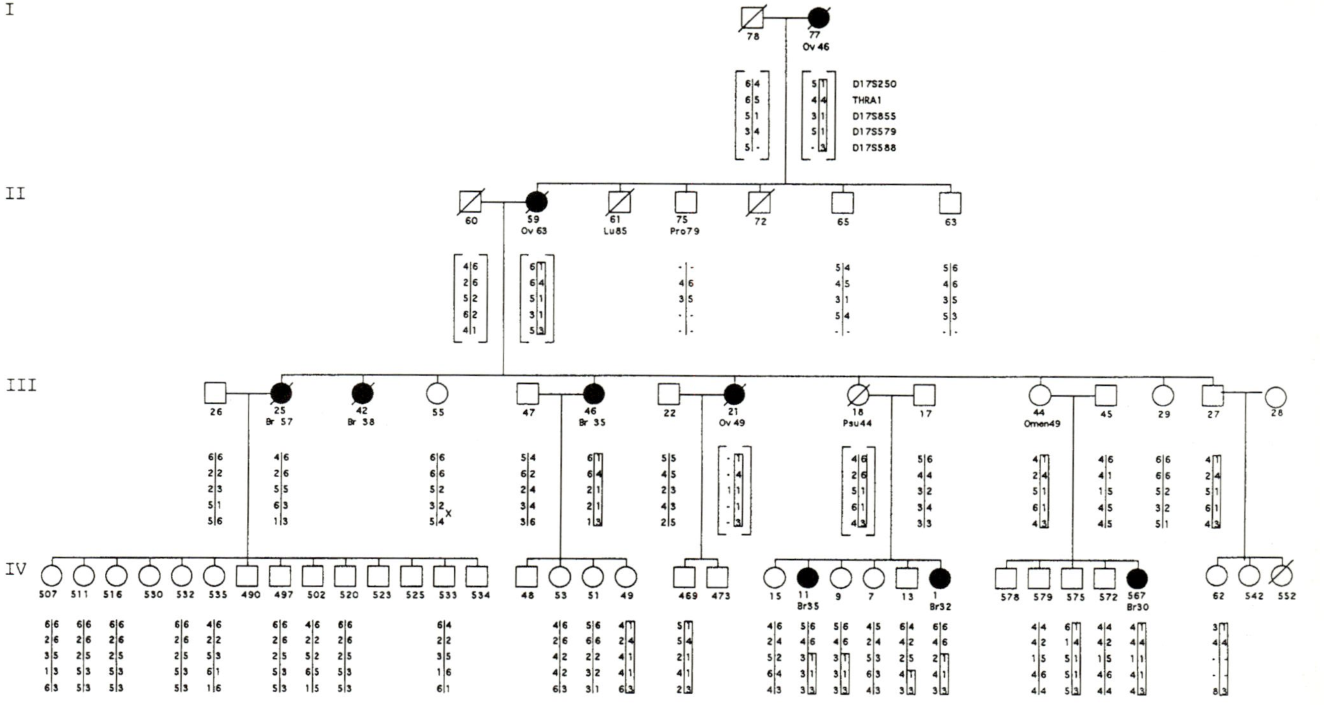

Fig. 7.3 Pedigree of family 1816. Five chromosome 17 markers have been used in this analysis. The haplotype enclosed in the vertical rectangle is associated with the BRCA1 gene mutation. Haplotypes in parentheses are inferred from spouses and children. Ov = ovarian cancer, Br = breast cancer, Omen = omental cancer, PSU = cancer, primary site unknown. Ages of diagnosis are given for affected women. Case 25 is a sporadic case of breast cancer. A crossover in case 18 places the BRCA1 gene below the THRA1 marker.

How screening tests perform when directed towards women with a family history of ovarian cancer: The King's College Hospital study

At King's College Hospital, in an attempt to draw together these different aspects of screening and pedigree analysis transvaginal ultrasound was used as a first stage screening test for ovarian cancer, and morphology assessment and colour Doppler were used to reduce the false positive rate.[17] The data were then assessed according to family pedigree to find out how this might alter the odds of finding cancer at operation in women with a positive screening result.

Two important groups of ultrasound markers of malignancy were introduced into the screening algorithm: firstly, a morphological assessment including the presence or absence of septae, and solid areas or irregularities in the cyst wall (giving a high morphology score); and secondly changes to the vascular pattern within a tumour showing low impedance to relatively high velocity blood flow.

The first 1000 women were screened according to the original algorithm[19] and the remainder according to the modified procedure, but the data on morphology and blood flow were stored for retrospective analysis. In the last 600 this information was incorporated and used for management purposes. The first 800 women were scanned both transabdominally and transvaginally. Interestingly the detection of ovarian lesions was the same using both approaches, but transvaginal ultrasonography proved to be more practical and more acceptable to patients as well as giving more detailed information about cystic lesions when one was present. Accordingly for subsequent scans only the transvaginal route was used.

The principal findings were as follows:

- Eight women (six who were premenopausal, one who had a hysterectomy, and one who was postmenopausal) had ovarian cancer (stage I (n = 6), stage II (n = 1), and stage III (n = 1)).
- Five of the eight were classified as serous cystadenocarcinoma (two borderline) and three as endometrioid (one borderline). Five occurred in the highest risk groups according to pedigree (mixed site cancer or site specific ovarian cancer).
- The detection rate was 89% (8/9) with a follow up of 12 months and 24 months.
- The false positive rate was 4.9% (54/1100) using transvaginal ultrasonography alone without any second stage tests, and 0.9% (15/1592) using a positive screen result based on an abnormal morphological score or Doppler indices.
- The odds of finding cancer at operation among women with a positive screen result were 1:13 after transvaginal ultrasonography alone and 1:2 after the application of the morphological score or Doppler indices (1:1 in the highest risk groups because of the prevalence effect). The newly

derived screening algorithm incorporating the new ultrasound endpoints worked well prospectively (the odds of a woman with a positive screening result having histological evidence of ovarian cancer were 3:5).

The false positive test results arose mainly from endometriosis, serous cystadenomas, and benign teratomas. Nine cancers have been reported at follow up to this study (serous cystadenocarcinomas all stage III (n=5); endometrioid, stage II (n=1); and peritoneal (n=3)). One ovarian cancer presented clinically nine months after the last scan, the remainder after at least 24 months had elapsed. The peritoneal cancers were detected 6–8 months after the last negative screen result.

Most true positive test results (primary ovarian cancer) occurred in premenopausal women. This is consistent with the view that in cancer families the cancers tend to develop at an earlier age than is usual. It also means that familial ovarian cancer screening programmes must include such women in their studies and not exclusively concentrate on postmenopausal women. Furthermore, by selecting this younger age group a greater number of years of life might be saved if an early stage tumour is detected. The fact that there was a similar false positive rate for both premenopausal women (28/954, 2.9%) and postmenopausal women (13/469, 2.8%) is perhaps surprising. This may in part be the result of careful adherence to a protocol that included repeat scans at an interval if any ovarian morphological abnormality was found. For women who had undergone hysterectomy the false positive rate of 8% (14/172) suggests that such women have an increased predisposition to ovarian disease. This further supports the view that when a hysterectomy is done for a woman with a family history of ovarian cancer, serious thought must be given to bilateral oophorectomy as a means of trying to reduce the chances of developing ovarian diseases in the future.

There seems little need for a second stage test when faced with a persistent ovarian mass in a woman with a family history suggestive of either mixed site cancer or site specific ovarian cancer, as the odds of finding cancer at operation are between 1:6 and 1:4 irrespective of any morphological or colour Doppler information. For the groups of women in whom screening may be most justified, morphological scores and colour Doppler seem least applicable (Fig. 7.4). Such techniques would, however, be necessary to improve test performance in families with no clear inheritance pattern or in other lower risk groups.[29]

Psychological impact of screening

A proportion of the women on the King's College Hospital ovarian screening study were interviewed by a clinical psychologist.[30 31] It is interesting that those whose style of coping was to ask for more information were most adversely affected by a positive scan result. This may reflect the way in which these women were dealt with by clinic staff. It may be

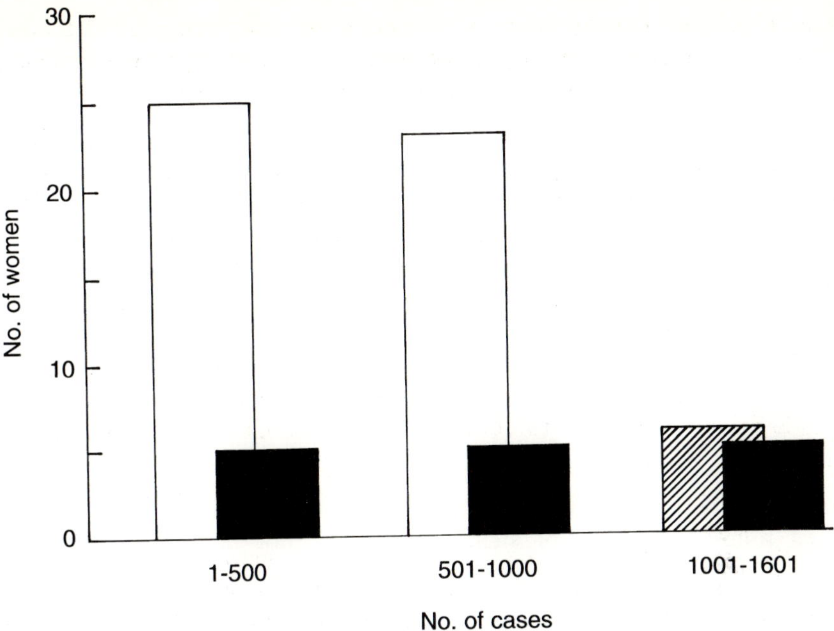

Fig. 7.4 The number of false positive screening results by number of cases. Open bars indicate at first stage, hatched bars after use of a high morphological score (≥ 5) or a low pulsatility index prospectively, and solid bars retrospectively.

beneficial to be able to select such women on entry into the programme for more intensive counselling. The result of the initial ultrasound had a clear impact on wellbeing. Women with negative scans experienced considerable relief and reassurance. Furthermore even when they were operated on for a benign lesion, the women associated this with a reduction in their risk of developing cancer. This is a concern to those of us involved in screening trials. Despite being told of the uncertain value of screening, these women were still reassured by various aspects of the screening procedure, and this persisted with time. This, however, may have been a reaction to the absence of a positive finding rather than feeling that a negative result meant that there could not be a cancer. Some of the women, however, showed an extreme commitment to the screening clinic, and travelled long distances to attend. They did this in the belief that if we could detect their ovarian cancer early then we would be in a position to cure them. Women who had bilateral oophorectomies were often happy, because having had both ovaries removed they felt certain that they had eradicated any risk of developing this form of cancer.

In the absence of a randomised trial of either ovarian cancer screening or oophorectomy such reassurance is not justified. All women participating in the study were made aware of the research nature of the project, but perhaps it is not surprising that some women still failed to grasp the concept of still being able to develop "ovarian" cancer in the absence of ovaries.

This highlights the problem of clinical trials where the line between "research" and "service" becomes blurred.

Implications for the relatives of women attending for cancer screening

Once the diagnosis of hereditary non-polyposis colorectal cancer or hereditary breast and ovarian cancer syndrome has been established, it should become obligatory that recommendations about cancer surveillance and management within the context of genetic counselling be extended to all available first degree relatives. Unfortunately, this is not common practice. For example, Arvanitis, et al[15] have shown that even in an obvious hereditary precancerous disorder such as familial adenomatous polyposis, where affected patients can easily be identified by proctosigmoidoscopy and colorectal cancer could be prevented by prophylactic colectomy, it has been found that 59% of such patients are not identified early and consequently die of metastatic colorectal cancer. Clearly, this means that the message about the risk of colorectal cancer among members of families with familial adenomatous polyposis is simply not being adequately put across. In hereditary non-polyposis colorectal cancer and hereditary breast and ovarian cancer in which reliable premonitory physical signs of cancer risk are lacking, the problem will admittedly be more difficult to deal with. Nevertheless, the solution of this problem should become more realistic once a simple blood test for hMSH2, hMLH1, and BRCA1 becomes available.

Much of the problem of neglect in providing high risk family members with genetic counselling, and thereby cancer control practices such as prophylactic colectomy in familial adenomatous polyposis or prophylactic oophorectomy in hereditary breast and ovarian cancer, relates to shortcomings in the medical education of physicians. Typically, physicians are concerned primarily with the medical problems of individual patients who are under their care, so there is a failure to focus attention on the disease risk in their patient's immediate relatives, and consequently, appropriate cancer control measures for them are not initiated.

Another problem is a presumed ethical concern about contacting high risk collateral relatives and how such an effort might be construed by the medical community. For example, in some medical circles this type of solicitation is disconcerting and may unjustifiably be construed as tantamount to attempts at economic gain on the part of the physician.

Other barriers to effective genetic counselling and its cancer control benefit involve patients' fear and denial which, in turn, may have a negative impact on compliance. There may also be socioeconomic and educational problems. Health insurance carriers may not pay for screening procedures and high risk patients might simply forego surveillance. Patients, based on past experiences of their relatives, may fear that if the insurance company is aware of the increased risk of cancer, they may cancel their policies or future policies may be refused. Some patients fear that insurance carriers' knowledge of their own cancer risk will translate into similar problems for their siblings and children and so they may prefer to keep this information from the insurance company and in turn not request coverage for surveillance.

Conclusions and recommendations

At present we have no data from randomised studies that screening for ovarian cancer is of practical value, but the increasing evidence about hereditary ovarian neoplasms has identified a group of women at high enough risk of the disease to warrant screening for their families and appropriate treatment.

Women with a family history of ovarian cancer are at increased risk of the disease. In the absence of an effective biological marker, a family pedigree should be taken to define risk more accurately. If there is one first degree relative only the risk may not be great, particularly if the index case was more than 55 years old. In site specific or mixed site cancer families the risk is high, and merits either screening or prophylactic oophorectomy.

Such high risk women should be seen in centralised clinics and consideration given to a comprehensive screening strategy involving mammography and ultrasound of the breast, ultrasound of the ovaries, and colonoscopy. A gynaecological oncologist should be affiliated to such a clinic so that consistent advice is given in the event of operation being necessary. The habit of collecting the odd high risk family in a practice is unhelpful in terms of research, and possibly irresponsible with regard to patient care.

Transvaginal ultrasound will detect most early stage cancers, and is the best method currently available to do this. If such scans are carried out, they probably need to be done six monthly in the highest risk groups. The scans should be done by someone who has considerable experience of what appearances are acceptable within normal ovaries. Clinicians not familiar with ultrasound should remember that this is not an exact science, and the results are totally dependent on the quality of the person carrying out the scan.

In the event of a persistently abnormal ovary the patient should be referred to a specialist gynaecological oncologist. Diagnostic laparoscopy will give no information about the internal structure of an ovarian mass and is illogical. Operative laparoscopy, on the other hand, might be a reasonable way of reducing the incidence of false positive results.

Some women with hereditary ovarian cancer and from mixed site cancer families may be offered prophylactic oophorectomy. Such women must be informed that this does not eradicate their risk of developing carcinoma. Concerns will often be voiced by these women regarding the use of hormone replacement therapy, particularly if they have a family history of breast carcinoma. There are few data on which to base any advice, but the relative risk of cardiovascular disease and bone loss in such young women probably outweighs the theoretical risk of breast cancer. The general view is that prophylactic oophorectomy may be offered to women over the age of 35, or once they have completed their families. Based on our data regarding the age of siblings developing the disease, we discuss oophorectomy with any woman from a high risk family whose sister has died within five years of her current age.

Questions will often be asked about the risk to the daughter of a woman attending the clinic. It is possible to point out that a genetic test of risk may be available to assess the child's risk. Otherwise it is reasonable to suggest that the daughter should consider taking the oral contraceptive pill to reduce her risk when she reaches a suitable age. Although there are no data about high risk women it is not unreasonable to give the contraceptive pill to any suitable women at increased risk of ovarian cancer in the hope that it may reduce the risk of developing the disease.

All patients attending clinics for these high risk women must be made aware of the limitations of both screening and prophylactic oophorectomy. The fact that the women derive benefit from having a point of contact to discuss their fears about the risk of cancer should not be underestimated, and may be as important as any of the more obviously practical aspects of the screening clinic. This means that the women must be seen by the person responsible for their care and not sent off to technicians. The care of these women is not restricted to sending them for a quick scan every six months.

References

1 Narod SA, Feunteun J, Lynch HT, *et al*. Familial breast/ovarian cancer locus on chromosome 17q12-q23. *Lancet* 1991; **338**: 82–3.
2 Miki Y, Swensen J, Shattuck-Eidens D, *et al*. A strong candidate for the breast and ovarian cancer susceptibility gene BRCA1. *Science* 1994; **266**: 66–71.
3 Lynch HT, Follett KL, Lynch PM, *et al*. Family history in an oncology clinic: Implications for cancer genetics. *JAMA* 1979; **242**: 1268–72.
4 Watson P, Lynch HT. Extracolonic cancer in hereditary nonpolyposis colorectal cancer. *Cancer* 1993; **71**: 677–85.
5 Lynch HT, Smyrk TC, Watson P, *et al*. Genetics, natural history, tumor spectrum, and pathology of hereditary nonpolyposis colorectal cancer: an updated review. *Gastroenterology* 1993; **104**: 1535–49.
6 Kinzler KW, Nilbert MC, Vogelstein B, *et al*. Identification of a gene located at chromosome 5q21 that is mutated in colorectal cancers. *Science* 1991; **251**: 1366–70.
7 Fusaro RM, Hoden RH, Johnsen LR, Egelston TG, Lynch HT. The clinical use of genealogical techniques in cancer investigations: a questionnaire survey. *J Cancer Educ* 1993; **8**: 217–25.
8 Fishel R, Lescoe MK, Rao MRS, *et al*. The human mutator gene homolog MSH2 and its association with hereditary nonpolyposis colon cancer. *Cell* 1993; **75**: 1027–38.
9 Leach FS, Nicolaides NC, Papadopoulos N, *et al*. Mutations of a MutS homolog in hereditary non-polyposis colorectal cancer. *Cell* 1993; **75**: 1215–35.
10 Papadopoulos N, Nicolaides NC, Wei Y-F, *et al*. Mutation of a mulL homolog in hereditary colon cancer. *Science* 1994; **262**: 1625–29.
11 Peltomäki P, Aaltonen L, Sistonen P, *et al*. Genetic mapping of a locus predisposing to human colorectal cancer. *Science* 1993; **260**: 810–12.

12 Bronner CE, Baker SM, Morrison PT, *et al.* Mutation in the DNA mismatch repair gene homologue hMLH1 is associated with hereditary non-polyposis colon cancer. *Nature* 1994; **368**: 258–61.

13 Nicolaides NC, Papadopoulos N, Liu B, *et al.* Mutations of two PMS homologues in hereditary nonpolyposis colon cancer. *Nature* 1994; **271**: 75–80.

14 David KL, Steiner-Grossman P. The potential use of tumor registry data in the recognition and prevention of hereditary and familial cancer. *NY State J Med* 1991; **91**: 150–2.

15 Arvanitis ML, Jagelman DG, Fazio VW, Lavery IC, McGannon E. Mortality in patients with familial adenomatous polyposis. *Dis Colon Rectum* 1990; **33**: 639–42.

16 Wooster R, Neuhausen SL, Mangion J, *et al.* Localization of a breast cancer susceptibility gene, BRCA2, to chromosome 13q12-13. *Science* 1994; **265**: 1088–2090.

17 Bourne TH, Campbell S, Reynolds K, *et al.* The potential role of serum CA 125 in an ultrasound-based screening program for familial ovarian cancer. *Gynecol Oncol* 1994; **52**: 379–85.

18 Woolas R, Jacobs IJ, Xu FJ, *et al.* Serum levels of CA 125, M-CSF, and OVX1 in stage I and preclinical ovarian cancer. *Proceedings of the American Society for Clinical Oncology* 1993; **12**: 110.

19 Bourne TH, Whitehead MI, Campbell S, Royston P, Bhan V, Collins WP. Ultrasound screening for familial ovarian cancer *Gynecol Oncol* 1991; **43**: 92–7.

20 Lynch HT, Harris RE, Guirgis HA, Maloney K, Carmody L, Lynch JF. Familial association of breast/ovarian cancer. *Cancer* 1988; **41**: 1543–8.

21 Campbell S, Bhan V, Royston P, Whitehead MI, Collins WP. Transabdominal ultrasound screening for early ovarian cancer. *BMJ* 1989; **299**: 1363–7.

22 Granberg S, Norstiom A, Wikland M. Tumours in the lower pelvis as imaged by transvaginal sonography. *Gynecol Oncol* 1990; **37**: 224–90.

23 Folkman J, Watson J, Ingber D, Hanahan D. Induction of angiogenesis during the transition from hyperplasia to neoplasia. *Nature* 1989; **339**: 58–61.

24 Bourne TH, Campbell S, Steer CV, Whitehead MI, Collins WP. Transvaginal colour flow imaging: a possible new screening technique for ovarian cancer. *BMJ* 1989; **299**: 1367–70.

25 Cramer DW, Hutchinson GB, Welch WR, Scully RE, Ryan KJ. Determinants of ovarian cancer risk. Reproductive experience and family history. *J Natl Cancer Inst* 1983; **71**: 711–16.

26 Ponder BAJ, Easton DF, Peto J. Risk of ovarian cancer associated with a family history: preliminary report of the OPCS study. In: Sharp F, Mason WP, Leake RE, eds. *Ovarian cancer. Biological and therapeutic challenges.* Cambridge: Chapman and Hall Medical, 1990: 3–6.

27 Houlston RS, Bourne TH, Collins WP, Whitehead MI, Campbell S, Slack J. Risk of ovarian carcinoma and genetic relationship to other cancers in families. *Hum Hered* 1993; **43**: 111–15.

28 Lynch HT, Watson P, Conway TA, *et al.* DNA screening for breast/ovarian cancer susceptibility based on linked markers. *Arch Intern Med* 1993; **153**: 1979–87.

29 Bourne TH, Campbell S, Reynolds K, *et al.* Screening for early familial ovarian cancer with transvaginal ultrasonography and colour blood flow imaging. *BMJ* 1993; **306**: 1025–9.

30 Pernet A, Wardle J, Bourne TH, Whitehead MI, Campbell S, Collins WP. A qualitative evaluation of the experience of surgery after false positive results in screening for early familial ovarian cancer. *Psycho-Oncology* 1992; **1**: 217–33.

31 Wardle J, Pernet A, Bourne TH, Collins WP, Campbell S, Whitehead MI. The psychological impact of screening for familial ovarian cancer. *J Natl Cancer Inst* 1993; **85**: 653–7.

8 Prognostic variables in ovarian cancer

PHILIP J BEALE, MICHAEL L FRIEDLANDER

Ovarian cancer results in higher mortality than any other gynaecological malignancy and is a leading cause of deaths from cancer in women in the western world.[1] In the absence of any recognised preventable causes and with no reliable means of early diagnosis any prospect for improved survival lies with the optimal management of patients after their initial presentation. There have been important advances in treatment strategies over the last three decades, but these have not significantly increased overall survival for most patients with advanced ovarian cancer. This may be attributed in part to a lack of understanding of the multiple prognostic variables—host, tumour, and treatment related—that contribute to the outcome; this has led to incorrect stratification of patients in randomised studies and consequently the publication of inconsistent reports. In some studies prognostic factors may be more important determinants of survival than the treatment under investigation. The establishment of prognostic criteria is therefore important in providing a framework for selection of treatment and allowing valid comparisons to be made among treatments.

Multiple prognostic factors have been identified in ovarian cancer and include clinicopathological variables such as stage, histological grade, and subtype; residual disease; and performance status. In addition a number of newer "biological" prognostic factors have been identified that may possibly complement or even replace some of the less objective clinical variables. In this chapter we review prognostic factors in ovarian cancer and highlight the recent attempts to develop prognostic indices that may serve better to predict the relative risk of relapse.

Tumour stage

The extent of anatomical spread, or stage, at the time of initial diagnosis is of major prognostic importance. The International Federation of Gynecology and Obstetrics (FIGO) staging system was initially introduced in 1964 and has subsequently been modified on a number of occasions. The staging is based on clinical and surgical findings, histopathological assessment and peritoneal cytology.[1] Survival correlates well with FIGO

96

Table 8.1 Five year survival of patients with ovarian cancer according to FIGO staging

Stage	FIGO	First author		
		Hogeberg[4]	Swenerton[10]	Unzelman[37]
I	—	80	80	—
a	72	87.4	—	—
b	62.5	80	—	—
c	57.4	76	—	—
II	—	63	61	—
a	52.5	66	—	—
b	37.5	63	—	—
c	37.5	62	—	—
III	10.8	17.5	16	—
a	—	50	—	47
b	—	13	—	49
c	—	15	—	5
IV	4.6	17	2	0
No of patients	4892	332	556	54

stage (table 8.1). Patients with FIGO stage I (50%–85% five year survival) or FIGO stage II (37%–79%) have a better prognosis than those with stage III (7%–18% five year survival) and stage IV (0–17%) tumours. Within each stage, however, there are often wide variations in prognosis. The often striking differences in survival among patients with similar FIGO stage suggests that inadequate or understaging is still common. In a recent study of patterns of care it was found that only 52% and 35% of cases operated on by obstetricians/gynaecologists and general surgeons, respectively, were evaluated correctly.[2] This is supported by a number of reports. In an important study the Ovarian Cancer Study Group restaged patients with stage I and II ovarian cancer by doing an additional laparotomy.[3] The stage was changed in 33% of patients with tumour commonly in unrecognised sites such as pelvic lymph nodes, pelvic peritoneum, para-aortic nodes, diaphragm, and omentum. Hogeberg *et al* found that only 20% of patients with stage I or II tumours were correctly staged.[4]

Tumour stage remains an important prognostic factor and is used in most centres to plan additional treatment. The importance of staging must be emphasised and standardised.

Pathology of ovarian cancer

Although the various histological subtypes of epithelial ovarian cancer differ in their biological behaviour there is still some debate about whether they are of independent prognostic importance. There is a relatively high degree of subjectivity in assigning histological subtype and this is highlighted in a recent study by Lund *et al* in which expert pathologists agreed on histological subtype in only 68% of cases.[5]

Table 8.2 Histological subtypes in epithelial ovarian cancer

Subtype	Percentage incidence	Percentage five year survival rate
Serous	40–70	20–35
Mucinous	5–20	40–60
Endometrioid	10–20	40–60
Clear cell	5–10	35–50
Undifferentiated	1	15–20
Brenner	1–5	—
Transitional cell	—	>50

Table 8.2 shows the relative frequency of the different histological subtypes and associated survival. Serous tumours are generally thought to be more aggressive but when grade and stage are taken into account the difference between serous tumours and other subtypes becomes less pronounced.[6-8] Mucinous tumours are commonly confined to the ovary at initial diagnosis and this contributes to the more favourable prognosis. Patients with stage III and IV mucinous tumours, however, often have poor survival.[4] Endometrioid tumours closely resemble those of the endometrium and are commonly associated with endometrial cancer. They constitute 20%–25% of epithelial ovarian cancers and are often confined to the ovary at the time of initial diagnosis; consequently they are associated with a favourable prognosis.[9 10] Clear cell carcinomas account for 5%–10% of ovarian malignancies.[9 11] Most patients with clear cell carcinomas have tumours confined to the ovaries at the time of initial diagnosis. There is still some controversy about the prognostic significance of clear cell cancer. Some studies have reported a 65%, five year survival rate in patients with stage I clear cell carcinoma of the ovary, in contrast to 80% with other histological subtypes.[12] A number of architectural patterns can be seen within clear cell tumours, and these are thought to relate to prognosis.[13 14] Patients with predominantly papillary or tubulocystic types have a better prognosis than those with a solid pattern, although this has not been a universal finding.[15 16]

Transitional cell carcinoma is a rare subgroup that has recently been found to have a relatively good prognosis even among patients with more advanced stages.[17-19] Robey *et al* initially reported prolonged disease free survival in 15 of 18 patients with advanced transitional cell carcinoma of the ovary.[18] In a subsequent larger study from the same institution comprising 62 patients with transitional cell carcinoma of the ovary, the favourable prognosis and good response to chemotherapy were confirmed.[19]

Borderline ovarian tumours

Borderline tumours are a well recognised subset of epithelial ovarian cancers that have an excellent prognosis.[21] They are also referred to as carcinomas of low malignant potential and are now included in the WHO classification.[22] They have an unusual degree of proliferation of the epithelial cells with no destructive stromal invasion. Borderline tumours are capable of implanting on peritoneal surfaces and the implants may be invasive in a small proportion of cases. The distinction between borderline tumours and truly invasive carcinomas is not always straightforward and has led to a wide variation in their reported incidence.[23 24] It is thought that they account for between 10% and 20% of all ovarian tumours.[11] Borderline ovarian tumours pursue an indolent course in most patients, and even patients with more advanced tumours have an excellent prognosis.[25 26] The wide variation in reported survival rates for patients with ovarian cancer in the past probably reflects the inclusion of borderline tumours as carcinomas in some series and their exclusion in others.

Tumour grade

The pattern of differentiation or amount of cellular anaplasia are the main criteria used to assess grade. Though histological typing according to the WHO classification has been generally accepted, there is no unanimously accepted grading system; histological grade is recognised as being of prognostic importance in ovarian cancer.[4 10 27–30] The five year survival for well differentiated (grade I tumours) is 70%–75%, moderately differentiated (grade II) 30%–46%, and poorly differentiated (grade III) 5%–24%.[4 30] Some studies have suggested that grade is a highly significant and independent prognostic factor while others do not. There are high interobserver and intraobserver differences associated with assigning grade.[31 32] A recent study from Denmark highlighted this problem and suggested that there was poor correlation among pathologists.[33] It is clear that a more objective method for assessing histopathological grade is needed and this may be possible with more quantitative techniques such as morphometry and DNA cytometry.

Residual tumour

The volume of tumour left after primary cytoreductive surgery is of important prognostic significance (table 8.3). Optimal cytoreduction is

Table 8.3 Relationship between survival and amount of residual tumour after primary cytoreduction in advanced ovarian cancer

First author	Median survival according to residual disease (months)						
	0	<1	<2	>1	>2	2–5	>5
Hogberg[4]	37.8	23.2	—	11.3	—	—	—
Lund[36]	Not reached	—	41	—	—	18	16
Gargano[38]	—	—	60	—	36	—	—
Unzelman[37]	58	—	34	—	9	—	—
Hacker[35]	—	40	18	—	6	—	—
Delgado[39]	—	—	45	—	16	—	—

achieved in at least 70% of cases in specialised centres and prognosis is associated with the volume of residual disease.[34 35] The most favourable subset are patients with either no macroscopic disease or tumour deposits of less than 1 cm in maximum diameter and they constitute the only subgroup of patients with advanced ovarian cancer who are treated with curative intent (30%–40% five year survival).[4 35 36] There is still some debate about whether the patients whose disease allows the optimal amount of tumour to be removed have tumours that are biologically different from those in which residual tumour is left. There never has been, and it is unlikely that there ever will be, a randomised trial of primary cytoreduction in patients with advanced ovarian cancer, but the vast body of data supports the concept of debulking surgery though this was questioned in a recent meta-analysis.[40] This meta-analysis suggested that surgery had only a small effect on outcome, but the study has been criticised on a number of grounds and the debate continues. Despite these questions, the volume of residual disease remains an important prognostic variable and correlates well with response to chemotherapy, median survival times and overall survival time.

Prognostive value of second-look laparotomy

Complete clinical response is achieved in about half of all patients with advanced ovarian cancer, but complete pathological responses occur in only about a quarter of patients.[41] The most accurate way of assessing response is second-look laparotomy, although there is still debate about the benefit of this procedure.[42 43] It is clear that between a quarter and half of patients in whom the second-look shows no evidence of recurrence will relapse at a later date.[41 42 44 45] Patients at high risk of developing recurrence after a negative second-look laparotomy include those with grade 3 tumours or with grossly visible disease at the end of the initial operation.[44–47] The question of whether additional treatment will influence the outcome of these high risk patients remains open.

Characteristics of patients

Both patient age and performance status correlate with outcome in patients with ovarian cancer.[10 11 48 49] Performance status has been shown to be an independent prognostic factor by several investigators but it is not always reported in studies.[48 49] In a meta-analysis by Voest et al performance status was reported in only 19 of 66 treatment groups.[50] Generally patients with a good performance status (Eastern Cooperative Oncology Group (ECOG) score 0–2 or Karnofsky score of over 70%) are more likely to respond to treatment, have less toxicity, and a more favourable outcome.

Age is also an important prognostic variable in a number of studies, with younger women having a better prognosis than older ones.[11] This may reflect different biological behaviour but it should be noted that older women often have more advanced disease at diagnosis and may be treated more conservatively than younger women.

Prognostic factors in non-metastatic (stage I) ovarian cancer

The reported five year survival rates for patients with stage I ovarian cancer vary considerably and range from 40% to more than 90% (table 8.1). This difference can be accounted for in part by understaging in some series, but there are also other intrinsic tumour characteristics that contribute to the variability in outcome. The FIGO classification subdivides stage I tumours into three prognostic subsets which include IA (unilateral tumours), IB (bilateral tumours), and IC (tumour spillage, capsular penetration, and positive peritoneal cytology).

The degree of differentiation, histological type, dense adhesions, large volume of ascites, and age have all been reported to be independent prognostic factors for stage I epithelial ovarian cancer. Dembo *et al* reported that patients with grade I tumours had a 98% five year survival compared with 81% for grade II, and 58% for those with grade III tumours.[51] Similar findings were reported from the Norwegian Radium Institute where patients with grade I tumours had a five year survival rate of 98% compared with 26% in patients with grade III densely adherent tumours. Kaern reported that DNA ploidy and FIGO substage were the only two independent prognostic variables in FIGO stage I disease and this supports findings of a number of other studies that suggested that ploidy analysis may be of particular value in predicting outcome in patients with stage I ovarian cancer.[52 53]

Multivariate analysis of prognostic factors in advanced ovarian cancer

Most patients with advanced ovarian cancer will die of their disease and have a median survival of 20–30 months. Only 20%–30% of patients will still be alive at five years. A recent overview of multivariate analyses in advanced ovarian cancer has been done in an attempt to identify important independent prognostic variables.[54] The findings were interesting and showed important differences between studies. Although there are pitfalls with such an analysis it is worthwhile noting that volume of residual tumour was identified as an independent prognostic factor in three of seven studies; tumour size was not a significant variable in the four studies in which it was considered; tumour grade was an independent prognostic variable in only one of six studies; stage was a significant factor in four of seven studies; histological subtype was of no prognostic value in any of the studies; and performance status did not appear as an independent variable in either of the two studies in which it was examined. The authors correctly caution over interpretation of their findings because of the many differences among the various studies, but it still raises questions.

In a meta-analysis of prognostic factors in roughly 3000 patients, Voest *et al* found that the factors that predicted survival were cisplatin-based chemotherapy, residual mass of less than 2 cm, FIGO stage II and III, and performance status.[50] After multivariate analysis only cisplatin and residual tumour were shown to be independent variables. In a similar meta-analysis of the effect of operation in almost 7000 patients with advanced ovarian cancer, Hunter *et al* found that cisplatin chemotherapy and FIGO stage were the only independent variables for predicting outcome.[40]

These studies all indicate inconsistencies and lack of agreement as to which factors are truly of independent prognostic importance. There is general acceptance that predicting the level of risk will best be achieved by the development of a prognostic index rather than by applying individual variables to each patient. Several groups have attempted to develop such an index (table 8.4). Swenerton *et al* stratified patients by grade, residual disease and performance status into subsets within each stage of good, intermediate, and poor prognosis.[10] Patients with stage III tumours had a 17% five year survival but a subset (comprising 30% of patients) had a 37% five year survival. Van Houwelingen *et al* managed to separate patients into six prognostic subsets and identified a group (comprising 10% of patients) with a 75% four year survival and a poor prognosis group (also comprising 10% of patients) who all died within four years.[49] Lund *et al* have also reported a prognostic index with good predictive power.[36]

There is much interest in developing a prognostic index that has a high predictive value, incorporates objective prognostic variables, and is relatively easy to use.

Table 8.4 Factors used to develop prognostic indexes based on the combined influence of independent prognostic variables

	Swenerton[10]	van Houwelingen[49]	Ansell[20]	Lund[36]
Performance status	Yes	Yes	—	Yes
FIGO stage	No	Yes	Yes	—
Grade	Yes	Yes	—	Yes
Residual tumour	Yes	Yes	Yes	Yes
Ascites	No	No	Yes	Yes
Histological type	No	No	No	Yes
Alkaline phosphatase activity	No	No	No	—
Rate of response	No	No	Yes	No
No of metastases	No	No	No	Yes

New prognostic factors

The general thrust of research into the biology of ovarian cancer has been to identify new and potentially more objective factors which predict outcome. A large number of these have been identified and some seem to be potentially clinically valuable. Though it is often tempting to accept these new prognostic factors, we need to subject them first to the same rigorous testing as the other clinicopathological variables. The ultimate value of any new prognostic factor is not only that it predicts outcomes, but also that it provides useful additional information.

Tumour ploidy and proliferative fraction

DNA ploidy expresses the nuclear DNA content as measured by cytometry. Friedlander *et al* showed that ovarian cancers were commonly aneuploid and that there was a close association between ploidy and subsequent outcome.[55] Diploid tumours represented 27% of advanced stage cancers, whereas 75% were aneuploid. Patients with diploid tumours had significantly longer survival (median 260 weeks) than those with aneuploid tumours (median 54 weeks) and multivariate analysis indicated that tumour ploidy was the most powerful determinant of survival in patients with advanced ovarian cancer. These results have now been confirmed by a large number of investigators.[56–58] Similar results have been reported in stage I ovarian tumours. Kaern *et al* investigated the prognostic significance of tumour ploidy in a cohort of 290 patients[52]; 51% of the tumours were diploid and 49% aneuploid. The five year disease free survival for patients with diploid tumours was 90% compared with 64% for patients with aneuploid tumours ($p<0.0001$). On multivariate analysis grade, ploidy, and FIGO stage were shown to have independent prognostic significance. Flow cytometric analysis of DNA ploidy has also been reported to be of prognostic value in patients with borderline ovarian tumours.[59] The largest study in 370 patients was reported from the Norwegian Radium Institute where DNA ploidy was shown to be the most powerful determinant of outcome.[60] It is possible that combining flow cytometric analysis with morphometric analysis may result in improved sensitivity and specificity in the detection of aneuploid cells and so improve their predictive value, but we need to confirm this in large studies.[61]

Flow cytometric analysis also provides information on cell cycle distribution. There are some data to suggest that S phase fraction may be of prognostic significance in ovarian cancer, but there are also a number of studies that contradict them.[56 62 63] There are still a number of technical problems in measuring S phase, particularly in paraffin-embedded tissue, and there are no international criteria for the calculation of S phase fraction.

Tumour associated antigens

CA-125

CA-125 is a glycoprotein antigen found in ovarian tumour cells which can be detected by a monoclonal antibody. CA-125 titres are raised in

75%–90% of patients with ovarian cancer.[64] It has a well established role in monitoring response to treatment and in detecting relapse. The prognostic value of CA-125 has been evaluated at four different times.

CA-125 before operation. Most authors have found that the preoperative CA-125 titre does not correlate with survival but probably reflects higher tumour burden and more advanced stage.[65 66]

CA-125 after operation and before chemotherapy. There is conflicting information about the titre of CA-125 after operation as a prognostic factor because it may reflect only the amount of residual tumour. However, Möbus *et al* found that among patients with minimal residual disease (less than 2 cm) all those with CA-125 titres of more than 65 U/ml were dead at 42 months, whereas 48% of those with a CA-125 concentration of less than 65 U/ml were alive at six years.[67] Makar *et al* in a study of 687 patients at all stages found that post operative CA-125 titres were of independent prognostic significance in all patients irrespective of the volume of residual disease.[68]

CA-125 during the first three cycles of chemotherapy. Measuring titres during the first three cycles of treatment seems to give the best prognostic information. Two criteria have been looked at—the absolute titre after one, two, or three cycles of chemotherapy, or the half life of CA-125. Mogensen found that if the CA-125 titre was 10 U/ml or less after three cycles of treatment than median survival was 60 months.[69] This compares with seven months' median survival if the CA-125 concentration was more than 100 U/ml. Other authors have found similar results.[70 71] Yedema *et al* noted that patients with a CA-125 half life of less than 20 days had a median survival of 28 months compared with a median survival of 19 months in those in whom the half life was greater than 20 days.[72]

The Medical Research Council Working Party on Gynaecological Cancer has recently analysed CA125 titre in 248 patients.[73] They found that the absolute value of CA125 after three cycles of treatment was the single most important factor for predicting progression at 12 months, but there was still a false positive rate of nearly 20% and they concluded that the prognostic information was not accurate enough to be used to manage individual patients.

CA-125 at second-look laparotomy. Patients with a raised serum CA-125 titre at second-look laparotomy have significantly shorter survival. Makar *et al* studied 208 patients and found that the titre of CA-125 before second-look laparotomy was an independent prognostic factor for survival.[74] Other authors have found this not to be true.

Sialyl Tn

Sialyl Tn is thought to be the antigen of the core region of mucin oligosaccharide and can be measured by a monoclonal antibody. Kobayashi *et al* found that concentrations were raised in 48% of patients with epithelial ovarian cancer of all stages before the initial operation.[75] The five year survival rate for patients with no antigen in their serum was 76% compared with 11% if it was present. On multivariate regression analysis the Sialyl Tn concentration was an important variable for predicting overall survival.

CA 54/61

These antigens are glycoproteins expressed on ovarian cancer cells and measured by monoclonal antibodies. Kobayashi *et al* looked at 45 patients with biopsy proven ovarian cancer (of all stages) and found the serum concentrations were raised in 44.4% of cases before operation.[76] Survival at three years for patients who did not have CA 54/61 in their serum was 64% compared with 35% for those that did. These findings were independent of CA-125 concentrations.

Steroid hormone receptor expression

The presence of oestrogen and progesterone receptors in ovarian tumours has been well documented and they are found in 40%–70% of patients.[77 78] Androgen receptors are found in 80%–90% of ovarian tumours.[79] Some investigators have found a correlation between steroid receptor concentrations and a number of tumour and individual characteristics such as histological type, stage, age, and histological grade, but others have not noted any association.[80–82] The association between steroid receptor expression and prognosis remains unclear.

Factors regulating transformation and growth

Oncogenes

The Her2-neu or C-erb-B2 is located on chromosome 17q and encodes a cell surface glycoprotein of 185 kDa. The function of the protein is unknown but it has homology for epidermal growth factor. Amplification of this proto-oncogene has been reported in ovarian cancer with a variable incidence ranging from 0–30%.[83–86] This may relate to the techniques used to assess amplification, as a recent study using strict criteria showed amplification in only 11% of malignant ovarian tumours.[83] Some studies have shown that amplification is a poor prognostic factor while others have not. Slamon *et al* reported that 31 out of 120 primary ovarian tumours had evidence of neu amplification and this was associated with an unfavourable outcome.[85] Patients with more than five copies of the neu gene had a poor prognosis. Berchuk *et al* found that 32% of 73 patients had increased neu expression and a median survival of 15.7 months.[86] Patients with normal neu expression had a median survival of 32.8 months and this difference was significant.

Other oncogenes have been studied including ras oncogene product p21, FGF-3(int 2), and the myc gene but overexpression or amplification of these has not been shown to be of prognostic significance.[87–89]

The FMS gene encodes a tyrosine kinase receptor for the macrophage colony stimulating factor (MCSF) or CSF-1. High levels of c-fms transcripts have been shown to correlate with advanced stage and high histological grade.[90] The prognostic significance of a raised serum concentration of macrophage colony stimulating factor is unknown, but in ascites raised concentrations of CSF-1 are associated with a significantly better prognosis.[91]

Growth factors

Ovarian cancers produce multiple peptide growth factors and their receptors. Overproduction of these could lead to autocrine stimulation. Factors that have been looked at include epidermal growth factors (EGF) and epidermal growth factor receptor, interleukin 1, tumour necrosis factor α, interleukin 6, and interleukin 2 receptor. Several studies have shown that expression of epidermal growth factor receptor is associated with a significantly worse prognosis although this has not been a consistent finding.[92–95] Expression of other growth factors correlates with the extent of disease but does not seem to be of independent prognostic importance.[96–98]

Heat shock proteins (HSP)

Heat shock genes encoding their protein are highly conserved through evolution. HSP 70 and HSP 90 are found in high concentrations in advanced stages of ovarian cancer.[99] HSP 60 is present in mitochondria of all cells and assists the correct folding of newly synthesised proteins. Kimura *et al* studied ovarian tumours and found on multivariate analysis that low expression of HSP 60 was associated with a highly significantly improved survival compared with high expression.[100]

Cathepsin D

Cathepsin D is a proteolytic enzyme which when secreted by cancer cells may facilitate tumour invasion and may also act as an autocrine mitogen. Scambia *et al* looked at 53 ovarian tumours and metastases and found that concentrations of cathepsin D were highest in metastases.[101] They suggest it could be an additional prognostic factor.

106

Tetranectin

Tetranectin is a plasminogen-binding protein that may have a role in tissue remodelling and cancer tissue growth. Hogdall *et al* in a small study of 37 patients with ovarian cancer (all stages) found that a low stromal tetranectin score and high plasma tetranectin concentration was associated with improved survival.[102]

Clonogenic growth in vitro

Clonogenic growth in semisolid media is thought to be related to the malignancy of cells. Colony formation has been reported to be related to prognosis in ovarian cancer. Bertoncello *et al* showed that there was a significant inverse relationship between clonogenicity and outcome in a small group of patients with ovarian cancer.[103] Dittrich *et al* reported a retrospective study of 84 patients with ovarian cancer and also found that clonogenic growth *in vitro* was associated with a worse prognosis.[104] Multivariate analysis identified clonogenic growth as a significant independent prognostic variable. Estimation of outcome was best accomplished by combining the residual tumour mass, age, and clonogenic growth.

References

1 Peterson F, Kolstad P, Ludwig H, Ulfelder H. *Annual report on the results of treatment in gynecological cancer.* Vol 20. Stockholm: International Federation of Gynecology and Obstetrics, 1988.
2 McGowan L. Patterns of care in carcinoma of the ovary. *Cancer* 1993; **71** (suppl 2): 628–33.
3 Young RC, Decker DG, Wharton JT, *et al.* Staging laparotomy in early ovarian cancer. *JAMA* 1983; **250**: 3072–6.
4 Högberg T, Carstensen J, Simonsen E. Treatment results and prognostic factors in a population-based study of epithelial ovarian cancer. *Gynecol Oncol* 1993; **48**: 38–49.
5 Lund B, Thomsen HK, Olsen J. Reproducibility of histopathological evaluation in epithelial ovarian carcinoma. Clinical implications *APMIS* 1991; **99**: 353–8.
6 Demopoulos RI, Bigelow B, Blaustein A, *et al.* Characterization and survival of patients with serous cystadenocarcinoma of the ovaries. *Obstet Gynecol* 1984; **64**: 557–63.
7 Malkasian GD, Decker DG, Webb MJ. Histology of epithelial tumours of the ovary: Clinical usefulness and prognostic significance of histologic classification and grading. *Semin Oncol* 1975; **2**: 191–201.
8 Silverberg SG. Prognostic significance of pathologic features of ovarian carcinoma. *Curr Top Pathol* 1989; **78**: 85–109.
9 Czernobilsky B, Silverman BB, Mikuta JJ. Endometrial carcinoma of the ovary. A clinicopathological study of 75 cases. *Cancer* 1970; **26**: 1141–52.
10 Swenerton KD, Hislop TG, Spinelli J, *et al.* Ovarian carcinoma: A multivariate analysis of prognostic factors. *Obstet Gynecol* 1985; **65**: 264–9.
11 Aure JC, Hoes K, Kolstad P. Clinical and histological studies of ovarian carcinoma. Long term follow up of 990 cases. *Obstet Gynecol* 1971; **37**: 1–9.
12 O'Brien MER, Schofield JB, Tan S, *et al.* Clear cell epithelial ovarian cancer (mesonephroid): Bad prognosis only in early stages. *Gynecol Oncol* 1993; **49**: 250–4.
13 Kennedy AW, Biscotti CV, Hart WR, Webster KD. Ovarian clear cell adenocarcinoma. *Gynecol Oncol* 1989; **32**: 342–9.

14 Crozier MA, Copeland CJ, Silva EG, *et al.* Clear cell carcinoma of the ovary: A study of 59 cases. *Gynecol Oncol* 1989; **35**: 199–203

15 Montag AG, Jenison EL, Griffiths CT, *et al.* Ovarian cell carcinoma: A clinicopathologic analysis of 44 cases. *Int J Gynecol Pathol* 1989; **8**: 85–96.

16 Imachi M, Tsukamoto N, Shimamoto T, *et al.* Clear cell carcinoma of the ovary: A clinicopathologic analysis of 34 cases. *International Journal of Gynaecological Cancer* 1991; **1**: 113–20.

17 Gersall DJ. Primary ovarian transitional cell carcinoma. Diagnostic and prognostic considerations. *Am J Clin Pathol* 1990; **93**: 586–8.

18 Robey SS, Silva EG, Gerhenson DG, *et al.* Transitional cell carcinoma in high-grade, high-stage ovarian carcinoma. An indicator of favorable response to chemotherapy. *Cancer* 1989; **63**: 839–47.

19 Gershenson DM, Silva EG, Mitchell MF, *et al.* Transitional cell carcinoma of the ovary: A matched control study of advanced study of advanced-stage patients treated with cisplatin-based chemotherapy. *Am J Obstet Gynecol* 1993; **168**: 1178–87.

20 Ansell SM, Rapoport BL, Falkson G, *et al.* Survival determinants in patients with advanced ovarian cancer. *Gynecol Oncol* 1993; **50**: 215–20.

21 Colgan TJ, Norris HJ. Ovarian tumours of low potential malignancy: A review. *International Journal of Gynecologic Oncology* 1983; **1**: 367–82.

22 Serov SF, Scully RE, Sobin LH. Histologic typing of ovarian tumours in: *International histological classification of tumours*. No 9. Geneva: World Health Organization, 1973.

23 Hopkins MP, Kumar NB, Morley GW. An assessment of pathologic features and treatment modalities in ovarian tumors of low malignant potential. *Obstet Gynecol* 1987; **70**: 923–9.

24 Kliman L, Rome RM, Fortune DW. Low malignant potential tumours of the ovary: A study of 76 cases. *Obstet Gynecol* 1986; **68**: 338–44.

25 Obel EB. A comparative study of patients with cancer of the ovary who have survived more or less than 10 years. *Acta Obstet Gynecol Scand* 1976; **55**: 429–39.

26 Kaern J, Trope CG, Abeler VM. A retrospective study of 370 borderline tumors of the ovary treated at the Norwegian Radium Hospital from 1970 to 1982. *Cancer* 1993; **71**: 1810–20.

27 Barber HRK, Sommers SC, Snyder R, Kwon TH. Histologic and nuclear grading and stromal reactions as indices for prognosis of ovarian cancer. *Am J Obstet Gynecol* 1975; **121**: 795–807.

28 Day TG Jr, Gallager HS, Rutledge FN. Epithelial carcinoma of the ovary: Prognostic importance of histologic grade. *Natl Cancer Inst Monogr* 1975; **42**: 15–18.

29 Jacobs AJ, Deligdisch L, Deppe G, Cohen CJ. Histologic correlations of virulence in ovarian adenocarcinoma. 1. Effects of differentiation. *Am J Obstet Gynecol* 1982; **143**: 574–80.

30 Sorbe B, Frankendal B, Veress B. Importance of histologic grading in the prognosis of epithelial ovarian carcinoma. *Obstet Gynecol* 1982; **59**: 576–82.

31 Hernandez E, Bhagavan BS, Parmley TH, Rosenshein NB. Interobserver variability in the interpretation of epithelial ovarian cancer. *Gynecol Oncol* 1984; **17**: 117–23.

32 Baak JP, Langley FA, Talerman A, Delemarre J. Interpathologist and intrapathologist disagreement in ovarian tumor grading and typing. *Analytical and Quantitative Histology* 1986; **8**: 354–7.

33 Bertelsen K, Holund B, Andersen E. Reproducibility and prognostic value of histologic type and grade in early epithelial ovarian cancer. *International Journal of Gynaecological Cancer* 1993; **3**: 72–9.

34 Heintz AM, Hacker NF, Berek JS, *et al.* Cytoreductive surgery in ovarian carcinoma: Feasibility and morbidity. *Obstet Gynecol* 1986; **67**: 783–7.

35 Hacker NF, Berek JS, Lagasse LD, *et al.* Primary cytoreductive surgery for epithelial ovarian cancer. *Obstet Gynecol* 1983; **61**: 413–20.

36 Lund B, Williamson P. Prognostic factors for overall survival in patients with advanced ovarian carcinoma. *Ann Oncol* 1991; **2**: 281–7.

37 Unzelman RF. Advanced epithelial ovarian carcinoma: Long-term survival experience at the community hospital. *Am J Obstet Gynecol* 1992; **166**: 1663–72.

38 Gargano G, Catino A, Correale M, *et al.* Prognostic factors in epithelial ovarian cancer. *Eur J Gynaecol Oncol* 1992; **13** [Suppl.]: 45–55.

39 Delgado G, Oram DH, Petrelli EG. Stage III epithelial ovarian cancer: The role of maximal surgical reduction. *Gynecol Oncol* 1984; **18**: 290–7.

40 Hunter RW, Alexander NDE, Soutter WP. Meta-analysis of surgery in advanced ovarian carcinoma is maximum cytoreductive surgery an independent determinant of prognosis? *Am J Obstet Gynecol* 1992; **166**: 504–11.

41 Creasman WT, Eddy GL. Prognostic factors in relation to second look laparotomy in ovarian cancer. *Ballieres Clin Obstet Gynaecol* 1989; **3**: 183–90.

42 Gershenson DM, Copeland LJ, Wharton JT, *et al.* Prognosis of surgically determined complete responders in advanced ovarian cancer. *Cancer* 1985; **55**: 1129–35.

43 Podratz KC, Kinney WK. Second-look operation in ovarian cancer. *Cancer* 1993; **71**: 1551–8.

44 Lippman SM, Alberts DS, Slymen DJ, *et al.* Second look laparotomy in epithelial ovarian cancer. Prognostic factors associated with survival duration. *Cancer* 1988; **61**: 2571–7.

45 Bertelsen K, Hansen MK, Pedersen PH, *et al.* The prognostic and therapeutic value of second look laparotomy in advanced ovarian cancer. *Br J Obstet Gynaecol* 1988; **95**: 1231–6.

46 Rubin SC, Hoskins WJ, Saigo PE, *et al.* Prognostic factors for recurrence following negative second-look laparotomy in ovarian cancer patients treated with platinum-based chemotherapy. *Gynecol Oncol* 1991; **42**: 137–41.

47 Lund B, Williamson P. Prognostic factors for outcome of and survival after second-look laparotomy in patients with advanced ovarian carcinoma. *Obstet Gynecol* 1990; **76**: 617–22.

48 Omura GA, Brady MF, Homesley HD, *et al.* Long term follow up and prognostic factor analysis in advanced ovarian carcinoma: The Gynaecologic Oncology Group Experience. *J Clin Oncol* 1991; **9**: 1138–50.

49 Van Houwelingen JC, Bokkel Huinink W, Van der Burg ATM, Neijt JP. Predictability of the survival of patients with ovarian cancer. *J Clin Oncol* 1989; **7**: 769–73.

50 Voest EE, van Houwelingen JC, Neijt JP. A metaanalysis of prognostic factors in advanced ovarian cancer with median survival and overall survival (measured with the log (relative risk)) as main objectives. *European Journal of Cancer and Clinical Oncology* 1989; **28A**: 1328–30.

51 Dembo AJ, Davy M, Stenwig AE, *et al.* Prognostic factors in patients with stage I epithelial ovarian cancer. *Obstet Gynecol* 1990; **75**: 263–73.

52 Kaern J. DNA ploidy in epithelial ovarian malignancies. Prognostic value and therapeutic implications. Norwegian Cancer Society 1993, PhD thesis.

53 Punnonen R, Kallioniemi OP, Mattila J, Koivula T. Prognostic assessment in stage I ovarian cancer using a discriminant analysis with clinicopathological and DNA flow cytometric data. *Gynecol Obstet Invest* 1989; **27**: 213–6.

54 Levin L, Lund B, Heintz APM. An overview of multivariate analyses of prognostic variables with special reference to the site of cytoreductive surgery. *Ann Oncology* 1993; 4 (4 suppl): S23–S29.

55 Friedlander ML, Taylor IW, Russell P, *et al.* Ploidy as a prognostic factor in ovarian cancer. *Int J Gynecol Pathol* 1983; **1**: 55–62.

56 Klemi PJ, Joensuu H, Kiiholma P, Maenpaa J. Clinical significance of abnormal nuclear DNA content in serous ovarian tumours. *Cancer* 1988; **62**: 2005–10.

57 Erba E, Ubezio P, Pepe S, *et al.* Flow cytometric analysis of DNA content in human ovarian cancers. *Br J Cancer* 1989; **60**: 45–50.

58 Kuhn W, Kaufman M, Feichter GE, *et al.* DNA flow cytometry, clinical and morphological parameters as prognostic factors for advanced malignant and borderline tumors. *Gynecol Oncol* 1989; **33**: 360–7.

59 Padberg B, Arps H, Franke U, *et al.* DNA cytophotometry and prognosis in ovarian tumors of borderline malignancy. *Cancer* 1992; **69**: 2510–14.

60 Kaern J, Trope CG, Kristensen GB, *et al.* DNA ploidy; the most important prognostic factor in patients with borderline tumors of the ovary. *International Journal of Gynaecological Cancer* 1993; **3**: 349–58.

61 Wils J, van Geuns H, Baak J. Proposal for therapeutic approach based on prognostic factors including morphometric and flow-cytometric features in stage III-IV ovarian cancer. *Cancer* 1988; **61**: 1920–5.

62 Kallioniemi OP, Punnonen R, Mattila J, *et al.* Prognostic significance of DNA index, multiploidy and S-phase fraction in ovarian cancer. *Cancer* 1988; **61**: 334–9.

63 Conte PF, Alama A, Rubagotte A, *et al.* Cell kinetics in ovarian cancer. Relationship to clinicopathologic features, responsiveness to chemotherapy and survival. *Cancer* 1989; **64**: 1188–91.

64 Bast RC, Klug TL, St John E, *et al*. A radioimmunoassay using a monoclonal antibody to monitor the course of epithelial ovarian cancer. *N Engl J Med* 1983; **309**: 883–7.

65 Sevelda P, Schemper M, Spona J. CA125 as an independent prognostic factor for survival in patients with ovarian cancer. *Am J Obstet Gynecol* 1989; **161**: 1213–6.

66 Cruikshank DG, Fullerton WT, Klopper A. The clinical significance of pre-operative serum CA125 in ovarian cancer. *Br J Obstet Gynaecol* 1987; **94**: 692–5.

67 Möbus V, Kreienberg R, Cromback G, *et al*. Evaluation of CA125 as a prognostic and predictive factor in ovarian cancer. *Journal of Tumour Markers Oncology* 1988; **3**: 251–8.

68 Makar A, Kirstensen GB, Bormer OP, Trope CG. Serum CA125 level allows early identification of nonrespondents during induction chemotherapy. *Gynecol Oncol* 1993; **49**: 73–9.

69 Mogensen O. Prognostic value of CA125 in advanced ovarian cancer. *Gynecol Oncol* 1992; **44**: 207–12.

70 Fisken J, Leonard RCF, Stewart M, *et al*. The prognostic value of early CA125 serum assay in epithelial ovarian carcinoma. *Br J Cancer* 1993; **68**: 140–5.

71 Redman CWE, Blackledge GR, Kelly K, *et al*. Early serum CA125 response and outcome in epithelial ovarian cancer. *Eur J Cancer* 1990; **26**: 593–6.

72 Yedema CA, Kenemans P, Voorhorst F, *et al*. CA125 half-life in ovarian cancer: a multivariate survival analysis. *Br J Cancer* 1993; **67**: 1361–7.

73 Fayers PM, Rustin GJS, Wood R, *et al*. The prognostic value of serum CA125 in patients with advanced ovarian carcinoma: An analysis of 573 patients by the Medical Research Council Working Party on Gynaecological Cancer. *International Journal of Gynaecological Cancer* 1993; **3**: 285–92.

74 Makar A, Kristensen GB, Bormer OP, *et al*. CA125 measured before second-look laparotomy is an independent prognostic factor for survival in patients with epithelial ovarian cancer. *Gynecol Oncol* 1992; **45**: 323–8.

75 Kobayashi H, Terao T, Kawashima Y. Serum sialyl Tn as an independent predictor of poor prognosis in patients with epithelial ovarian cancer. *J Clin Oncol* 1992; **10**: 95–101.

76 Kobayashi H, Ohi H, Sugimura M, Shinohara H, Terao T. Monoclonal antibodies MA54 and MA61 as potential reagents in the prognosis of patients with ovarian cancer. *Gynecol Oncol* 1993; **49**: 80–5.

77 Bizzi A, Codegoni AM, Landoni F, *et al*. Steroid receptors in epithelial ovarian carcinoma. Relation to clinical parameters and survival. *Cancer Res* 1988; **48**: 6222–6.

78 Creasman WT, Sasso RA, Weed JC, McCarty KS. Ovarian carcinoma. Histologic and clinical correlation of cytoplasmic estrogen and progesterone binding. *Gynecol Oncol* 1981; **12**: 319–27.

79 Kühnel R, DeGraafü, Rao BR, Stolk JG. Androgen receptor predominance in human ovarian carcinoma. *Journal of Steroid Biochemistry* 1987; **26**: 393–7.

80 Rose PJ, Reale FR, Longcope C, Hunter RE. Prognostic significance of estrogen and progesterone receptors in epithelial ovarian cancer. *Obstet Gynecol* 1990; **76**: 258–63.

81 Sevelda P, Denison U, Schemper M, *et al*. Oestrogen and progesterone receptor content as a prognostic factor in advanced epithelial ovarian carcinoma. *Br J Obstet Gynaecol* 1990; **97**: 706–12.

82 Slotman BJ, Narta JJP, Rao BR. Survival of patients with ovarian cancer. Apart from stage and grade, tumor progesterone receptor content is a prognostic indicator. *Cancer* 1990; **66**: 740–4.

83 Leary JA, Edwards BG, Houghton CRS, *et al*. Amplification of HER-2/neu oncogene in human ovarian cancer. *International Journal of Gynaecological Cancer* 1992; **2**: 291–4.

84 Rubin SC, Finstad CL, Wong GY, *et al*. Prognostic significance of HER-2/neu expression in advanced epithelial ovarian cancer: A multivariate analysis. *Am J Obstet Gynecol* 1993; **168**: 162–9.

85 Slamon DJ, Godolphin W, Jones LA, *et al*. Studies of the HER-2/neu protooncogene in human breast and ovarian cancer. *Science* 1989; **244**: 707–12.

86 Berchuck A, Kamel A, Whitaker R, *et al*. Overexpression of HER-2/neu is associated with poor survival in advanced epithelial ovarian cancer. *Cancer Res* 1990; **50**: 4087–91.

87 Baker VV, Borst MP, Dixon D, *et al*. c-myc amplification in ovarian cancer. *Gynecol Oncol* 1990; **38**: 340–2.

88 Rosen A, Sevelda P, Klein M, *et al*. First experience with FGF-3 (INT-2) amplification in women with epithelial ovarian cancer. *Br J Cancer* 1993; **67**: 1122–5.

89 Yaginuma Y, Yamashita K, Kuzumaki N, *et al*. ras Oncogene product p21 expression and prognosis of human ovarian tumors. *Gynecol Oncol* 1992; **46**: 45–50.

90 Kacinski BM, Carter D, Mittal K, *et al.* Ovarian adenocarcinomas express fms-complementary transcripts and fms antigen, often with coexpression of CSF-1. *Am J Pathol* 1990; **137**: 135–47.

91 Price FV, Chambers SK, Chambers JT, *et al.* Colony-stimulating factor-1 in primary ascites of ovarian cancer is a significant predictor of survival. *Am J Obstet Gynecol* 1993; **168**: 520–7.

92 Scambia G, Panici PB, Battaglia F, *et al.* Significance of epidermal growth factor receptor in advanced ovarian cancer. *J Clin Oncol* 1992; **10**: 529–35.

93 Foekens JA, van Putten WLJ, Portengen H, *et al.* Prognostic value of pS2 protein and receptors for epidermal growth factor (EGF-R), insulin-like growth factor-1 (IGF-1-R) and somatostatin (SS-R) in patients with breast and ovarian cancer. *J Steroid Biochem Molec Biol* 1990; **37**: 815–21.

94 Berchuck A, Rodriguez GC, Kamel A, *et al.* Epidermal growth factor receptor expression in normal ovarian epithelium and ovarian cancer. *Am J Obstet Gynecol* 1991; **164**: 669–74.

95 van der Burg M, Henzen-Logmans SC, Foekens JA, *et al.* The prognostic value of epidermal growth factor receptors, determined by both immunohistochemistry and ligand binding assays, in primary epithelial ovarian cancer: A pilot study. *Eur J Cancer* 1993; **29A**: 1951–7.

96 Owens OJ, Taggart C, Wilson R, Walker JJ, *et al.* Interleukin-2 receptor and ovarian cancer. *Br J Cancer* 1993; **68**: 364–7.

97 Berek JS, Chung C, Kaldi K, *et al.* Serum interleukin-6 levels correlate with disease status in patients with epithelial ovarian cancer. *Am J Obstet Gynecol* 1991; **164**: 1038–43.

98 Moradi MM, Carson LF, Weinberg J, *et al.* Serum and ascitic fluid levels of interleukin-1, interleukin-6, and tumor necrosis factor-alpha in patients with ovarian epithelial cancer. *Cancer* 1993; **72**: 2433–40.

99 Mileo AM, Fanuele M, Battaglia F, *et al.* Selective overexpression of mRNA coding for 90KDa stress protein in human ovarian cancer. *Anticancer Res* 1990; **10**: 903–6.

100 Kimura E, Enns RE, Alcaraz JE, *et al.* Correlation of the survival of ovarian cancer patients with mRNA expression of the 60-kD heat-shock protein HSP-60. *J Clin Oncol* 1993; **11**: 891–8.

101 Scambia G, Benedetti P, Ferrandina G, *et al.* Cathepsin D assay in ovarian cancer: correlation with pathological features and receptors for oestrogen, progesterone and epidermal growth factor. *Br J Cancer* 1991; **64**: 182–4.

102 Hogdall CK, Christensen L, Clemmensen I. The prognostic value of tetranectin immunoreactivity and plasma tetranectin in patients with ovarian cancer. *Cancer* 1993; **72**: 2415–22.

103 Bertoncello I, Bradley TR, Campbell JJ, et al. Limitations of the clonal agar assay for the assessment of primary human ovarian tumour biopsies. *Br J Cancer* 1982; **45**: 803–11.

104 Dittrich C, Dittrich E, Sevelda P, *et al.* Clonogenic growth in vitro: An independent biologic prognostic factor in ovarian carcinoma. *J Clin Oncol* 1991; **9**: 381–8.

9 Ovarian tumours of low malignant potential: borderline epithelial ovarian carcinoma

PAUL J HOSKINS

Taylor, in 1929, was the first to report a subset of patients with a variant of epithelial ovarian carcinoma in whom the outlook was good despite the presence of peritoneal implants. The histopathological features were intermediate between benign and frankly malignant and therefore initially he called them "semimalignant" and later on "borderline".[1] [2] They were further defined in the 1950s by Kottmeier, and Woodruff and Novak,[3] [4] but were not formally recognised until 1971.[5] In 1973 the World Health Organization produced standard diagnostic criteria and terminology for this intermediate form of epithelial ovarian cancer.[6] More recently some authors have proposed that this grouping is wrong, and that while some of these cancers are malignant with the ability to invade and metastasise, others are benign with no malignant potential.

The histological criteria used for making the diagnosis are: stratification of the epithelial lining of the papillae; nuclear atypia; mitotic activity; intracystic clusters of free floating cells; and a lack of stromal invasion. We have to analyse a certain minimum number of sections of the tumour to rule out invasion, and the empiric recommendation was for one section for every 1–2 cm of the tumour's maximum diameter.[7] This recommendation has been shown to be adequate.[8] Only two of 51 invasive tumours were missed using these guidelines. Pragmatically, these infrequent underdiagnoses are unimportant as the prognosis for surgically staged patients with stage I, grade 1 invasive cancers is the same as for those with borderline cancers.[9] Despite these standardised criteria there are considerable differences among pathologists when assessing these tumours. Four pathologists reviewed a representative section of 137 borderline and invasive tumours on two occasions at least six months apart, and the results were presented by comparing histological diagnosis with survival. Within the borderline group, as defined by each pathologist, five year survival rates varied from 80%–100% and the intraobserver change in five year survival

"

Table 9.1 Histological types of epithelial ovarian cancers of low malignant potential (based upon the original WHO definitions but with some latter modifications)[6 11 12 15–20]

Serous
Mucinous
  Endocervical
  Intestinal type

Mixed
Endometrioid
  Adenofibromatous
  Papillary adenoma
  Mixed pattern

Clear cell
Intermediate Brenner tumour of low malignant potential*

* Not all authors agree with subdividing intermediate Brenner tumours into the three distinct groups of metaplastic, proliferating, and low malignant potential.

after their second reading of the slides ranged from 0 to 12%. The number of cases diagnosed as borderline ranged from 25 to 35.[10]

The different histological types of borderline cancer are listed in table 9.1. There are certain provisos—the diagnosis is based on the ovarian primary only: the presence of invasive peritoneal implants does not change the diagnosis to invasive cancer; in the absence of invasion the mucinous subtype is called "invasive cancer" if the cellular stratification exceeds three cells in thickness.[7] More recently two variants of mucinous borderline tumours have been recognised.[11] The rarer "endocervical" or "Mullerian" type, which comprises roughly 15% of mucinous tumours, commonly has exuberant tufting and marked stratification of the epithelium but this does not have an adverse effect on outcome. True invasion is needed for the diagnosis of ovarian cancer with this type of mucinous tumour.

In contrast, the outcome for the "intestinal" type is worse if there is stratification which is more than three cells deep, or a mitotic rate of greater than 10 mitoses/10 high power fields, or nuclear enlargement greater than three times the size of a normal stromal cell nucleus. The survival for patients with any of these indices falls from 90–100% to 75%. If they are present therefore the tumour should be regarded as invasive.[12] Not all investigators agree with this broadening of the definition of invasive cancer. Their objection is that the data on which it is based comes from small numbers of patients, and in addition they do not think that it is worthwhile to increase the diagnostic complexity because the prognostic precision remains unchanged.[13 14] An excellent review of this progressively more complex topic is that by Bell.[11] The relative percentages of the different types are: serous (55%), mucinous (40%), mixed (2%), endometrial (2%), clear (<1%), and Brenner (<1%) (table 9.2).

Table 9.2 Number of patients by histological subtype

First author	Year	Reference No	Histological subtype					
			Serous	Mucinous	Endometrioid	Clear cell	Brenner	Mixed
Russell	1979	13	70	52	14	3	1	4
Nikrui	1981	21	36	25	1	0	0	0
Creasman	1982	22	14	13	2	0	0	0
O'Quinn	1985	23	10	1	2	0	0	0
Barnhill	1985	24	77	17	0	0	0	0
Tasker	1985	25	22	39	0	0	0	0
Bostwick	1986	26	73	30	—	—	—	—
Nation	1986	27	28	36	0	0	0	1
Kliman	1986	28	29	39	3	0	0	5
Hopkins	1987	29	61	33	0	0	0	0
Chambers	1988	30	51	11	3	0	0	3
Yazigi	1988	31	25	11	0	0	0	0
Yoonessi	1988	32	10	14	1	0	0	1
Chien	1989	33	35	6	0	0	0	0
Nakashima	1989	34	12	52	2	0	1	4
Fort	1989	35	32	14	1	0	1	0
Rice	1990	36	39	37	0	0	0	1
Massad	1991	37	23	6	1	0	0	1
Manchul	1992	38	58	23	0	0	0	0
Kaern	1993	39	174	178	7	2	0	8
Total (n = 1589)			879	637	37	5	3	28
Percentage			55	40	2	Rare	Rare	2

Epidemiology

To get an accurate idea of how common a tumour is it is necessary to use population based statistics as this avoids the problem of referral bias which confounds the data from hospital series. Two reports fulfill this criterion. That from Katsube *et al* relates to the Denver Metropolitan area and details the number of new cases diagnosed over a six month period in 1979.[40] A total of 383 ovarian neoplasms were diagnosed of which 220 (57%) were of epithelial origin. The percentages of benign, borderline, and invasive epithelial neoplasms were 60%, 8%, and 32% respectively. These percentages changed when the different histological types were analysed independently. For serous tumours 33% were invasive, 17% borderline, and 50% benign. In contrast only 5% of mucinous tumours were invasive and 79% benign. The percentages also vary with age with benign tumours predominating before 40 years of age and malignant thereafter (table 9.3).

The second study was from Harlow *et al* and quoted the incidence from 13 counties in the state of Washington over an eight year period[41]; the annual incidence was 24.7/million women, which is about 30% of the incidence of invasive epithelial ovarian cancer. In the study by Katsube *et al* 20% of all epithelial malignancies were borderline.[40] The percentage in hospital based series tends to be lower, usually about 10% with a range from 9% to 20%.[21 27 39 42] This reflects the trend for patients with borderline

Table 9.3 Population-based incidence of ovarian neoplasms by age[40]

Type of tumour	Actual No of patients by age (years)				Total
	<20	20–39	40–49	>50	
Benign					
Epithelial	9	54	47	21	131
Non-epithelial	15	85	41	16	157
Borderline	0	8	8	2	18
Malignant					
Epithelial	0	8	35	28	71
Non-epithelial	1	5	0	0	6
Total	25	160	131	67	383

tumours not to be referred to tertiary centres. The lifetime risk of developing a borderline cancer for any women in the USA is 1:200 to 1:300. There are about 3000 new cases a year. Table 9.4 lists the incidence/year within different age groups in western Washington. The incidence shows a continuous increase, though at a slower rate than that of invasive cancer, until at least the age of 79. The incidence has also been increasing with time, from 16.6/million women/year in 1978 to 390 in 1983.[41] Whether this is a true increase in incidence or merely reflects an increase in diagnosis because of the routine use of ultrasound scans or a change in pathologists' diagnostic patterns is unknown.

Table 9.4 Incidence/million women/year of serous and mucinous borderline and invasive ovarian cancers by age in western Washington[41]

Age (years)	Borderline	Invasive
20–34	11.3	12.6
35–49	25.8	51.0
50–64	31.6	152.7
65–79	45.9	186.3

The mean age at diagnosis of women with borderline tumours is about 45 years, which is 10 years younger than that of invasive cancer.[28 29 42 43] The annual incidence is higher among whites than among non-whites, being 26.2/million and 16.5/million, respectively, in the state of Washington.[41]

Aetiology

The risk profile for developing a borderline tumour is similar to that of invasive ovarian cancer. An analysis of 12 case control studies in the USA showed that parity is inversely related to risk with an odds ratio of 0.54 for women who had ever had a pregnancy that went to term. The greater the number of pregnancies, the lower the risk. A younger age at the time of the first live birth, an increasing number of failed pregnancies, and a history of breast feeding reduced the risk, although not significantly so. A history of infertility increased the risk with an odds ratio of 3.8 for nulliparous

115

women.[43] There are several case reports that also implicate infertility as a potential risk factor.[44 45] Increasing duration of oral contraceptive use was protective as was ever having used them at all which had an odds ratio of 0.8. A body mass index (weight (kg)/height (m)^2) over 28 increased the risk twofold. Oestrogen replacement therapy, age at menarche, and age at natural menopause had no effect on risk. There are several possible aetiological hypotheses that have been proposed to account for these facts. Fathalla proposed that "incessant ovulation" in which the repeated replication of the epithelium after ovulation with the switching on and off of cell growth had the potential to cause unchecked growth.[46] Cassagrande *et al* modified this further by adding the concept of "ovulatory age"—that is, the more ovulations, the greater the risk.[47] In contrast it could be that the endocrine disorders that lead to infertility are themselves risk factors.[48] These two theories are not mutually exclusive.

As with invasive epithelial ovarian cancer, perineal exposure to talc may have a role.[49 50] A case control study, with all its limitations, showed a 2.8 times increased risk for women who used talc on the perineum.[49] In contrast the use of baby powder or cornstarch did not increase the risk. The likeliest mechanism is that mineral talc is structurally similar to asbestos which is known to induce cancer of mesothelial origin.

Viruses have been implicated in some gynaecological malignancies, but human papillomavirus types 6,11,16 and 18 were not found in 24 borderline cancers by polymerase chain reaction technology.[51]

Are borderline tumours precursors of invasive cancer?

Are borderline epithelial cancers like cervical intraepithelial neoplasms in being precursors of invasion? There are few data, but indirect evidence suggests that they share a common cell of origin, the ovarian epithelial cell; the risk factors for borderline and invasive cancers are the same; and that their age at diagnosis is different with a progressive increase from benign to borderline to invasive neoplasms.

Other more direct, though not conclusive evidence is available. There is histological evidence for all processes occurring within the same specimen. Plaxe *et al* studied the epithelium adjacent to 50 stage I invasive cancers and 50 normal control ovaries. Nuclear and cellular atypia, features of borderline neoplasia, were found in 49 of the invasive and none of the normal ovaries.[52] It has been shown on electron and light microscopy that there is a benign-borderline-invasive continuum in some serous tumours.[53 54] A change from a borderline tumour at diagnosis to invasive cancer at relapse is occasionally seen, but this could be the result of undersampling of the initial specimen with the diagnosis of invasion being missed.

Another line of evidence comes from the knowledge that the malignant state arises from the accumulation of a critical number of mutations within the regulatory genes, both the growth inducers (oncogenes) and the tumour suppressor genes. Evidence of genetic changes in common would support,

116

but not prove, the concept of the continuum. The K-ras oncogene has been studied but with conflicting results. It was found in 30%–47% of all borderline tumours with a higher incidence in mucinous tumours.[55 56] Two studies, however, reported it in invasive cancers[56 57] and two did not.[55 58] It is likely that assay sensitivity explains this discrepancy. Mutation of the P53 tumour suppressor gene was found only in invasive cancers.[55] This implies that either there is no continuum or that if there is a continuum this mutation is a late event in the progression to the invasive genotype. In situ hybridisation for Neu proto-oncogene transcripts and immuno-histochemical tests for Neu antigen showed that it was present in only one of six borderline tumours but in 30 of 68 invasive cancers.[59] Neu expression is therefore not needed for the borderline phenotype to occur. Gallion *et al* studied allelic loss on chromosomes from benign, borderline, and invasive cancers.[60] Loss of heterozygosity on chromosome 11 was seen only in the invasive cohort and in only 53% of them implying that it is not involved in the origin of benign or borderline tumours. In contrast an abnormality on chromosomes 13 and 17 was common to all three processes. The individual tumours, however, did not show all the abnormalities.

The data provide some support for the idea of a continuum but are far from conclusive. The issue is confounded by the fact that it is likely that different genetic pathways exist with the common end point being the malignant phenotype, so that benign tissue can become malignant either directly or through an intermediate "borderline" step.

Preoperative diagnosis of borderline tumours

There are no specific signs, symptoms, tumour markers or imaging techniques, with one possible exception, which can be used to make the diagnosis preoperatively. The age of the patient can make one suspicious, but no more than that.

The signs and symptoms, as with invasive ovarian cancers, are non-specific.[21 26–29 31 34 61] Abdominal distension (in 20%–92% of cases) and abdominal discomfort (in 10%–58%) are the most common findings. Abnormal bleeding (6%–44%), urinary symptoms (6%–13%), and alteration in bowel habit (5%) are also found. From 8%–30% of the time borderline tumours are either an incidental finding at surgery or found on routine examination of women with no symptoms.

Data about preoperative tumour markers are scanty; only two studies present the results by stage. In the first, 10 of 25 stage I patients and 12 of 13 stage II patients (or higher) had increased concentrations of CA125, with the highest being 318 U/l.[62] In the other there was no difference by stage, with eight of 10 stage I patients and all 10 stage II and III patients having increased concentrations; the highest was 272 U/l[63] and the highest reported anywhere was 900 U/l.[64 65] How applicable these results are is not clear as only 21% of the patients in these studies had preoperative markers assayed. Indirect evidence that selection bias may have been operating is that raised CA125 concentrations were found in 54% and 100%, respectively, of the stage I mucinous tumours in the two studies. For invasive mucinous tumours of all stages the figures would be 40%–50%.[66]

Current imaging techniques have produced conflicting results. Some authors have shown that borderline tumours have features similar to those of their invasive counterparts with thick, irregular cyst walls covered with vegetations.[67 68] In contrast, others have shown that they are indistinguishable from benign tumours.[69 70] There is one intriguing early report which concerned blood flow velocity waveforms analysed with transvaginal Doppler ultrasound.[70] The three sonographically "benign" borderline tumours had high diastolic flows and low pulsatility similar to those seen in invasive tumours. The explanation for these findings is that the tumour vasculature lacks muscle.

Role of surgery in borderline tumours

Surgery has several distinct roles. Tissue can be removed for diagnosis and for analysis of pathological prognostic variables such as DNA content. There is a caveat about obtaining an histological diagnosis: frozen section is not 100% accurate.[71] In a study from Holland frozen section analysis was done 19% of the time and the diagnosis of borderline tumour was made in 13 women. On formal histological review four of these 13 (31%) had invasive cancer and one (8%) had a benign tumour. Six of the 98 "benign" tumours were upgraded to borderline and two to malignant. In contrast there were no false positives if the frozen section diagnosis was of invasive cancer. Staging information can also be obtained which in turn directs treatment strategy and provides an assessment of the likely outcome for the patient. Finally it is therapeutic in its own right in that the cancer is removed.

Traditional surgical staging and treatment has followed the guidelines used for invasive ovarian cancer which includes bilateral salpingo-oophorectomy, hysterectomy, omentectomy, washings, diaphragmatic palpation, and lymph node palpation. Not all groups regard this degree of surgery as adequate for women with invasive cancer but would also do nodal biopsy or dissection, diaphragmatic scraping, and take random biopsy specimens. The data suggest that the more limited surgical approach is sufficient and that even more restricted resection may be appropriate in some circumstances.

Surgical staging

Tables 9.5 and 9.6 list the stage distribution for mucinous and serous tumours. The degree of "surgical thoroughness" varies and this explains some of the variations reported in the relative percentages by stage. There is an instructive report by McGowan *et al* which deals with the issue of what actually occurs at the laparotomy.[74] He reviewed retrospectively the

118

Table 9.5 Distribution by stage of serous tumours of borderline malignancy

First author	Year	Reference No	FIGO stage			
			I	II	III	IV
Nikrui	1981	21	23	6	4	0
Barnhill	1985	24	36	12	28	1
Bostwick	1986	26	57	8	14	0
Nation	1986	27	19	4	5	1
Kliman	1986	28	18	4	7	0
Hopkins	1987	29	20	13	14	0
Chambers	1988	30	43	8	10	0
Chien	1989	33	19	4	12	0
Nakashima	1989	34	10	1	1	0
Rice	1990	36	26	4	9	0
Massad	1991	37	12	4	7	0
Leake	1992	61	135	24	41	0
De Nictolis	1992	72	25	5	14	0
Kaern	1993	39	148	13	13	0
Total (n = 882)			591	110	179	2
Percentage			67	12	20	<1

Table 9.6 Distribution by stage of mucinous tumours of borderline malignancy

First author	Year	Reference No	FIGO stage			
			I	II	III	IV
Nikrui	1985	21	21	0	1	1
Barnhill	1985	24	11	1	5	0
Chaitin	1985	74	21	1	10	1
Bostwick	1986	26	30	0	0	0
Nation	1986	27	34	1	2	0
Kliman	1986	28	31	1	7	0
Hopkins	1987	29	9	0	2	0
Chambers	1988	30	30	1	2	0
Chien	1989	33	4	0	2	0
Nakashima	1989	34	41	2	9	0
Rice	1990	36	37	0	0	0
Massad	1991	37	4	1	1	0
Kaern	1993	39	147	6	25	0
Total (n = 502)			420	14	66	2
Percentage			84	3	13	<1

operative reports of women staged for ovarian cancer between 1978 and 1981, (which may underestimate what actually took place), and in only one third were all staging components carried out when general surgeons were involved. This improved to 50% for general gynaecologists, and was close to 100% for gynaecological oncologists.

Mucinous tumours are significantly more likely to be stage I than serous tumours (84% compared with 67%). The rate of bilateral ovarian

Table 9.7 Incidence of bilateral involvement of the ovaries

First author	Year	Reference No	Histological type	
			Mucinous	Serous
Hart	1973	7	7/97	—
Katzenstein	1978	75	—	22/50
Russell	1979	13	3/52	24/70
Chaitin	1985	74	0/15	—
Nikrui	1985	21	—	9/33
Bostwick	1986	26	3/30	38/79
Kliman	1986	28	5/39	14/29
Hopkins	1987	29	0/11	23/51
Nakashima	1989	34	5/54	8/12
Massad	1991	37	0/6	12/23
Leake	1992	61	—	72/200
Total			23/304	222/547
Percentage			8	41

involvement also changes depending on the histological picture (table 9.7). For serous tumours the rate is 41% (range 28%–66%) whereas for mucinous tumours it is 8% (range 0–13%). Germane to the question of "How little resection is needed for accurate staging?" is the question "Is the other ovary involved macroscopically or microscopically?" Information is limited, as most reports do not address how bilaterality was documented. In three studies the cancer was always macroscopically evident[28 31 61]: one study found that one of 27 normal looking ovaries was involved,[76] and in the final report all mucinous bilateral tumours were macroscopically obvious.[39] By inference, some of the serous tumours must have been microscopic, so for a mucinous tumour removal of a normal looking opposite ovary is not indicated. For a woman past her reproductive years, however, there is no real benefit in leaving the second ovary. In contrast, for serous tumours (given the increased risk of bilaterality, both macroscopic and microscopic), unless preservation of fertility is an issue, the other ovary should be removed. There is no role for "bivalving" the normal ovary in a woman who wants to maintain her fertility. Tazelaar *et al* did this and found no ovarian involvement in the seven women they treated.[77] The incidence of infertility after this type of procedure is 14% and given the low diagnostic rate this negates its value.[78]

Does the more rigorous surgical approach advocated by some authorities provide more information about unsuspected disease? For mucinous tumours it does not, but for serous tumours it does. Leake *et al* sampled retroperitoneal nodes in 34 (20%) women.[79] In six of the 33 (18%) who underwent para-aortic sampling the nodes contained tumour. Two of the 12 who had pelvic node sampling had nodal involvement. None of the seven patients with mucinous tumours had involved nodes. For the serous tumours the risk of nodal involvement was 19% for the apparent stage I tumours and 33% for those in stage II. Yazigi *et al* documented the detection rate from various elements of the staging process.[31] Washings showed involvement in two of 27, omental biopsy in three of 23, diaphragmatic biopsy in one of 15, sampling of pelvic nodes in four of 15 and of para-aortic nodes in one of 15. Four (16%) of the apparent stage I tumours were upgraded as were three (75%) of the stage II tumours. Snider *et al* reported the outcome of restaging with nodal biopsy, diaphragmatic scraping, and peritoneal biopsy in a group of 27 patients and, again, none of the patients with mucinous tumours were upgraded. Five patients in total were upgraded: four with omental or peritoneal disease, and one with a tumour in the other ovary. No nodes were documented as involved. It is of note that only one of the four involved peritoneal biopsy specimens contained invasive implants.[76] The operative morbidity was 7.4%. This rate of upgrading of serous tumours is low and also probably irrelevant given that non-invasive implants do not affect outcome.

Another common event is where malignancy is not suspected at the initial operation and so the abdomen is not explored. Should this patient be re-operated on? Given the normal staging distribution one would expect up to 30% to have more disease. Seven of 15 such patients re-operated on by Hopkins and Morley were found to have disease.[80] No breakdown by histological type was given. In such cases repeat operation would be reasonable.

Surgery as treatment

The ultimate test of the utility of the more rigorous surgical procedure which undoubtedly uncovers unsuspected disease in patients with serous tumours, is its effect on outcome. The Norwegians routinely do more limited staging with hysterectomy, bilateral salpingo-oophorectomy, infracolic omentectomy, and diaphragmatic and nodal palpation. Some of these patients must be "understaged" yet the 15 year survival of their patients with stage I and II tumours was 99% with only three of 234 women relapsing.[80] It would be difficult to improve on this outcome, so the more limited staging and therapeutic approach should be regarded as the standard. One reason why unsuspected disease does not alter outcome is that it may not have malignant potential, as in the example of non-invasive implants. It is likely that borderline tumours are a mixed population, some of which are truly malignant with metastatic potential and others, the majority, being benign.

For some patients conservative surgery is justifiable. A total of 30% of patients are under 40 years at the time of diagnosis,[9] and for some of them preserving their fertility is important. For patients with mucinous tumours and apparent stage I disease, upstaging as a result of more thorough surgery is rare and so unilateral oophorectomy is safe. For those with serous tumours unilateral salpingo-oophorectomy, if the other ovary is macroscopically normal and provided that the abdomen has been explored thoroughly, is a reasonable approach with no obvious detrimental effect on survival. The duration of follow-up in the studies on which this statement is based is relatively short and so one cannot be dogmatic about the long term outcome. The recurrence rate for stage IA patients treated by unilateral salpingo-oophorectomy alone ranged from 0 to 30%.[26 30 61 75 77] All who developed recurrences were re-operated on and remained disease free for the duration of follow-up, which was up to four years. All relapses were local and all but one occurred in women with serous tumours. In contrast, the recurrence rate for those treated by bilateral salpingo-oophorectomy was 5%.[77]

Even more limited surgery (excision of the cyst), can be done. Lim-Tan *et al* reported on 33 women treated by excision alone, although 60% of them had additional operations relatively soon afterwards which reduced the potential recurrence rate. Four tumours recurred in the ovary(s). All 33 women were alive three to 18 years later.[82] Eight of 16 who wished to become pregnant did so. Predictors of relapse were resection margins containing tumour cells and the necessity to remove more than one cyst. The recurrence rates after excision of cysts in the other two studies in which data were presented were 12.5% and 40%.[30 61] Such excision does not obviously improve the chances of maintaining fertility compared with unilateral salpingo-oophorectomy, and given the higher rate of recurrence it is not worthwhile.

All women who are treated conservatively need careful follow-up with some form of imaging as pelvic examination is relatively inaccurate. Implicit in this recommendation is the idea that early intervention when the relapse is asymptomatic is more beneficial than later intervention. There is, however, no evidence for or against this. Some authors have recommended completing formal surgery once the woman has completed her family. Given that the relapse rate is low, that relapse is most common within the first five years and that it may take this long to complete the family, and provided that the patient is followed up closely, further operation is not necessary.

Outcome

Survival rates by stage and histology are shown in tables 9.8–9.11. The extent of operation and the use of adjuvant treatment varied from series to series as did the method of calculating survival. Despite this the results are remarkably similar. The most striking feature is that most patients are either cured or live a long time. For those with mucinous tumours 85% or more were alive at 15 years. The figures are similar, 87%–92%, for those with serous tumours. When stage is taken into account the 15 year figures from Norway are stage I—94.7%, stage II—66.7%, and stage III—57.4% for mucinous, and 100%, 76.9%, and 64%, respectively, for serous tumours.[39] In the report by Manchul *et al* the 10 year disease free survival, irrespective of stage and histology, was 78%.[38]

The actual number of relapses by stage and histology are given in table 9.12. For stage I mucinous tumours there was only one relapse in the two

Table 9.8 Percentage survival by stage[1]

First author	Reference No	Survival time (years)		
		5	10	15
Stage 1				
Kaern	39	98	97	96
Kliman	28	—	94	—
Manchul	38	100	100	—
Manchul[2]	38	94	87	—
Russell	13	92	—	—
Stage II				
Kaern	39	80	73	73
Kliman	28	—	94	—
Stage III				
Kaern	39	80	73	73
Kliman	28	—	72	—
Barnhill	24	88	—	—
Russell	13	64	—	—
Hopkins	29	82	60	—

[1] All are cause specific except
[2] Manchul—disease free survival

Table 9.9 Percentage survival for patients with mucinous borderline tumours

First author	Reference No	FIGO stage				All patients regardless of stage
		I	II	III	IV	
Five years						
Kaern	39	—	—	—	—	95
Aure	43	—	—	—	—	92
Hart	7	98	—	—	—	—
Chaitin	73	95	—	—	—	—
Nakashima	34	93	—	—	—	69
Barnhill	24	—	—	—	—	78
Nikrui	21	—	—	—	—	81
10 years						
Kaern	39	—	—	—	—	91
Aure	42	—	—	—	—	88
Hart	7	96	—	—	—	—
Chaitin	73	90	—	—	—	—
Nakashima	34	—	—	—	—	62
Nikrui	21	—	—	—	—	73
15 years						
Kaern	39	95	67	58	—	88
Aure	42	—	—	—	—	85
20 years						
Aure	42	—	—	—	—	85

Table 9.10 Percentage survival for patients with serous borderline tumours

First author	Reference No	FIGO stage				All patients regardless of stage
		I	II	III	IV	
Five years						
Kaern	39	100	—	—	—	96
Aure	42	—	—	—	—	96
Leake	61	—	—	—	—	97
Barnhill	24	—	—	—	—	97
Nikrui	21	—	—	—	—	92
10 years						
Kaern	39	100	—	—	—	95
Aure	42	—	—	—	—	90
Leake	61	—	—	—	—	95
Nikrui	21	—	—	—	—	83
15 years						
Kaern	39	100	76.9	64	—	95
Aure	42	—	—	—	—	87
Leake	61	—	—	—	—	92
20 years						
Aure	42	—	—	—	—	76
Leake	61	99	96	45*	—	89

* In the text the mortality rate for stage III was given as 26.8%, so 45% may be an error.

Table 9.11 Percentage survival, irrespective of stage or histological type

	Reference No	Survival (years)	
		5	10
Petterson	9	89	—
Chambers	30	83	—
Barnhill	24	95	87
Manchul	38		
Disease free		—	78
Overall		—	85
Cause specific		—	96

Table 9.12 Relapse rate by histology and stage

	Reference No	Stage			Duration of follow up (years)
		I	II	III	
Serous/mixed					
Bostwick	26	8/57	0/8	9/14	3.0—21
Leake	61	8/135	4/24	22/41	3.0—27
Chambers	30	4/43	0/8	2/10	0.1—10
Mucinous					
Bostwick	26	0/30			3.0—21
Chambers	30	1/30	0/1	0/2	0.1—10

series that presented data that could be interpreted. This reinforces the concept that it is wrong to call such mucinous tumours "borderline" or of "low malignant potential". Instead they should be regarded as of "no malignant potential".[83] Most relapses develop relatively soon after diagnosis.[84] The cumulative rate reported by Kaern *et al* was 15 within two years, 23 in five years, and 27 within 10 years.[39] The figures given by Nikrui were four within two years and seven within 10 years.[21] Bostwick *et al* gave no figures but reported that most were within the first five years.[26] One report, however, documented 18 recurrences 14 of which occurred more than five years after diagnosis.[85] Late relapse can occur. Hopkins *et al* reported two at 23 and 25 years after diagnosis.[29] Data for relapse rates after 10 years are not available. Prolonged survival after relapse is possible with a median of five years (range one and a half to six years) reported by Hopkins *et al*,[29] and one patient in Nikrui's series was still alive 14 years later.[21]

Role of additional treatment after surgery

The standard treatment for borderline tumours is surgical removal. Additional non-surgical treatment may be needed in some instances. Two questions need to be answered. Is the risk of relapse/recurrence and eventual death sufficient to merit using chemotherapy or radiotherapy? If the answer is yes, is there treatment that will alter such an outcome?

For stage I patients the answer to the first question is "no". The 15 year survival is 95% or more. We would have to treat 100 potentially to benefit five, with all the attendant acute and chronic toxicity. Furthermore, to show that a treatment helped would require a worldwide collaborative trial as thousands of patients would have to be treated to achieve a statistically significant result given the small number of adverse events.

The three randomised studies, all with small numbers of patients, provide no hint of a beneficial effect of treatment.[22 81 86] Creasman randomised 55 stage IA and B patients to observation, pelvic irradiation, or melphalan.[27] The one relapse was in the irradiation group. In the Norwegian trial 66 patients with stage I and II tumours were randomised to receive thiotepa or to have no additional treatment. The only patient who died of the cancer was in the thiopeta group. In addition the group compared the survival of their treated patients from their randomised studies with that of a control group (not randomised) who had been treated by operation alone. The survival rate was the same. Young *et al* compared the use of melphalan with no adjuvant treatment in stage I invasive ovarian cancer. On review, 27 had borderline tumours and of these only one patient died.

For patients with stage II tumours the 15 year survival is about 75%. At this level there is a role for adjuvant treatment but unfortunately there are no randomised trial data about efficacy. Chambers summarised the data from the individual series and the recurrence/persistence/death rates for those treated by surgery alone, or surgery plus chemotherapy or radiotherapy, or chemotherapy plus radiotherapy were 4.4%, 11.2%, 6.7%, and 11.1%, respectively.[87] This suggests that additional treatment has no beneficial effect.

The 15 year survival rate for patients with stage III disease is only 60%,[39] so good adjuvant treatment is needed. The results of treatment with either alkylating agents or cisplatin-based combinations as assessed at repeat operations are shown in Table 9.13. There is no doubt that some of the cancer cells were killed in a small percentage (30%) of those treated. There are, however, no data available to show if this is enough to prolong survival or alter the quality of life.

Table 9.13 Chemotherapy for stage III tumours with macroscopic residual tumour after initial operation: response assessed at second-look laparotomy

First author	Reference No	No of patients	No with no sign of recurrence at second-look laparotomy
Fort	35	6	3
Nation	27	2	0
O'Quinn	23	5	0
Chambers	30	4	3
Yazigi	88	8	1
Sutton	89	Unknown	2
Kliman	28	5	0

It is not surprising that treatment has so little effect. Radiotherapy and chemotherapy require cells to be in the cell cycle to kill them. Borderline tumours are indolent and only a small fraction of the cells in the tumour will be going through the cell cycle. To alter the outcome an altered approach which bypasses the need for cells to be in cycle is needed—for example, using differentiating agents, or restoring growth control by hormonal manoeuvres or by manipulation of oncogenes or their products.

Currently there is no adjuvant treatment that can be regarded as the standard to be used routinely in patients at risk. Any treatment should be regarded as experimental. An eminently reasonable approach is to use repeat surgical debulking while it is technically feasible for relapses, and then to change to chemotherapy or radiotherapy when debulking is no longer possible.

Prognostic factors

Prognostic factors have a great potential value in a disease such as borderline ovarian cancer in which most of the patients are cured by surgery alone. Their use enables us to offer additional treatments, with their attendant acute and chronic toxicities, only to those who are at risk of relapse. Unfortunately it is not as simple as this, because the "adverse prognostic factors" do not invariably predict relapse and death. A combination of factors may be more accurate but so far I know of only one group that has studied this.[39] A further problem is that most patients do not relapse, and relapse if it occurs can be delayed. The available studies contain small numbers of patients with relatively short follow-up. The number of adverse events is therefore small and this may lead to a statistically insignificant result despite there being a clinically relevant effect.

Stage is an important prognostic factor (table 9.8). Using the Norwegian experience as a representative example, the 10-year survival rates for stage I, II, and III tumours were 97%, 73%, and 68%, respectively.[39] It is not quite as simple as this, however. The type of peritoneal implant seems to alter outcome, at least when the results from all studies are combined (table 9.14). Overall, if there were invasive implants 32% died as opposed to 9% if they were not invasive. The individual studies gave contradictory results, though the outcome for non-invasive implants was constant among the studies. It is the outcome for the invasive implants that varied. As discussed by Bell *et al* this may be the result of different histological interpretations about what is invasive,[93] or differences in duration of follow-up or in the thoroughness of the sampling of the peritoneal implants. Additionally, other biological characteristics may be important. The degree of nuclear atypia and mitotic activity were important with regard to outcome for non-invasive implants.[93] Similarly, only patients with aneuploid implants died.[72] It is still not clear whether all implants are metastases or whether they merely indicate a field change. The excellent outcome for non-invasive implants favours the latter interpretation.

Table 9.14 Influence of type of peritoneal implant on survival in patients with serous borderline tumours

First author	Year	Reference No	Type of implant**	
			Non-invasive*	Invasive
McCaughey	1984	90	2/13	4/5
Russell	1984	91	2/6	3/6
Michael	1986	92	1/6	1/7
Bell	1988	93	3/50	4/6
Kliman	1988	28	1/8	0/2
Gershenson	1990	94	2/37	1/13
DeNictolis	1992	72	0/10	4/9
Manchul	1992	38	1/7	0/5
Total			12/127	17/53
Percentage			9	32

* Benign (endometriosis, endosalpingiosis) not included; ** Denominator is the total number of patients, numerator is the number who died.

Another clinical characteristic that affects outcome is the presence of macroscopic residuum. Four of 12 women with such residuum died compared with none of the 26 who had none in the series reported by Bell *et al.*[93] Leake *et al* reported a 66% recurrence rate and 31% mortality in their patients who had residual disease compared with 19% and 8%, respectively, among those who did not.[61] The Norwegian experience was similar.[39]

Histological type may alter outcome.[39] The 10 year survival rate was better for serous or mucinous tumours compared with any other type—95% compared with 75%. The same study also identified size of the primary (if it was under 10 cm in diameter, 4% died; if it was more than 10 cm, 14% died) and the presence of surface extension (21% compared with 6% died) as adverse factors. In contrast, three other studies found no correlation with surface extension.[29 38 61] They also showed that surgical rupture of the tumour had no effect. In addition, cytology indicating malignant cells, adhesions,[38] and the presence of psammoma bodies[29] did not alter outcome.

The only multivariate analysis is that by Kaern *et al.*[39] They identified three independent adverse factors: stage II or III tumours; histological type other than serous or mucinous, and age over 70 years. They identified a low risk group in which all patients survived and a high risk group with a survival rate of 25% or less using these criteria. The low risk group had stage 1A serous or mucinous tumours and were aged 40 or less, and the high risk group were over 70 with mucinous or serous tumours that were stage II or III, or they were any age with stage III non-serous or non-mucinous tumours.

Even using combinations of clinical factors, patients cannot accurately be put into pigeon holes with guaranteed outcomes. This is where non-clinical factors such as DNA content, grading, and morphometry may have a role.

Conflicting results have been reported for the rate of aneuploidy in borderline ovarian cancer with rates ranging from 0 to 52% (table 9.15). There are several explanations for this variation. Different techniques are used to analyse DNA content, and flow cytometry[97–99 104] produces lower rates than image analysis.[72 96 107] Two distinct G_O/G_I peaks are needed before a tumour can be described as aneuploid. To be able to discriminate between these a 10% difference in DNA content is needed. Smaller differences cannot be reliably detected by flow cytometry. If cell size is analysed as well, however, then smaller differences can be detected.[105] The number of samples analysed, the technique for preparing single cells, and the representative nature of the sample are also important. The subject has been reviewed in two papers.[105 108] The likelihood of aneuploidy seems to increase with increasing stage, being 12% for stage I and 36% for stage III (table 9.15). The outcome is worse for women with aneuploid tumours. Only 7% of diploid tumours compared with 23% of aneuploid tumours relapsed.[72 96 99 100 102 104 107] The true difference is not clear given the discriminatory abilities of the different techniques, variations in follow-up, and patient selection.

These results do not include the stage I result from Norway. The results from this large study are instructive, but its size would overwhelm the smaller studies. In it, 321 borderline tumours were analysed and 17 of 28 aneuploid (61%) and 9 of 293 (3%) diploid patients died of cancer.[95] If this is confirmed from long term follow-up in other series then aneuploidy would indicate the need for additional treatment.

128

Table 9.15 Number of aneuploid tumours by stage

First author	Reference No	Stage				All stages (%)
		I	II	III	IV	
Vergote	95	28/321	—	—	—	—
Drescher	96	←5/15→		4/6	—	9/21 (43)
Lage	97	0/17	0/4	0/6	—	0/27
De Nictolis	72	12/25	3/5	8/14	—	23/44 (52)
Iversen	98	—	—	—	—	0/6
Klemi	99	—	—	—	—	7/43 (16)
Padberg	100	—	—	—	—	17/80 (21)
Kuhn	101	—	—	←1/8→		1/8 (12)
Seidman	102	12/30	—	4/14	—	16/44 (36)
Freidlander	103	—	—	—	—	2/7 (29)
Freidlander	104	0/30	—	2/14	—	2/44 (5)
Eriksen	105	—	—	—	—	15/45 (33)
Griffiths	106	0/14	0/2	—	—	0/21
Fu	107	←0/9→		4/7	—	4/16 (25)

Conversely, the diploid tumours do not need additional postoperative treatment.

The degree of nuclear atypia is important. Two of 44 (5%) and two of 12 (42%) women with implants with mild/moderate atypia and severe atypia died.[93] Interestingly, severe atypia was an adverse factor in non-invasive implants. Hopkins *et al* studied the ovarian tumour and reported 10 year survivals of 98% and 52% for tumours with low and high atypia.[29] In contrast, Kaern *et al* found no significant difference although the 10 year rates were 93% and 80%, respectively.[39] Hopkins *et al* also graded the primary tumour based on the amount of stratification, number of nucleoli, degree of nuclear atypia, and presence of a cribriform pattern. The long term survival rates for low grade and high grade tumours were 100% and 38%.[29] Bell *et al* showed that the presence of mitoses within implants indicated a poor prognosis with three deaths in the nine patients with mitoses and only four in the 47 without.[93]

Morphometry (measurement of sizes and shapes of cells and nuclei) gives prognostic clues. Baak *et al* showed that more than 30 mitoses/25 high power fields, an increased ratio of epithelial to epithelial/stromal volume, and more irregular nuclei, were adverse findings. None of the 15 in the favourable group died while all three in the poor group did.[109]

All these techniques probably measure interrelated properties. No multivariate analysis has been done to identify if they are of independent prognostic value or which is the superior discriminant. DNA content and morphometry are more reproducible as they are not subjective, unlike grading and degree of atypia, so they are probably the more valuable indices.

Pseudomyxoma peritonei

Pseudomyxoma peritonei was originally described by Werth in 1884. It describes the syndrome in which gelatinous material fills the peritoneal cavity and is associated with mucinous tumours of the ovary and appendix. If careful sectioning of the "primary" is carried out, rupture will almost inevitably be found. Mucinous epithelium growing on the visceral and parietal peritoneum is always found although it can be difficult to show. There are two theories about its histogenesis. The first is that implantation arises as a result of the spillage, and the second is that the mucin is an irritant which causes metaplasia of the mesothelial cells.[110] Coexisting ovarian and appendiceal lesions are common and it has been postulated that the appendix is the original site of the problem.[111][112] An appendiceal tumour is not always found on routine sampling, however.[39] Interestingly, pseudomyxoma peritonei is not found with the Mullerian variant of mucinous tumour but only with the intestinal type.[113] It is not particularly common, and is found in association with about 16% of mucinous borderline tumours.[28][34] The survival rates vary depending on when the report was published. Long *et al* reported a 40% 10 year survival rate in 1969,[114] and Campbell *et al* found only seven deaths in 26 patients when they reviewed published reports in 1973.[116] In contrast, more modern series have a 10 year survival rate of only 18% and death is the norm, often within five years.[28][34][110][113] The standard course is a pattern of recurring bowel obstruction leading to eventual death. Distant metastases are unusual.

The standard treatment is repeated laparotomies with removal of the viscous material. Paracentesis is ineffective as the material is too thick to drain. Various other treatments including chemotherapy, both intravenous and intraperitoneal; radiotherapy, again intraperitoneal and external beam; mucolytics; and 5% dextrose have been tried.[28][116–118] There have been occasional responses, but the small numbers treated make the relevance unclear.

References

1 Taylor HC. Malignant and semi-malignant tumors of the ovary. *Surg Gynecol Obstet* 1929; **48**: 204–30.

2 Taylor HC, Alsop WE. Spontaneous regression of peritoneal implantations from ovarian papillary cystadenoma. *Am J Cancer* 1932; **16**: 1305–25.

3 Kottmeir HL. The classification and treatment of ovarian tumors. *Acta Obstet Gynecol Scand* 1952; **31**: 313–63.

4 Woodruff JD, Novak ER. Papillary serous tumors of the ovary. *Am J Obstet Gynecol* 1954; **67**: 1112–26.

5 Ingleman-Sundberg A. Classification and staging of malignant tumors in the female pelvis. *Acta Obstet Gynecol Scand* 1971; **50**: 1–7.

6 Serov SF, Scully RE, Sobin LH. *Histological typing of ovarian tumors.* International histologic classification of tumors, No 9. Geneva: World Health Organization, 1973.

7 Hart WR, Norris HJ. Borderline and malignant mucinous tumors of the ovary. Histologic criteria and clinical behaviour. *Cancer* 1973; **31**: 1031–45.

8 Gramlich T, Austin RM, Lutz M. Histologic sampling requirements in ovarian carcinoma: a review of 51 tumors. *Gynecol Oncol* 1990; **38**: 249–56.

9 Pettersson F. *Annual report of the results of treatment in gynecological cancer*. Vol 21. International Federation of Obstetrics and Gynecology. Stockholm: Elsevier, 1990.

10 Baak JPA, Langley FA, Talerman A, Delemarre JFM. The prognostic variability of ovarian tumor grading by different pathologists. *Gynecol Oncol* 1987; **27**: 166–72.

11 Bell DA. Ovarian surface epithelial-stromal tumors. *Hum Pathol* 1991; **22**: 750–62.

12 Sumithran E, Susil BJ, Looi L-M. The prognostic significance of grading in borderline mucinous tumors of the ovary. *Hum Pathol* 1988; **19**: 15–18.

13 Russell P, Merkur H. Proliferating ovarian "epithelial" tumors: A clinicopathologic analysis of 144 cases. *Aust NZ J Obstet Gynaecol* 1979; **9**: 45–51.

14 The Ovarian Tumor Panel of the Royal College of Obstetricians and Gynaecologists. Ovarian epithelial tumors of borderline malignancy: pathologic features and current status. *Br J Obstet Gynaecol* 1983; **90**: 743–50.

15 Bell DA, Scully RE. Atyical and borderline endometrioid adenofibromas of the ovary. A report of 27 cases. *Am J Surg Pathol* 1985; **9**: 205–14.

16 Rutgers JL, Scully RE. Ovarian mixed-epithelial papillary cystadenomas of borderline malignancy of Mullerian type. A clinico-pathologic analysis. *Cancer* 1988; **61**: 546–54.

17 Snyder RR, Norris HJ, Tavassoli F. Endometrioid proliferative and low malignant potential tumors of the ovary. A clinicopathologic study of 46 cases. *Am J Surg Pathol* 1988; **12**: 661–71.

18 Bell DA, Scully RE. Benign and borderline clear cell adenofibromas of the ovary. *Cancer* 1985; **56**: 2922–31.

19 Roth LM, Langley FA, Fox H, Wheeler JE, Czernobilsky B. Ovarian clear cell adenofibromatous tumors: benign, of low malignant potential, and associated with invasive clear cell carcinoma. *Cancer* 1984; **53**: 1156–63.

20 Roth LM, Dallenbach-Hellweg G, Czernobilsky B. Ovarian Brenner tumors. 1. metaplastic, proliferating and of low malignant potential. *Cancer* 1985; **56**: 582–91.

21 Nikrui N. Survey of the clinical behaviour of patients with borderline epithelial tumors of the ovary. *Gynecol Oncol* 1981; **12**: 107–19.

22 Creasman WT, Park R, Norris H, Disaia PJ, Morrow PA, Hreshchyshyn MM. Stage I borderline ovarian tumors. *Obstet Gynecol* 1982; **59**: 93–6.

23 O'Quinn AG, Hannigan EV. Epithelial ovarian neoplasms of low malignant potential. *Gynecol Oncol* 1985; **21**: 177–85.

24 Barnhill D, Heller P, Brzozowski P, Advani H, Gallup D, Park R. Epithelial ovarian carcinoma of low malignant potential. *Obstet Gynecol* 1985; **65**: 53–9.

25 Tasker M, Langley AFA. The outlook of women with borderline epithelial tumors of the ovary. *Br J Obstet Gynaecol* 1985; **92**: 969–73.

26 Bostwick DG, Tazelaar HD, Ballon SC, Hendrickson MR, Kempson RL. Ovarian epithelial tumors of borderline malignancy. A clinical and pathologic study of 109 cases. *Cancer* 1986; **58**: 2052–65.

27 Nation JG, Krepart GV. Ovarian carcinoma of low malignant potential: Staging and treatment. *Am J Obstet Gynecol* 1985; **154**: 290–3.

28 Kliman L, Rome RM, Fortune DW. Low malignant potential tumors of the ovary. A study of 76 cases. *Obstet Gynecol* 1988; **72**: 775–81.

29 Hopkins MP, Kumar NB, Morley GW. An assessment of pathologic features and treatment modalities in ovarian tumors of low malignant potential. *Obstet Gynecol* 1987; **70**: 923–9.

30 Chambers JT, Merino MJ, Kohorn EI, Schwartz PE. Borderline ovarian tumors. *Am J Obstet Gynecol* 1988; **159**: 1088–94.

31 Yazigi R, Sandstad J, Munoz AK. Primary staging in ovarian tumors of low malignant potential. *Gynecol Oncol* 1988; **31**: 402–8.

32 Yoonessi M, Crickard K, Celik C, Yoonessi S. Borderline epithelial tumors of the ovary: Ovarian intraepithelial neoplasia. *Obstet Gynecol Surv* 1988; **43**: 435–44.

33 Chien RTY, Rettenmaier MA, Micha JP, Disaia PJ. Ovarian epithelial tumors of low malignant potential. *Surg Gynecol Obstet* 1989; **169**: 143–6.

34 Nakashima N, Nagasaka T, Oiwa N, *et al.* Ovarian epithelial tumors of borderline malignancy in Japan. *Gynecol Oncol* 1990; **38**: 90–8.

35 Fort MG, Piene VK, Saigo PE, Hoskins WJ, Lewis JL. Evidence for the efficacy of adjuvant therapy in epithelial ovarian tumors of low malignant potential. *Gynecol Oncol* 1989; **32**: 269–72.

36 Rice LW, Berkowitz RS, Mark SD, Yauner DL, Lage JM. Epithelial ovarian tumors of borderline malignancy. *Gynecol Oncol* 1990; **39**: 195–8.

37 Massad LS, Hunter VJ, Szpak CA, Clarke-Pearson DL, Creasman WT. Epithelial ovarian tumors of low malignant potential. *Obstet Gynecol* 1991; **78**: 1027–32.

38 Manchul LA, Simm J, Levin W, *et al.* Borderline epithelial ovarian tumors: a review of 81 cases with an assessment of the impact of treatment. *Int J Radiat Oncol Biol Phys* 1992; **22**: 867–74.

39 Kaern J, Trope CG, Abeler VM. A retrospective study of 370 borderline tumors of the ovary treated at the Norwegian Radium Hospital from 1970–1982. A review of clinico-pathologic features and treatment modalities. *Cancer* 1993; **71**: 1810–20.

40 Katsube Y, Berg JW, Silverberg SG. Epidemiologic pathology of ovarian tumors: A histopathologic review of primary ovarian neoplasms diagnosed in the Denver standard metropolitan statistical area, 1 July–31 December 1969, and 1 July–31 December 1979. *Int J Gynecol Pathol* 1982; **1**: 3–16.

41 Harlow BL, Weiss NS, Lofton S. Epidemiology of borderline ovarian tumors. *J Natl Cancer Inst* 1987; **78**: 71–4.

42 Aure JC, Hoeg K, Kolstad P. Clinical and histologic studies of ovarian carcinoma. Long term follow up of 990 cases. *Obstet Gynecol* 1971; **37**: 1–9.

43 Harris R, Whittemore AS, Hnyte J. Characteristics relating to ovarian cancer risk: collaborative analysis of 12 US case-control studies. *Am J Epidemiol* 1992; **136**: 1204–11.

44 Nijman HW, Burger CW, Baak JPA, Schats R, Vermorken JB, Kenemans P. Borderline malignancy of the ovary and controlled hyperstimulation, a report of 2 cases. *Eur J Cancer* 1992; **28**: 1971–3.

45 Goldberg GL, Runowicz CD. Ovarian carcinoma of low malignant potential, infertility, and induction of ovulation—is there a link? *Am J Obstet Gynecol* 1992; **166**: 853–4.

46 Fathalla MF. Incessant ovulation—a factor in ovarian neoplasia? *Lancet* 1971; **ii**: 163.

47 Cassagrande JT, Louie EW, Pike MC, Roy S, Henderson BE. Incessant ovulation and ovarian cancer. *Lancet* 1979; **ii**: 170–3.

48 Whittemore AS, Wu ML, Paffenbarger RS, *et al.* Epithelial ovarian cancer and the ability to conceive. *Cancer Res* 1989; **49**: 4047–52.

49 Harlow BL, Weiss NS. A case control study of borderline ovarian tumors: the influence of perineal exposure to talc. *Am J Epidemiol* 1989; **130**: 390–4.

50 Cramer DW, Welch WR, Scully RE, Wojciechowski CA. Ovarian cancer and talc. *Cancer* 1982; **50**: 372–6.

51 McLellan R, Buscema J, Guerrero E, Shah KV, Woodruff JD, Currie JL. Investigation of ovarian neoplasia of low malignant potential for human papillomavirus. *Gynecol Oncol* 1990; **38**: 383–5.

52 Plaxe SC, Deligdisch L, Dottino PR, Cohen CJ. Ovarian intraepithelial neoplasia demonstrated in patients with stage I ovarian carcinoma. *Gynecol Oncol* 1990; **38**: 367–72.

53 Steinback F. Benign, borderline and malignant serous cystadenomas of the ovary. *Pathol Res Pract* 1981; **172**: 58–72.

54 Puls E, Powell DE, DePriest PD, *et al.* Transition from benign to malignant epithelium in mucinous and serous ovarian cystadenocarcinoma. *Gynecol Oncol* 1992; **47**: 53–7.

55 Teneriello G, Ebina M, Linnoila RI, *et al.* P53 and Ki-Ras mutations in epithelial ovarian neoplasms. *Cancer Res* 1993; **53**: 3103–8.

56 Mok SC, Bell DA, Knapp RC, *et al.* Mutation of K. Ras proto-oncogene in human ovarian epithelial tumors of borderline malignancy. *Cancer Res* 1993; **53**: 1489–92.

57 Van't Veer LJ, Hermens R, Van den berg-Bakker LAM, *et al.* Ras oncogene activation in human ovarian carcinoma. *Oncogene* 1988; **2**: 157–65.

58 Enomoto T, Weghorst CM, Inoue M, Tanizawa O, Rice JM. K-ras activation occurs frequently in mucinous adenocarcinomas and rarely in other common epithelial tumors of the human ovary. *Am J Pathol* 1991; **139**: 777–85.

59 Kacinski M, Mayer AG, King BL, Carter D, Chambers SK. Neu protein overexpression in benign, borderline, and malignant ovarian neoplasms. *Gynecol Oncol* 1992; **44**: 245–53.

60 Gallion HH, Powell DE, Morrow JK, *et al.* Molecular genetic changes in human epithelial ovarian malignancies. *Gynecol Oncol* 1992; **47**: 137–42.

61 Leake JF, Currie JL, Rosenshein NB, Woodruff TD. Long term follow up of serous ovarian tumors of low malignant potential. *Gynecol Oncol* 1992; **47**: 150–8.

62 Rice LW, Lage JM, Berkowitz RS, *et al.* Preoperative serum CA125 levels in borderline tumors of the ovary. *Gynecol Oncol* 1992; **46**: 226–9.

63 Makar AP, Kaern J, Kristensen GB, Vergote I, Bormer OP, Trope CG. Evaluation of serum CA125 level as a tumor marker in borderline tumors of the ovary. *International Journal of Gynecological Cancer* 1993; **3**: 299–303.

64 Mogensen O, Mogensen B, Jakobsen A, Sell A. Preoperative measurement of cancer antigen 125 (CA125) in the differential diagnosis of ovarian tumors. *Acta Oncol* 1989; **28**: 471–3.

65 Einhorn N, Bast RC, Knapp RC, Tjernberg B, Zurawski VR. Preoperative evaluation of serum CA125 levels in patients with primary epithelial ovarian cancer. *Obstet Gynecol* 1986; **67**: 414–16.

66 Brioschi PA, Irion O, Bischof P, Bader M, Forni M, Krauer F. Serum CA125 and epithelial ovarian cancer. A longitudinal study. *Br J Obstet Gynaecol* 1987; **94**: 196–201.

67 Hermann U, Locher GW, Golhirsch A. Sonographic patterns of ovarian tumors: prediction of malignancy. *Obstet Gynecol* 1987; **69**: 777–81.

68 Buy JN, Ghossain MA, Sciot C, *et al.* Epithelial tumors of the ovary: CT findings and correlation with US. *Radiology* 1991; **178**: 811–88.

69 Achiron R, Schejfer E, Malinger G, Zakut H. Observation on the ultrasound diagnosis of ovarian neoplasms. *Arch Gynecol Obstet* 1987; **241**: 183–90.

70 Itata K, Itata T, Manabe A, Kitao M. Ovarian tumors of low malignant potential: transvaginal doppler ultrasound features. *Gynecol Oncol* 1992; **45**: 259–64.

71 Twaalfhoven FCM, Peters AAW, Trimbos JB, Hermans J, Fleuren GJ. The accuracy of frozen section diagnosis of ovarian tumors. *Gynecol Oncol* 1991; **41**: 189–92.

72 de Nictolis M, Montironi R, Tommasoni S, *et al.* Serous borderline tumors of the ovary. A clinicopathologic, immunohistochemical and quantitative study of 44 cases. *Cancer* 1992; **70**: 152–60.

73 Chaitin BA, Gershenson DM, Evans HL. Mucinous tumors of the ovary. A clinicopathologic study of 70 cases. *Cancer* 1985; **55**: 1958–62.

74 McGowan L, Lesher LP, Norris HJ, Barnett M. Misstaging of ovarian cancer. *Obstet Gynecol* 1985; **65**: 568–72.

75 Katzenstein ALA, Mazur MT, Morgan TE. Proliferative serous tumors of the ovary: Histological features and prognosis. *Am J Surg Pathol* 1978; **2**: 339–55.

76 Snider DD, Stuart GCE, Nation JG, Robertson DI. Evaluation of surgical staging in stage I low malignant potential ovarian tumors. *Gynecol Oncol* 1991; **40**: 129–32.

77 Tazelaar HD, Bostwick DG, Ballon SC, Hendrickson MR, Kempson RL. Conservative treatment of borderline ovarian tumors. *Obstet Gynecol* 1985; **66**: 417–22.

78 Weinstein D, Polishuk Z. The role of wedge resection of the ovary as a cause for mechanical sterility. *Surg Gynecol Obstet* 1975; **141**: 417–18.

79 Leake JF, Rader JS, Woodruff JD, Rosenshein NB. Retroperitoneal lymphatic involvement with epithelial ovarian tumors of low malignant potential. *Gynecol Oncol* 1991; **42**: 124–30.

80 Hopkins MP, Morley GW. The second-look operation and surgical re-exploration in ovarian tumor of low malignant potential. *Obstet Gynecol* 1989; **74**: 375–8.

81 Trope C, Kaern J, Vergote IB, Kristensen G, Abeler V. Are borderline tumors of the ovary overtreated both surgically and systematically? A review of four prospective randomized trials including 253 patients with borderline tumors. *Gynecol Oncol* 1993; **51**: 236–43.

82 Lim-Tan SK, Cajigas HE, Scully RE. Ovarian cystectomy for serous borderline tumors: A follow study of 35 cases. *Obstet Gynecol* 1988; **72**: 775–81.

83 Siriaunkjul S, Robbins K, McGowan L, Silverberg SG. Mucinous borderline tumors of the ovary: tumors of no malignant potential? *International Journal of Gynecologic Oncology* 1993; **3**: 271.

84 Julian CG, Woodruff D. The biologic behaviour of low grade papillary serous carcinoma of the ovary. *Obstet Gynecol* 1972; **40**: 860–7.

85 Minyi T, Lijuan L, Tonghua L. The characteristics of ovarian serous tumours of borderline malignancy. *Chin Med J* 1980; **93**: 459–64.

86 Young RC, Walton LA, Ellenberg SS, *et al.* Adjuvant therapy in stage I and stage II epithelial ovarian cancer. *N Engl J Med* 1990; **322**: 1021–7.

87 Chambers JT. Borderline ovarian tumors: a review of treatment. *Yale J Biol Med* 1989; **62**: 351–65.

88 Yazigi R, Munoz AK, Sandstad J, Litshitz S, Choi DJ. Cisplatin based combination therapy in the treatment of stage III ovarian epithelial tumors of low malignant potential. *Eur J Gynaecol Oncol* 1991; **12**: 451–5.

89 Sutton GP, Bundy BN, Omura GA, Yordan EL, Beecham JB, Bonfiglio T. Stage III ovarian tumors of low malignant potential treated with cisplatin combination therapy (a gynecologic oncology group study). *Gynecol Oncol* 1991; **41**: 230–3.

90 McCaughey WTE, Kirk ME, Lessler W, Dardick I. Peritoneal epithelial lesions associated with proliferative serous tumors of ovary. *Histopathology* 1984; **8**: 195–208.

91 Russell P. Borderline epithelial tumors of the ovary: a conceptual dilemma. *Clin Obstet Gynecol* 1984; **11**: 259–77.

92 Michael H, Roth LM. Invasive and non-invasive implants in ovarian serous tumors of low malignant potential. *Cancer* 1986; **57**: 1240–7.

93 Bell DA, Weinstock MA, Scully RE. Peritoneal implants of ovarian serous borderline tumors. Histological features and prognosis. *Cancer* 1988; **62**: 2212–22.

94 Gershenson DM, Silva EG. Serous ovarian tumor of low malignant potential with peritoneal implants. *Cancer* 1990; **65**: 578–85.

95 Vergote I, Kaern J, Trope C. Adjuvant treatment of stage I ovarian cancer: How can we prevent overtreatment. *Proceedings of the American Society for Clinical Oncology* 1992; **11**: 715.

96 Drescher CW, Flint A, Hopkins MD, Roberts JA. Prognostic significance of DNA content and nuclear morphology in borderline ovarian tumors. *Gynecol Oncol* 1993; **48**: 242–6.

97 Lage JM, Weinberg DS, Huettner PC, Mark SD. Flow cytometric analysis of nuclear DNA content in ovarian tumors. Association of ploidy with tumor type, histologic grade, and clinical stage. *Cancer* 1992; **69**: 2668–75.

98 Iversen O-E, Skaarland E. Ploidy assessment of benign and malignant ovarian tumors by flow cytometry. *Cancer* 1987; **60**: 82–7.

99 Klemi PJ, Joensuu H, Kiilholma P, Maenpaa J. Clinical significance of abnormal nuclear DNA content in serous ovarian tumors. *Cancer* 1988; **62**: 2005–10.

100 Padberg B-C, Arps H, Franke U, *et al*. DNA cytophotometry and prognosis in ovarian tumors of borderline malignancy. *Cancer* 1992; **69**: 2510–14.

101 Kuhn W, Kaufman M, Feichter GE, Rummel HH, Schmid H, Heberling D. DNA flow cytometry, clinical and morphological parameters as prognostic factors for advanced malignant and borderline ovarian tumors. *Gynecol Oncol* 1989; **33**: 360–7.

102 Seidman JD, Norris HJ, Griffith JL, Hitchcock CL. DNA flow cytometric analysis of serous ovarian tumors of low malignant potential. *Cancer* 1993; **71**: 3947–51.

103 Friedlander ML, Hedley DW, Taylor IW, Russell P, Coates AS, Tattersall MHN. Influence of cellular DNA content on survival in advanced ovarian cancer. *Cancer Res* 1984; **44**: 397–400.

104 Friedlander ML, Russell P, Taylor IW, Hedley DW, Tattersall MHN. Flow cytometric analysis of cellular DNA content as an adjunct to the diagnosis of ovarian tumors of borderline malignancy. *Pathology* 1984; **16**: 301–6.

105 Eriksen B, Miller DS, Murad TM, Lurain JR, Bauer KD. Dual parameter flow cytometric analysis coupling the measurements of forward-angle light scatter and DNA content of archival ovarian carcinomas of low malignant potential. *Anal Quant Cytol Histol* 1991; **13**: 45–53.

106 Griffiths AP, Cross D, Kingston RE, Harkin P, Wells M, Quirke P. Flow cytometry and Ag NORs in benign, borderline, and malignant mucinous and serous tumors of the ovary. *Int J Gynecol Pathol* 1993; **12**: 307–14.

107 Fu YS, Ro J, Reagan JW, Hall TL, Berek J. Nuclear deoxyribonucleic acid heterogeneity of ovarian borderline malignant serous tumors. *Obstet Gynecol* 1986; **67**: 478–82.

108 Braly PS, Klevecz RR. Flow cytometric evaluation of ovarian cancer. *Cancer* 1993; **71**: 1621–8.

109 Baak JPA, Chan KK, Stolk JG, Kenemans P. Prognostic factors in borderline and invasive ovarian tumors of the common epithelial type. *Pathol Res Pract* 1987; **182**: 755–74.

110 Sandenbergh HA, Woodruff JD. Histogenesis of pseudomyxoma peritonei. *Obstet Gynecol* 1977; **49**: 339–45.

111 Young RH, Gilks CB, Scully RE. Mucinous tumors of the appendix associated with mucinous tumors of the ovary and pseudomyxoma peritonei. A clinicopathologic analysis of 22 cases supporting an origin in the appendix. *Am J Surg Pathol* 1991; **15**: 415–29.

112 Kahn MA, Demopoulos RI. Mucinous ovarian tumors with pseudomyxoma peritonei: a clinicopathologic study. *Int J Gynecol Pathol* 1992; **11**: 15–23.

113 Rutgers JL, Scully RE. Ovarian mullerian mucinous papillary cystadenomas of borderline malignancy. A clinicopathologic analysis. *Cancer* 1988; **61**: 340–8.

114 Long RTL, Spratt JS, Dowling E. Pseudomyxoma peritoneii. New concepts with a report of seventeen patients. *Am J Surg* 1969; **117**: 162–9.
115 Campbell JS, Lou P, Ferguson JP, *et al*. Pseudomyxoma peritoneii et ovarii with occult neoplasms of appendix. *Obstet Gynecol* 1973; **42**: 897–902.
116 Piver SM, Lele SB, Patsner B. Pseudomyxoma peritoneii: possible prevention of mucinous ascites by peritoneal lavage. *Obstet Gynecol* 1984; **64**: 95–6.
117 Green N, Gancedo H, Smith R, Bernett G. Pseudomyxoma peritonei—nonoperative management and biochemical findings. A case report. *Cancer* 1975; **36**: 1834–7.
118 Nasr MF, Kemp GM, Given FT. Pseudomyxoma peritonei: treatment with intra-peritoneal 5-fluorouracil. *Eur J Gynaecol Oncol* 1993; **14**: 213–17.

10 Management of stage I and II ovarian cancer: the value of prognostic factors in therapeutic decisions

CLAES G TROPÉ, AMIN MAKAR

The incidence of ovarian cancer in Norway is 19.3/100 000 and there are about 400 deaths annually.[1][2] Despite important advances in surgery, chemotherapy, and radiotherapy, the overall survival for patients—even those with early stage disease—has not improved appreciably with five year survival of 73% and 46%, respectively, for patients with stage I and II disease over the period 1979–1981.[1][2]

Survival might be improved if high risk patients among those with early stage disease could be identified and treated with more intensive regimens while those at low risk could be spared over treatment. In some cases operations to preserve fertility might be possible.

Patterns of spread in patients with ovarian malignancy

It has been postulated that neoplasms originating in the ovary have two major routes of spread, one by migration of exfoliated cells within the normal circulation of peritoneal fluid, reaching the domes of the diaphragm and omentum through the para-colic gutters, and the other by lymphatic permeation.[3-14] Six to eight lymphatic channels originate on the ovarian surface and drain by three main routes: along the infundibulopelvic ligament to the supracaval and intercavoaortic nodes, along the broad ligament to the interiliac and upper gluteal nodes, and by the round ligament to the external iliac and inguinal nodes. Haematogenous spread is rare.

Prognostic factors in early ovarian cancer

The influence of traditional prognostic factors (histological type, tumour grade, International Federation of Gynecology and Obstetrics (FIGO) stage at presentation) has been evaluated thoroughly.[15-18] The best prognosis is

136

associated with borderline and grade I tumours of mucinous or endometrioid type and with early stage disease. Other prognostic factors have been identified and may be of use in defining risk categories further, particularly in patients with stage I disease.

The large studies by Kaern *et al* from the Norwegian Radium Hospital have shown that DNA ploidy is the most important factor in patients with borderline tumours.[19,20] Aneuploidy is associated with older age, advanced stage, and non-serous histology, and patients with aneuploid tumours have a 19-fold increased risk of dying of their disease compared with those with diploid tumours. The authors concluded that a young patient with a stage IA diploid tumour could undergo fertility-conserving surgery.

The high incidence of relapse in stage I, grade 2 or 3 disease has repeatedly been reported, although even in grade 1 disease there will be a small but significant relapse rate.[15,17,21–23]

The excellent prognosis for patients with early stage, completely excised, mucinous or endometrioid tumours has encouraged the recommendation that no adjuvant therapy is necessary.[21,22,24] On the other hand, large studies reported by Young *et al* and Guthrie *et al* confirmed that clear cell tumours have a poor prognosis.[21,25] Results from our hospital confirm this. Half of all patients with no residual disease after primary operation but with clear cell cancer were dead within 24 months, whilst all such patients with residual cancer were dead by the same time.[22] Patients with stage IC disease have the worst prognosis.[26,27] It seems that the presence of ascites, rather than rupture of the cyst during operation, however, has the worst impact on survival.[26] This report also suggested that the presence of adhesions was a poor prognostic factor, although this remains controversial[27] and is not judged to be of prognostic significance by FIGO.

Preoperative measurement of serum CA125 concentrations in our hospital is of no prognostic significance, but in patients with no residual cancer a raised concentration four to six weeks after the primary operation seems to indicate a poor prognosis, presumably reflecting the presence of microscopic disease.[22]

Staging

Staging of ovarian cancer is based on the findings at the initial operation and on histological examination, and the problems of understaging are well documented.[28–32] Up to 30% of patients presumed to have stage I disease as a result of an inadequate primary operation are upstaged after a second, thorough exploration.[29] The most common sites of occult cancer are within peritoneal fluid or washings, the pelvic peritoneum or omentum, or in the subdiaphragmatic areas or nodes.[29] Staging should be done through a vertical midline incision to allow palpation and biopsy of all peritoneal surfaces. It does not seem to be necessary to sample the subdiaphragmatic areas routinely.[33] In patients with stage II disease and peritoneal extension we recommend total excision of the pelvic peritoneum. The role of node sampling or lymphadenectomy in patients with early stage disease is discussed later.

The five year survival for patients with inadequately staged "stage IA" disease is only 60%,[31] but conservative surgery can be offered to some patients with FIGO stage IA disease. Candidates for such an approach can be summarised as follows: those with diploid, borderline tumours and those with invasive grade 1 diploid disease of non-clear cell type, in whom an increased preoperative serum CA125 concentration returned to the reference range after operation. The extent of staging when conservation of fertility is being contemplated, however, should be limited. A staging laparotomy that includes lymphadenectomy may lead to extensive adhesions which will reduce the chances of conception. We think that for patients who fulfil the above selection criteria, staging in association with conservative surgery should include only analysis of peritoneal cytology, infracolic omentectomy, and peritoneal biopsies. In serous tumours the opposite ovary could be sampled as they are commonly bilateral. The endometrium should be curetted, as endometrial lesions coexist in about 5% of cases of ovarian cancer.

Role of lymphadenectomy

The integration of lymphadenectomy in the management of early stage ovarian cancer has been hampered by the belief that once there are retroperitoneal metastases, the first group of involved nodes will be those in the para-aortic region, away from the pelvis and therefore beyond the reach of conventional surgery and radiation. Many groups, however, have shown the importance of the pelvic nodes as sites for metastases, and that the external iliac and obturator nodes are more likely to be involved than the common iliac nodes. Table 10.1 shows the incidence of para-aortic nodal metastases in patients with apparent stage I or II disease. This means that a considerable number of patients with apparent stage I or II disease would be upstaged to IIIC as a result of lymphadenectomy. Studies have also shown that the incidence of nodal disease is highest in grade 3 tumours, while the chance of nodal disease in mucinous tumours is extremely small.[4-14]

Table 10.1 Para-aortic metastases in apparent stage I and II epithelial ovarian cancer

First author	Year	Reference No	No of patients	Percentage with involved nodes
Knapp	1974	5	2*	19.5
Delgado	1977	11	10	20
Chen	1984	8	21	19
Knipscheer	1982	9	20*	25
Lanza	1988	14	29	34

* Only patients with stage I disease.

The benefits that might be expected from lymphadenectomy include appropriate treatment for upstaged patients, the excision of retroperitoneal disease if intraperitoneal treatment is being considered, and possible survival benefits associated with the removal of occult disease.[6 10 13 14] It has been suggested that platinum treatment may be less effective against nodal disease.[6 34 35]

Wu *et al* showed that microscopic disease was found in a quarter of apparently normal nodes,[6] while Burghardt *et al* showed that in 81 patients with involved nodes, most had only two or three individual nodes involved.[13] These data imply that lymph node sampling is insufficient to find nodal disease and a complete lymphadenectomy should be carried out. Retroperitoneal exploration can, however, carry an appreciable morbidity. Operative problems include injury to the third part of the duodenum, the vena cava and mesenteric vessels, and the ureters. Postoperative morbidity includes lymph cyst formation, infection, prolonged ileus and, in up to 10% of cases, unilateral or bilateral leg oedema. When done by a gynaecological oncologist the operative mortality should be less than 3%.[10 36] As the diagnosis of stage I ovarian cancer is often unexpected, and therefore often made by a general gynaecologist or surgeon, lymphadenectomy at a gynaecological cancer centre must be considered. Tables 10.2 and 10.3 list the criteria which may be used to decide whether reoperation is necessary.

Table 10.2 Tumours in FIGO stage I for which lymphadenectomy is not indicated

Borderline diploid tumours
Grade 1 mucinous tumours
Grade 1 diploid serous or endometrioid tumours

Table 10.3 Tumours in FIGO stage I for which lymphadenectomy is indicated

Borderline aneuploid tumours
Grade 1 aneuploid tumours of serous, mixed, endometrioid or undifferentiated types
Grade 2 and 3 tumours
Clear cell adenocarcinomas

In cases of stage II disease, we recommend that lymphadenectomy should be carried out because up to a half of these patients will have nodal disease and their overall five year survival is only about 60%.

Despite all these data there is a lack of convincing evidence that a survival benefit is gained from lymphadenectomy. The limited number of patients with stage I disease in these reports calls for a multicentre randomised study to elucidate the problem.

Chemotherapy and radiotherapy

There is little argument about the role of platinum-based chemotherapy after primary surgery in all patients with FIGO stage II disease.[37 38] In addition, because many of these cases are understaged FIGO stage III disease, it seems appropriate to recommmend chemotherapy rather than pelvic irradiation, particularly as there are data to show that systemic chemotherapy may be less hazardous than radiotherapy after lymphadenectomy.[27 39]

Giving adjuvant chemotherapy or radiotherapy to all patients with FIGO stage I disease remains controversial, however, and randomised studies have failed to show that cure rates might be improved by such treatment.[27 40-45] Kolstad *et al*, in a study including 418 patients with stage I and II disease, compared the results of postoperative external beam irradiation with the intraperitoneal instillation of radioactive gold (^{198}Au). There was no significant difference in five year survival for patients with stage I disease, although that for patients with stage II disease was 54.1% for those treated with ^{198}Au compared with 40% for those treated with radiotherapy. The morbidity in the former group was, however, considerably worse.[15]

Workers at the MD Anderson Hospital randomised 149 patients with stage I-III disease to receive 12 cycles of melphalan (0.2 mg/kg daily for five days every 28 days) or moving strip radiotherapy with a pelvic boost. Five year survival rates were 78% and 72%, respectively, and were not significantly different, although the study was criticised because the latter group received a reduced dose of radiation to the upper abdomen and because a number of patients with borderline tumours were included.[42] Dembo *et al* in Toronto randomised 190 patients with stage IB-III (completely resected disease) to pelvic irradiation alone, pelvic irradiation plus oral chlorambucil for 24 months, or moving strip whole abdomen irradiation.[43] The first treatment arm was discontinued prematurely because of a high recurrence rate. Five and 10 year survival rates were 78% and 64% compared with 51% and 40% for strip irradiation and irradiation/chlorambucil, respectively. This study was also criticised because of poor surgical staging between the groups and the low dose of chlorambucil that was given. Importantly, the apparent superiority of whole abdominal radiotherapy was not confirmed in subsequent trials.[44 45]

Later studies ensured that subgroups of patients were better controlled. The Gynecologic Oncology Group published a study comparing oral melphalan (0.2 mg/kg day for five days every four to six weeks for 12 months) compared with no postoperative treatment in 81 patients with stage IA/B cancer and with a longer than seven year median follow-up showed that the survival was over 90% in both arms for patients with grade 1 or 2 cancer.[21] In addition a further 141 patients with grade 3 IA-II cancer were randomised to receive melphalan as before or a single dose of intraperitoneal radioactive chromic phosphate. Once again, there was no significant difference in five year disease free survival or overall rates, both predicted to be over 80%.

Sell *et al* confirmed these observations when they reported that there was no difference in recurrence free interval for four year survival among

118 patients with IB-IIB cancer randomised to receive either abdominal irradiation or pelvic irradiation plus cyclophosphamide.[45]

The substantial improvement in long term survival after treatment with cisplatinum in patients with advanced disease has encouraged its use in patients with early stage cancer. In an Italian study 90 patients with grade 2 or 3 stage IA/B disease were randomised to receive cisplatinum (50 mg/m^2 every 28 days for six cycles) or observation only. There was no survival difference between the two arms.[46] One hundred and eighty two patients with stage IC disease were treated with either cisplatinum as before or intraperitoneal ^{32}phosphorus. The five year disease free survival was 82% and 70% respectively (p = 0.006). A large randomised study of 347 patients from our hospital has, however, shown that cisplatinum is no better than ^{32}P.[27] Crude and disease free survival was similar for both groups with estimated five year survival overall and for those with stage I disease being 82% and 86%, respectively. Toxicity in the form of late bowel complications was commoner in the ^{32}P group. One of the shortcomings of the study was the high proportion of patients (51%) with borderline or grade 1 tumours.

Conclusion

In conclusion there have been no reported randomised studies that show that either postoperative chemotherapy or irradiation improve cure rate in FIGO stage I disease. Morbidity, however, seems higher after radiotherapy or treatment with intraperitoneal radioactive isotopes. The prognosis for subgroups of low risk patients appears to be excellent without adjuvant treatment and the integration of flow cytometry and postoperative serum CA125 measurement can identify these patients.

Patients considered to be at high risk of relapse may benefit from adjuvant treatment, but, without an observation arm to a study, this will be impossible to prove. At the Norwegian Radium Hospital a study has been instituted in which patients with grade 2 or 3 tumours, grade 1 aneuploid tumours, or clear cell tumours are randomised to treatment with carboplatin or observation only. The place of lymphadenectomy in these patients will not be investigated in this study: a multicentre study will be required to shed further light on this problem.

References

1 Incidence of cancer in Norway 1989. Oslo: The Cancer Registry of Norway; Institute for epidemiological cancer research, 1990.
2 Makar A. Prognostic studies in cancer of the ovary and fallopian tube with emphasis on the CA125 antigen and c-erbB-2 oncogene. Thesis. Oslo, Norway 1993.
3 Bergman F. Carcinoma of the ovary: a clinico-pathological study of 86 autopsied cases with special reference to mode of spread. *Acta Obstet Gynecol Scand* 1966; **45**: 211–32.
4 Burghardt E, Pickel H, Lahousen M, Stettner H. Pelvic lymphadenectomy in operative treatment of ovarian cancer. *Am J Obstet Gynecol* 1986; **155**: 315–19.

5 Knapp RC, Friedman EA. Aortic lymph node metastases in early ovarian cancer. *Am J Obstet Gynecol* 1974; **119**: 1013–17.

6 Wu PC, He Lang J, Huang R, Yi Qu Lian LJ. Lymph node metastasis and retroperitoneal lymphadenectomy in ovarian cancer. *Bailliere's Clin Obstet Gynaecol* 1989; **3**: 143–55.

7 Chen SS, Lee L. Incidence of para-aortic and pelvic node metastases in epithelial carcinoma of the ovary. *Gynecol Oncol* 1983; **16**: 95–100.

8 Chen SS, Lee L. Prognostic significance of morphology of tumor and retroperitoneal lymph nodes in epithelial carcinoma of the ovary. *Gynecol Oncol* 1984; **18**: 84–99.

9 Knipscheer RJL. Para-aortal lymph node dissection in 20 cases of primary epithelial ovarian carcinoma stage I: influence on staging. *Eur J Obstet Gynecol Reprod Biol* 1982; **13**: 303–7.

10 Di Re F, Fontanelli R, Raspagliesi F, Di Re E. Pelvic and para-aortic lymphadenectomy in cancer of the ovary. *Bailliere's Clin Obstet Gynaecol* 1989; **3**: 131–42.

11 Delgado G, Chun B, Caglar H, Bepko F. Para-aortic lymphadenectomy in gynecologic malignancies confined to the pelvis. *Obstet Gynecol* 1977; **50**: 418–23.

12 Creasman WT, Abu-Gazaleh S, Schmidt HJ. Retroperitoneal metastatic spread of ovarian cancer. *Gynecol Oncol* 1987; **6**: 447–51.

13 Berghardt E, Girardi F, Lahousen M, Tamussino K, Stettner H. Patterns of pelvic and para-aortic lymph node involvement in ovarian cancer. *Gynecol Oncol* 1991; **40**: 103–6.

14 Lanza A, D'addato F, Valli M, *et al*. Pelvic and para-aortic lymph node positivity in ovarian carcinoma: it's prognostic significance. *Eur J Gynecol Oncol* 1988; **9**: 36–9.

15 Kolstad P, Davy M, Hoeg K. Individualised treatment of ovarian cancer. *Am J Obstet Gynecol* 1977; **128**: 617–25.

16 Einhorn N, Nilsson B, Sjovall K. Factors influencing survival in carcinoma of the ovary. Study from a well defined Swedish population. *Cancer* 1985; **55**: 2019–25.

17 Bjorkholm E, Pettersson F, Einhorn N, Krebs I, Nilsson B, Tjernberg B. Long term follow-up and prognostic factors in ovarian carcinoma. The Radiumhemmet series. *Acta Radiologica Oncologica* 1982; **21**: 413–19.

18 Swenerton KD, Hislop TG, Spinelli J, LeRiche JC, Yang N, Boyes DA. Ovarian carcinoma: a multivariate analysis of prognostic factors. *Obstet Gynecol* 1985; **65**: 264–70.

19 Kaern J, Trope C, Kjorstad KE, Abeler VM, Petterson EO. Cellular DNA content as a new prognostic tool in patients with borderline tumors of the ovary. *Gynecol Oncol* 1990; **38**: 452–7.

20 Kaern J, Trope CG, Abeler VM, Pettersen EO. DNA ploidy, the most important prognostic factor in patients with borderline tumors of the ovary. *International Journal of Gynecological Cancer* [in press].

21 Young RC, Walton LA, Ellenbergh SS, *et al*. Adjuvant therapy in stage I and stage II epithelial ovarian cancer. Results of two prospective randomised trials. *N Engl J Med* 1990; **322**: 1021–7.

22 Makar A. Kristensen GB, Kaern J, Bormer OP, Abeler VM, Trope CG. Prognostic value of pre- and postoperative serum CA125 levels in ovarian cancer: New aspects and multivariate analysis. *Obstet Gynecol* 1992; **72**: 1002–10.

23 Vergote IB, Kaern J, Abeler VM, Petterson EO, Devos L, Trope CG. Analysis of prognostic factors in stage I ovarian carcinoma. Importance of degree of differentiation and DNA ploidy in predicting relapse. *Am J Obstet Gynecol*, 1993; **169**: 40–52.

24 Makar A, Kristensen GB, Baeklandt M, Trope CG. Size of residual disease after debulking surgery in the main prognostic factor in ovarian cancer FIGO stage III. *Gynecol Oncol* 1995; **56**: 175–80.

25 Guthrie D, Davy ML, Philips PR. A study of 656 patients with early ovarian cancer. *Gynecol Oncol* 1984; **17**: 363–9.

26 Dembo AJ, Davy M, Stenwig AE, Berle EJ, Bush RS, Kjorstad K. Prognostic factors in patients with stage I epithelial ovarian cancer. *Obstet Gynecol* 1990; **75**: 263–73.

27 Vergote IB, Vergote-De Vos LN, Abeler VM, *et al*. Randomised trial comparing cisplatin with radioactive phosphorus or whole abdominal irradiation as adjuvant treatment of ovarian cancer. *Cancer* 1992; **69**: 741–9.

28 Piver MS, Barlow JJ, Lele SB. Incidence of subclinical metastases in stage I and II ovarian carcinoma. *Obstet Gynecol* 1978; **52**: 100–4.

29 Young RC, Decker DG, Wharton JT, *et al*. Staging laparotomy in early ovarian cancer. *JAMA* 1983; **250**: 3072–6.

30 McGowan L, Lesher LP, Norris HJ, *et al*. Mis-staging of ovarian cancer. *Obstet Gynecol* 1985; **65**: 568–72.

31 Sevelda P, Vavra N, Schemper M, Salzer H. Prognostic factors for survival in stage I epithelial ovarian carcinoma. *Cancer* 1990; **65**: 2349–52.

32 Mayer AR, Chambers SK, Graves E, *et al.* Ovarian cancer staging: does it require a gynecologic oncologist? *Gynecol Oncol* 1992; **47**: 223–7.

33 Bertelson K. Tumor reduction surgery and long-term survival in advanced ovarian cancer: a DACOVA study. *Gynecol Oncol* 1990; **38**: 203–9.

34 Podratz KC, Malkasian GD, Wieland HS, *et al.* Recurrent disease after negative second-look laparotomy in stages III and IV ovarian carcinoma. *Gynecol Oncol* 1988; **29**: 274–82.

35 Burghardt E, Winter R. The effect of chemotherapy on lymph node metastases in ovarian cancer. *Bailliere's Clin Obstet Gynaecol* 1989; **3**: 167–71.

36 Conte M, Beneditti P, Guariglia L, Scambia G, Gregi S, Mancuso S. Pelvic lymphocele following radical para-aortic and pelvic lymphadenectomy for cervical carcinoma: incidence rate and percutaneous management. *Obstet Gynecol* 1990; **76**: 268–71.

37 Walton LA, Yadusky A, Rubinstein L, Roth LM, Young R. Stage II carcinoma of the ovary: an analysis of survival after comprehensive surgical staging and adjuvant therapy. *Gynecol Oncol* 1992; **44**: 55–60.

38 Piver MS, Malfetano J, Hempling RE, Baker TR, Driscoll DL. Cisplatin based chemotherapy for stage II ovarian adenocarcinoma. A preliminary report. *Gynecol Oncol* 1990; **39**: 249–52.

39 Sigurdsson K, Johansson JE, Trope C. Carcinoma of the ovary stage I and II. A prospective, randomized study of the effects of post-operative chemotherapy and radiotherapy. *Ann Chir Gynaecol* 1982; **71**: 321–9.

40 Hreshchyshyn MM, Park RC, Blessing JA, *et al.* The role of adjuvant therapy in stage I ovarian cancer. *Am J Obstet Gynecol* 1980; **138**: 139–45.

41 Hammond R, Bull C, Houghton CRS. Primary adjunctive whole abdominal radiotherapy in epithelial ovarian cancer: Results of 10 years experience. *Aust NZ J Obstet Gynaecol* 1999; **32**: 267–9.

42 Smith JP, Rutledge FN, Delclos L. Post-operative treatment of early cancer of the ovary: a random trial between postoperative irradiation and chemotherapy. *National Cancer Institute Monographs* 1975; **42**: 149–53.

43 Dembo AJ, Bush RS, Beale FA, Bean HA, Pringle JF, Sturgeon JF. The Princess Margaret Hospital study of ovarian cancer: stage I, II and asymptomatic III presentations. *Cancer Treatment Reports* 1979; **63**: 54.

44 Klaassen D, Shelly W, Starreveld A, Kirk M, *et al.* Early stage ovarian cancer. A randomized clinical trial comparing whole abdominal radiotherapy, melphalan and intraperitoneal chromic phosphate: A National Cancer Institute of Canada Clinical Trials Group Report. *J Clin Oncol* 1988; **6**: 1254–63.

45 Sell A, Bertelsen K, Andersen JE, Styer I, Panduro J. Randomized study of whole abdomen irradiation versus pelvic irradiation plus cyclophosphamide in the treatment of early ovarian cancer. *Gynecol Oncol* 1990; **37**: 367–73.

46 Bolis G, Torri V, Babilionti L. Multicenter controlled trial in patients with stage I epithelial ovarian cancer. *Proceedings of the International Gynecologic Cancer Society* 1991; **3**: 16.

11 Surgical management of advanced ovarian cancer

NEVILLE F HACKER

Epithelial ovarian cancer is relatively slow growing and usually asymptomatic until it has disseminated around the peritoneal cavity, so about 70% of patients have advanced disease when they present. Resection of all the tumour is usually impossible for these patients, so the surgical options are either biopsy only, limited resection including the primary tumour, or aggressive cytoreduction.

Cytoreduction or debulking is a procedure whereby surgically incurable tumour is partially removed to improve the effectiveness of subsequent treatment, usually chemotherapy but occasionally radiation. It contravenes the traditional principles of cancer surgery because clear resection margins are usually not obtained, so it has remained controversial. Even the nomenclature is not universally agreed on, which adds to the confusion in published reports. For the purposes of this report, debulking surgery is considered under three headings: primary debulking (which is done before the initiation of chemotherapy); intervention debulking (which is done after at least one cycle but before the completion of a prescribed primary course of chemotherapy); and secondary debulking (which is done after the primary course of chemotherapy has been completed).

The concept of cytoreductive surgery for advanced ovarian cancer has been discussed for over half a century. Meigs was the first to suggest that as much tumour as possible should be removed to increase the effect of postoperative irradiation.[1] In 1968 Munnell reported an overall increase in survival of patients with ovarian cancer from 28% to 40% and concluded that this improvement was related to more frequent use of postoperative irradiation and more aggressive resection of the tumour.[2] He introduced the concept of "the maximum surgical effort," but Griffiths was the first to quantify the surgical objective when he suggested that all tumour nodules should be reduced to 1.5 cm in diameter or less.[3] Subsequently, both my group and van Lindert *et al* reported further improvement of survival if all metastatic masses larger than 5 mm could be resected (fig. 11.1).[4][5] More recently, in an analysis of data from the Royal Hospital for Women in Sydney, we reported that the only patients with a better than 50% chance of survival for longer than four years were those with no residual disease,

144

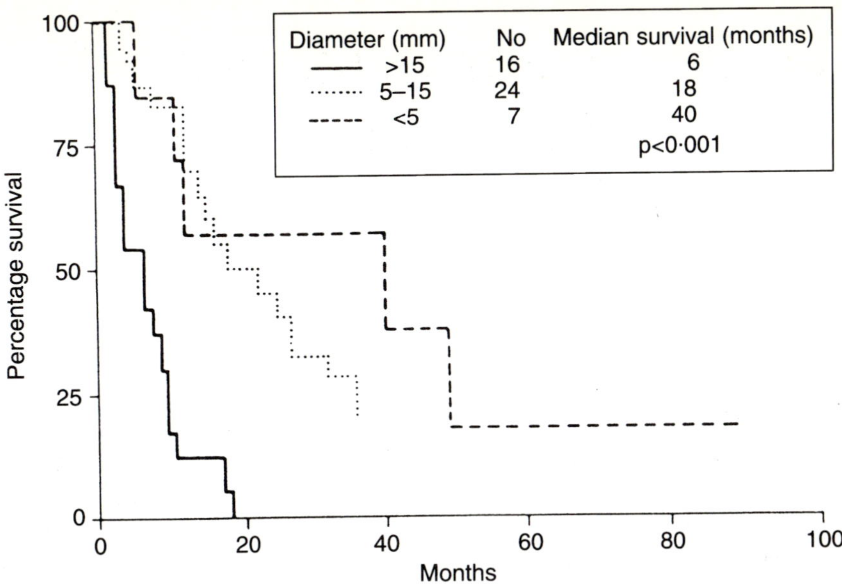

Fig. 11.1 Influence of amount of residual disease on survival.

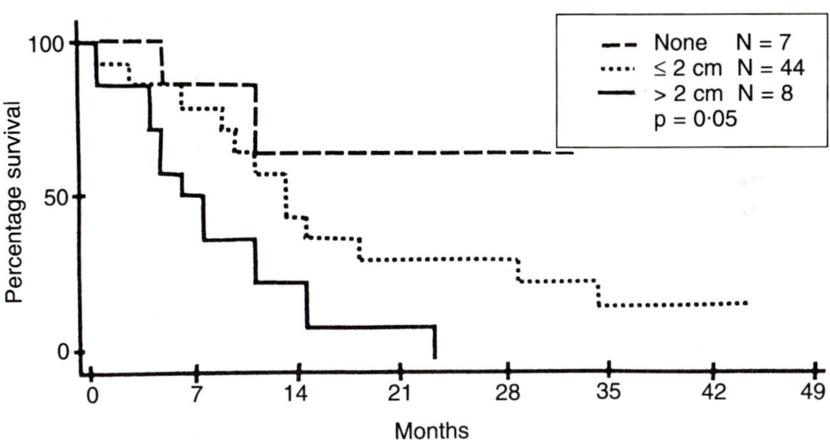

Fig. 11.2 Influence of presence of residual disease on survival.

and this should be the aim of cytoreductive surgery (fig. 11.2).[6] Schwartz *et al* in a study of 100 patients, also found that women with no residual disease survived significantly longer than those with any residual disease.[7] Patients with residual disease less than 2 cm in diameter had no advantage over those with larger residual nodules unless the former group also included patients with no macroscopic residuum.

It has been conventional to refer to an "optimal" cytoreductive operation as one in which no tumour nodules larger than 1.5 cm in diameter remain. This convention allows comparison among different studies. Where centres have a particular interest and expertise in cytoreductive surgery optimal cytoreduction is possible in about 80% of patients.[6 9 10] Feasibility increases with the experience of the surgical team.[9]

From the results of three national cooperative ovarian cancer trials it must be concluded that most patients with advanced ovarian cancer are not afforded the benefits of aggressive cytoreduction. In the Italian study of 531 patients, only 31% had optimal cytoreduction[11]; in the Australian study, only 45% of 284 patients had appropriate surgery[12]; and in the Dutch study only 48% of 191 patients had optimal cytoreduction.[13] This partly reflects the difficulty in making a preoperative diagnosis of ovarian cancer in some patients, but also probably reflects some degree of scepticism about the need for such extensive surgery.

Prerequisites for cytoreduction

There are two prerequisites for the proper surgical management of patients with advanced ovarian cancer.[14] Firstly, units must be established with medical and nursing personnel who have the necessary expertise to deal with the preoperative, operative, and postoperative care of these patients. The surgeon should be thoroughly conversant with the relevant gynaecological, gastrointestinal, and urological procedures, and knowledgeable about the disease. Secondly, patients with a diagnosis of probable ovarian cancer must be referred to these units, preferably before any surgical intervention. As the diagnosis of ovarian cancer is usually not made before laparotomy it will not be possible to refer all patients preoperatively, but the following should be referred: patients with a pelvic mass and an upper abdominal mass; patients with a pelvic mass and ascites; and patients with a pelvic mass and a raised serum CA125 concentration.

The CA125 concentration may be falsely raised, particularly in premenopausal patients, so an alternative in such patients is laparoscopy to exclude such conditions as fibroids or endometriosis. If ovarian cancer is suspected at laparoscopy and the disease is confined to the pelvis it is preferable not to biopsy the ovary or aspirate a cystic mass as this may encourage dissemination of tumour by the slow leakage of fluid. Cytology of the cyst fluid is inaccurate, false negative rates of greater than 50% being reported.[15 16]

If an inexperienced oncological surgeon inadvertently does a laparotomy on a patient with advanced ovarian cancer, the safest and wisest thing to do is to biopsy the tumour and retreat. Serious vascular, gastrointestinal or urological injury may result from inappropriate attempts to debulk it. The patient should then be referred to a gynaecological cancer unit for appropriate cytoreduction before chemotherapy.

Rationale for cytoreduction

Cytoreductive surgery has been advocated for a number of tumours, including those in the testis, kidney, adrenal gland, and central nervous system, but evidence suggests that there is a significant benefit only for ovarian carcinoma and Burkitt's lymphoma.[17] Highly chemosensitive tumours such as gestational trophoblastic neoplasms and ovarian germ cell tumours respond regardless of tumour bulk, while tumours of low chemosensitivity such as cervical carcinomas and sarcomas are not curable by chemotherapy regardless of small initial tumour bulk. Epithelial ovarian tumours are of intermediate chemosensitivity, and it is therefore logical to attempt to overcome as many barriers as possible to drug resistance before starting chemotherapy. The rationale for cytoreductive surgery for tumours of intermediate chemosensitivity[18] is that: pharmacological sanctuaries are eliminated by removal of poorly perfused, bulky tumour masses; there is a higher growth fraction in the better perfused small residual tumour masses, and this increases the number of tumour cells killed by chemotherapy; small tumour masses require fewer cycles of chemotherapy, so there is less likelihood of induced drug resistance; host immunocompetence is increased by removal of large tumour masses; and clones of phenotypically resistant cells may possibly be removed, particularly if all macroscopic tumour is removed.

Temporary resistance to chemotherapy may be related to either physiological barriers or to alterations in the tumour's cellular kinetics.[19] For solid tumours the major physiological barrier is poor perfusion of the tumour. Adequate diffusion of the drug to all cells is fundamental for cytotoxicity. As tumours grow the blood supply to the central region becomes particularly tenuous, ultimately resulting in tumour necrosis, but there are adjacent viable though poorly perfused areas and these may act as pharmacological sanctuaries. This poor perfusion of the tumour is reflected in the low content of free glucose and high concentrations of lactic acid in the interstitial fluid of solid tumours.[20]

The second cause of temporary resistance to chemotherapy is alterations in the cellular kinetics associated with large tumour masses. As large tumours outstrip their blood supply there is a reduction in the growth fraction of the tumour.[21] The growth fraction is the proportion of proliferating cells in a population, and it can be estimated by an autoradiographic method after parenteral injection of tritiated thymidine.[22] The reason a high growth fraction is important is that non-proliferating cells (G_o) are not susceptible to cell-cycle specific chemotherapeutic agents, whereas aklylating agents and DNA binders (such as cisplatin) react with or bind to DNA regardless of the phase of the cell cycle during exposure, but apparently kill only those neoplastic or normal cells that attempt DNA replication before repair.[23 24] For all chemotherapeutic agents, therefore, a high growth fraction is necessary to ensure the optimal number of cells are killed, and this phenonenon is also exploitable by cytoreductive surgery.

Exposure to chemotherapeutic agents alone may produce drug resistance. Skipper *et al* calculated that for an exponentially growing tumour the killing effect of chemotherapeutic agents was a logarithmic function—that is, a given dose killed a constant fraction not a constant number of cells regardless of the cell population initially present.[25] A tumour nodule with a volume of 1 ml will contain about a billion cells.[24] The larger the tumour volume before the start of treatment, the greater the number of cycles that will be needed to eradicate the disease, so increasing the likelihood of induced resistance.

Aggressive cytoreductive surgery seems to increase the immuno-competence of the patient. Morton regarded surgery as a form of immunotherapy, pointing out that in many patients the major role of the surgeon was to remove the bulk of the tumour to lower the level of immunosuppression that had been induced by the neoplasm.[26] Cell mediated immunity[27] and blocking factor activity[28] have both been shown to have a direct relationship to tumour burden in cancers of the female genital tract.

The main obstacle to the chemotherapeutic cure of epithelial ovarian cancer is the development of permanent drug resistance. This is caused mainly by spontaneous mutation to phenotypic drug resistance, and occurs as an intrinsic property of genetically unstable malignant cells. Goldie and Coldman developed a mathematical model which relates the drug sensitivity of tumours to their spontaneous mutation rate.[29] The development of a resistant clone by the tumour is a random event that is dependent on the growth curve of the tumour and the mutation rate. As the tumour increases in size the probability of resistant clones increases so the expectation of cure might be expected to depend not only on the amount of the residual disease, but also on the amount of the original disease, particularly the metastatic disease, before the start of surgery and chemotherapy.

Criticisms of cytoreduction

One early criticism of cytoreductive surgery was that it carried a high morbidity, but several studies have refuted this.[9 30] The two most compelling criticisms are the lack of randomised prospective studies, and the issue of whether the improved prognosis for patients with small residual tumour nodules is related to the surgical resection of the bulky disease or to the surgical resectability of the disease.

The Gynecologic Oncology Group in the United States did initiate a randomised study of primary cytoreduction compared with six cycles of primary chemotherapy for patients with stage III epithelial ovarian cancer who had an "open-and-close" laparotomy at a community hospital before referral to a gynaecological cancer centre, but the study was stopped because patients were not being entered. Both ethical and medicopolitical considerations prevented its successful completion, and it is unlikely that such a study could be undertaken successfully in the future given the subsequent accumulation of evidence supporting a role for surgery. This evidence now includes a prospective randomised study of intervention cytoreduction (after three cycles of chemotherapy) which showed a significant survival advantage with a reduced risk of progression and of death for patients in the surgical arm of the study.

With respect to the issue of resection as opposed to resectability, Griffiths initially claimed that patients whose tumours could be reduced to "optimal" status had the same prognosis as those with "optimal" metastatic disease in the first place.[31] In 1983, however, we reported that patients with extensive metastatic disease (>10 cm diameter) had a worse prognosis than patients with a small initial tumour burden regardless of the extent of cytoreduction.[4] Recently, Hoskins *et al* confirmed this observation in a retrospective analysis of some data from the Gynecologic Oncology Group.[32]

These observations indicate that intrinsic factors within the tumour itself are of prognostic importance, but they do not negate the role of cytoreduction. Indeed, Heintz *et al* reported that for patients with extensive metastatic disease treated with cisplatin-based combination chemotherapy, optimal cytoreduction conferred a 12 month survival advantage over patients having suboptimal resections (35 compared with 23 months). Patients with little metastatic disease who underwent optimal cytoreduction had a median survival of longer than five years.[33]

The presence of peritoneal carcinomatosis is also a poor prognostic factor.[33] Farias-Eisner *et al* analysed 66 patients with minimal residual disease (nodules less than 5 mm) after cytoreductive surgery for advanced ovarian cancer,[34] and reported a median survival of 48 months for patients with no residual nodules, 35 months for those with less than 10 nodules, 21 months for 10–40 nodules, and 16 months for extensive carcinomatosis (more than 40 nodules) (p<0.01).

A recent report by Eisenkop *et al* suggested that removal of all small metastatic nodules can significantly improve survival.[35] They used carbon dioxide laser, electrocautery, the cavitron ultrasonic surgical aspirator, the argon beam coagulator, or sharp dissection to remove all peritoneal implants from a group of 21 patients. Results were compared with those of a group of 30 patients who had macroscopic disease of less than 1 cm remaining exclusively on peritoneal surfaces after primary cytoreduction. No attempt was made in this latter group to remove the peritoneal implants, but both groups were comparable with respect to age, volume of ascites, diameter of largest metastasis, and histological type. A third group was rendered macroscopically free of disease by total abdominal hysterectomy, bilateral salpingo-oophorectomy and omentectomy, without the need for excision of peritoneal implants. Results indicate a significantly worse survival for the group without peritoneal implants that were not resected before initiation of chemotherapy (table 11.1).

A recent meta-analysis purported to show that resection had only a small effect on survival for advanced ovarian cancer, with the type of chemotherapy being a more important variable.[36] The report looked at 58 studies involving

Table 11.1 Effect of excision of peritoneal implant on survival in epithelial ovarian cancer

Group	No of patients	Median survival (months)	Four-year survival (%)
No peritoneal nodules, no residuum	7	39	60
Small implants (≤ 1 cm) removed	21	37	52.9
Small implants (≤ 1 cm) not removed	30	26	18.2

6962 patients with "predominantly" stage III and IV ovarian cancer. Optimal cytoreduction was variously defined, but in 5.2% of cohorts it meant residual nodules less than 3 cm diameter and in 10.3% of cases there was only a verbal description such as "no palpable residual tumour". Therefore in 15% of cohorts residual tumour was too large to show any advantage for cytoreduction. More disconcerting was the fact that optimal cytoreduction was achieved in a much lower proportion of cases than would be expected if aggressive surgery had been undertaken, so optimal cytoreduction was more truly related to the biology of the disease than the skill of the surgeon in most cases. In 23 of the cohorts optimal cytoreduction was achieved in 20% or less of cases making it impossible to examine the true role of cytoreduction. In addition, only half the patients were treated with cisplatin-based chemotherapy, undoubtedly reducing the beneficial effect of surgery on survival.

About a quarter of patients with ovarian cancer have chemoresistant tumours. Such patients will not be helped by cytoreduction, but they cannot be identified preoperatively. They will make it difficult to show a major advantage for cytoreduction in a meta-analysis, but to conclude that cytoreduction is unimportant is to deny an appreciable survival advantage to the group with chemosensitive tumours.

The issue of quality of life after cytoreductive surgery has been inadequately studied. In the only report addressing this question Blyth and Wahl reported significantly improved quality of life in patients who had aggressive cytoreductive surgery.[37] Most of them (84%) were able to continue their normal activities, including employment. In addition, 79% of them were able to enjoy their regular diet, compared with 40% of patients who had suboptimal initial surgery.

Primary cytoreduction

Numerous reports have indicated that the diameter of the largest residual tumour nodule before chemotherapy is a significant prognostic indicator in patients with advanced ovarian cancer. It is the only prognostic factor over which the treating team has any control, all other factors such as histological grade, age, chemosensitivity, and DNA ploidy being predetermined.

This was well reflected in a study by Dutch workers who attempted to predict the outcome of patients treated for ovarian cancer by constructing a prognostic index based on information from two clinical trials done by the Netherlands Joint Study Group for Ovarian Cancer.[38] All patients in these trials received cisplatinum-based chemotherapy. Factors found to have independent prognostic significance were: the Karnofsky index, International Federation of Gynecology and Obstetrics (FIGO) stage, Broder's grade, presence of ascites, treating hospital, and residual tumour size. Factors not found to have independent prognostic significance included the size of the primary tumour and of the metastatic disease.

In the absence of a dedicated attempt to remove all small residual implants of tumour, only about 12% of patients with advanced disease will be rendered free of tumour,[6] but current evidence suggests that a serious attempt should be made to remove all residual tumour nodules before the start of chemotherapy.[6 7 35] Primary cytoreductive surgery should include removal of the uterus, tubes, ovaries and omentum in all cases. Additional procedures such as bowel resection, urological resections, resection of peritoneal nodules (and occasionally splenectomy or resection of diaphragmatic peritoneum and muscle) may be required in selected cases.

The value of routine pelvic and para-aortic lymphadenectomy in patients who have undergone optimal cytoreduction is unresolved. Burghardt and colleagues reported involved pelvic or para-aortic lymph nodes, or both, in 74% of patients with stage III disease (n=114) and 73% of those with stage IV (n=15).[39] Actuarial five-year survival rates with none, one, or more than one involved node were 69%, 58%, and 28%, respectively. An interesting finding from their data,[40] also reported by Wu et al,[41] was that chemotherapy seemed to be ineffective in eradicating ovarian cancer deposits from lymph nodes, the same incidence of involved nodes being found whether or not the lymphadenectomy was done at the time of the primary operation or at the time of second-look laparotomy.

To find out whether pelvic and para-aortic lymphadenectomy is of any therapeutic value we have initiated a randomised prospective study in collaboration with oncology units in Europe, the United States and Australia. Patients with stage III disease who have residual tumour nodules of 1 cm in diameter or more will be randomised to either systematic lymphadenectomy or resection of bulky nodes only, before six cycles of cisplatin-based chemotherapy. Roughly 170 patients will be required in each arm of the study, and the final results will not be known for some years.

Technique of cytoreductive surgery

Preoperative investigations

All patients with a presumptive diagnosis of ovarian cancer should have a full blood count, measurement of serum creatinine and electrolyte concentrations, liver function tests, clotting studies, assays of tumour markers, including CA125 concentration, a chest radiograph and an intravenous pyelogram. Baseline arterial blood gas tensions should be measured on patients with pulmonary problems. As about 5% of ovarian cancers are metastatic from other primaries, a barium enema examination should be seriously considered for all patients with any change in bowel habits or a family history of bowel cancer. A barium meal examination should be done if there are any symptoms referable to the stomach, and bilateral mammograms if there are any suspicious breast lumps. Pelvic ultrasonography may be useful in the evaluation of a mass 8 cm in diameter or less in a woman of reproductive age. Computed tomograms are not usually helpful, and neither are liver, bone, or brain scans unless there are specific symptoms or signs to suggest involvement of that particular organ.

Preoperative management

When laboratory results are available, appropriate measures should be taken to prepare the patient preoperatively. These should include: blood transfusion if the haemoglobin is less than 100 g/l, correction of any acid base or electrolyte imbalance, parenteral nutrition through a central venous line for 10 days if the patient has lost more than 10% of her body weight, cessation of smoking for at least a week and intensive physiotherapy for patients with a history of respiratory disease, and thorough mechanical and antibiotic bowel preparation (table 11.2). Prophylactic antibiotics are given (for example, cefoxitin 1 g two hours preoperatively and 1 g every 8 hours for three doses postoperatively), and pneumatic calf compressors are used to reduce the risk of venous thrombosis. It is not our policy to use subcutaneous heparin before or during the operation because of the increased oozing that is sometimes encountered.

Table 11.2 Timetable for 48 hour bowel preparation

Day 1:	Admit to hospital
	Clear liquid diet
	Glycoprep (polyethylene glycol) orally
Day 2:	Neomycin tablets 1 g four hourly × 4
	Erythromycin base tablets 1 g four hourly × 4
	Tap water enemas until effluent is clear

Surgical objectives

A realistic appraisal of the likelihood of removing all or most of the macroscopic disease must be made after exploring the abdomen. With judicious use of the end-to-end stapling device (EEA), it is usually possible to clear the pelvis of the bulk of the disease without the need for a colostomy, and even the largest of omental "cakes" can be resected safely. Inability to achieve optimal cytoreduction therefore will result from extension of the disease to involve the diaphragm, liver, porta hepatis, or lesser omentum. Occasionally a matted mass of retroperitoneal lymph nodes that are invading the underlying vessels will be unresectable.

If exploration suggests that optimal cytoreduction is not possible, the surgical goals should be limited to removal of the primary disease and omentum to improve the patient's comfort. Bowel should be resected only to relieve obstruction and urological resections would not be indicated. Removal of a large omental cake in patients with ascites will often prevent or reduce the accumulation of ascitic fluid postoperatively.

Surgical technique

The patient should be examined under general anaesthesia and if there is fixed disease in the cul-de-sac suggesting the possible need for rectosigmoid resection, she should be placed in the low lithotomy (or "ski") position to facilitate subsequent bowel anastomosis with the EEA stapler. Otherwise, the operation is done with the patient supine. A lower midline or paramedian incision is made initially and subsequently extended as far upwards as necessary. The following description assumes that optimal cytoreduction is considered feasible.

On opening the abdomen, there is often an omental cake present, which must be dissected off the parietal peritoneum of the anterior abdominal wall. The inferior surface of the omental cake is often adherent to loops of small bowel, which must be dissected free to allow the bowel to be packed out of the pelvis. If the omental cake is massive it will be necessary to remove it primarily to gain adequate access to the pelvis and upper abdomen. Usually, however, the pelvic dissection is done first.

It may be possible to do a total abdominal hysterectomy and bilateral salpingo-oophorectomy in the standard manner, or with only minor modifications. Often, however, the tumour mass is densely adherent to the pelvic organs (fig. 11.3) and cannot be separated from the pelvic peritoneum which grossly distorts the anatomy. When the pelvic anatomy is grossly distorted, removal of the pelvic tumour is facilitated by using a retroperitoneal approach.[42] The dissection is started by incising the parietal peritoneum just lateral to the external iliac vessels, and opening the retroperitoneal space bilaterally. Each ureter is identified as it crosses the bifurcation of the common iliac artery and kept under direct vision throughout the pelvic dissection. Each infundibulopelvic ligament is isolated and ligated early in the dissection to help reduce blood loss.

Removal of the pelvic tumour is facilitated by extending the peritoneal incisions anteriorly until they meet over the bladder, distal to the tumour mass. Posteriorly the peritoneal incisions meet anterior to the rectum, beyond the tumour mass. Tumour bearing peritoneum can usually be dissected off the bladder or rectal muscle. The ureters are dissected to the

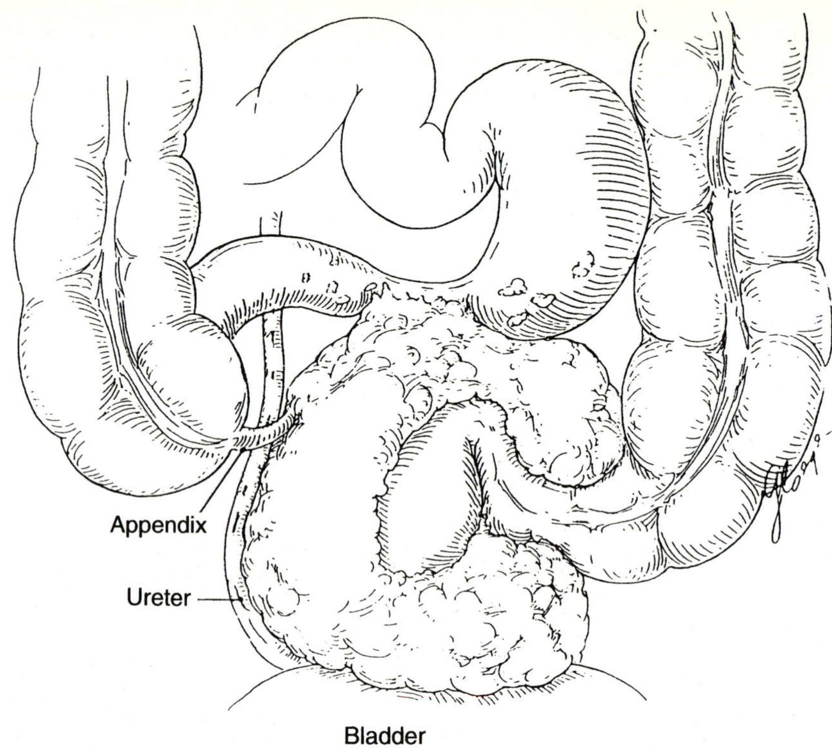

Fig. 11.3 Extensive pelvic tumour that is invading the bladder, rectosigmoid, and iliocaecal region.

level of the uterine arteries which are ligated, and then the hysterectomy can be done in continuity with the tumour mass.

If the tumour is growing into the rectum, sigmoid colon, or sigmoid mesentery, low anterior resection of the colon in continuity with the pelvic mass will be required. It is desirable to transsect the sigmoid colon early in the dissection with the GIA (gastrointestinal anastomosis) stapler, and then ligate and divide the vessels in the sigmoid mesentery to gain access to the presacral space. In the presacral space it is possible to get below the pelvic tumour with blunt dissection and it can then be raised after transection of the rectal pillars between the presacral and pararectal spaces. This allows access to the distal rectum. As much distal rectum as possible should be spared when the bowel is transsected below the tumour. Primary bowel anastomosis should be done with the EEA stapler.

Occasionally, the pelvic tumour will be invading the lower urinary tract. If the bladder is invaded, partial cystectomy with primary repair is required. If the distal ureter is involved, it should be resected in continuity with the mass, and the lower urinary tract reconstructed. If only the distal 2–3 cm are resected, a ureteroneocystostomy may be done, the bladder being suspended from the pelvic sidewall by a psoas hitch. If a larger segment of

154

ureter is resected, transureteroureterostomy is required.[43] At the end of the pelvic dissection, any enlarged pelvic lymph nodes are removed. No attempt should be made to close the pelvic peritoneum, and no drains are placed in the pelvis.

Attention is then turned to the upper abdomen. If there is no gross involvement of the omentum, infracolic omentectomy may be satisfactory, but if an omental "cake" is present, total omenectomy is required. Although an omental "cake" may seem to be invading the transverse colon it can usually be dissected free to expose the lesser sac and transverse mesocolon. The omentum is then taken off the greater curvature of the stomach, care being taken to ligate all short gastric vessels. Traction downwards and medially on the omental cake facilitates its removal from the fundal region of the stomach and gastrosplenic ligament. At times the tumour will invade close to the hilar region of the spleen necessitating careful dissection around the splenic pedicle and (rarely) splenectomy (fig. 11.4).

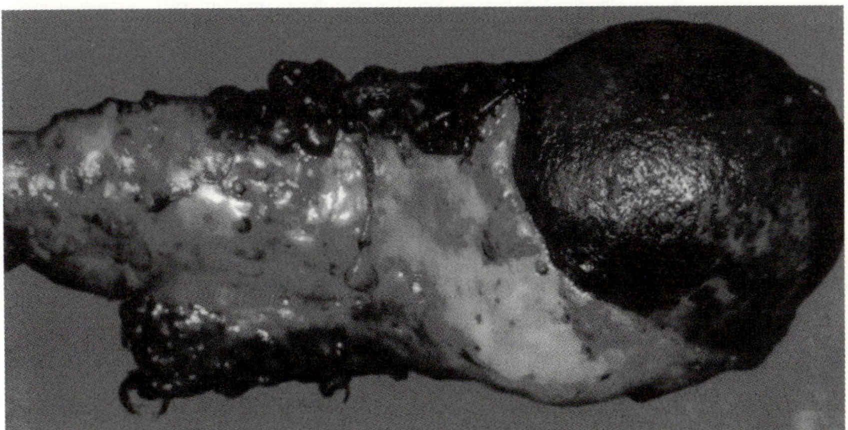

Fig. 11.4 An omentum "cake" that has been resected in continuity with the spleen because of dense adherence.

After the pelvic tumour and omentum have been removed, any grossly involved para-aortic lymph nodes should be excised, together with gross nodules on the bowel serosa or parietal peritoneum. Ovarian cancer usually "cakes" on to serosal surfaces rather than invading deeply, so lines of cleavage can usually be found. If there are large tumour masses involving the small bowel mesentery or invading beyond the serosa, however, partial small bowel resection should be done. To save time during an otherwise long operation, stapling devices may be used to expedite the anastomoses.

Metastases to the diaphragm are usually diffuse and not resectable, but isolated diaphragmatic nodules may be resected after mobilising the liver by transsecting the triangular ligament. It may be possible to dissect the nodule off the diaphragmatic muscle, but if the muscle has to be resected, the defect can be closed with interrupted sutures. A chest tube may be necessary in such circumstances.[44] Resection of disease from the diaphragm as well as from the bowel or its mesentery may be facilitated by using the cavitron ultrasonic surgical aspirator[45] or the argon beam coagulator.[46]

After resection of the tumour, the peritoneum should be irrigated with normal saline and the wound closed using a continuous, mass-closure technique with polydioxanone or polygluconate.

Postoperative care

Some of these patients require intensive care nursing postoperatively. All patients need careful monitoring of fluid and electrolyte balance and urine output, particularly during the first 72 hours. Urine output from the catheter should be measured hourly and maintained at more than 30 ml/hour. Incentive spirometry should be started on the day of operation and be supervised every four hours when the patient is awake. An occasional patient with chronic lung disease will require intubation and ventilation for the first 24 to 48 hours postoperatively. Pneumatic calf compression should continue until the patient is fully mobilised, and our practice is to start calcium heparin (Calciparine) 5000 units subcutaneously every eight hours immediately postoperatively.

Antibiotics are discontinued after 24 hours unless a specific infection is being treated. Early ambulation is encouraged, but oral fluids are withheld until the patient has passed flatus. Early oral intake will only prolong the postoperative ileus. If a colonic resection has been done suppositories and enemas are contraindicated and the patient is given nothing by mouth for seven days. Total parenteral nutrition may be required in such patients if they were malnourished preoperatively.

Morbidity

Despite earlier concerns, morbidity after aggressive cytoreductive surgery is tolerably low. In a recent report of 264 patients who underwent primary cytoreductive surgery in Finland, Venessmaa and Ylikorkala reported that only 26% of patients lost more than a litre of blood.[30] Postoperative fever developed in 5%, urinary tract infections in 19%, bowel complications (mainly paralytic ileus) in 9%, wound complications (mainly infection or haematoma) in 3%, and thromboembolism in 3%. There were four postoperative deaths (2%). One patient died of a pulmonary embolism four days after operation, and three died of cancer and obstructive ileus seven, eight, and 26 days respectively, after operation.

Survival after primary cytoreduction

Before any discussion about survival it should be stated that most patients with advanced ovarian cancer will eventually die of their disease, so the five-year survival rate is not the yardstick by which the value of cytoreduction should be measured. Rather, the two most important indices are the

156

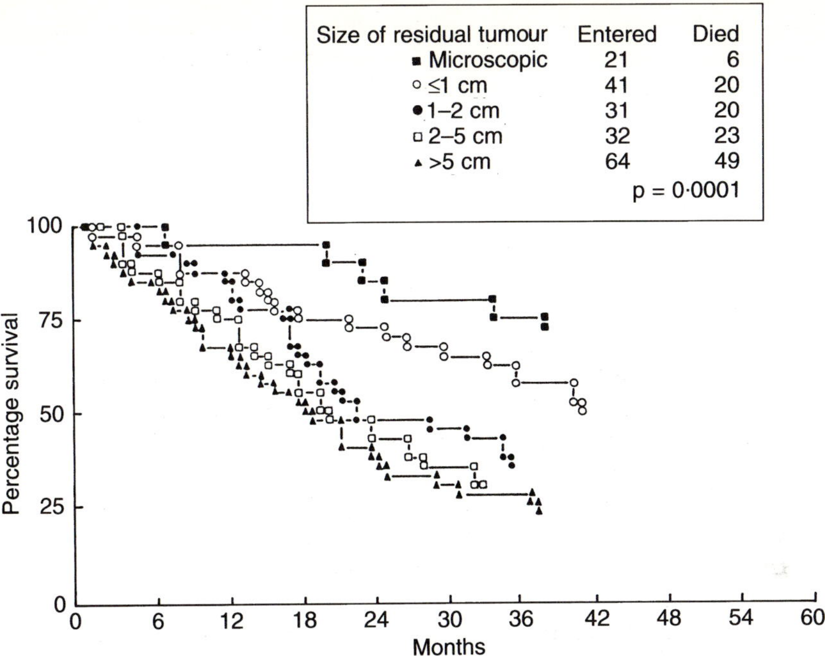

Fig. 11.5 Survival after primary cytoreduction.

quantity of the patient's life (progression-free interval or survival) and the quality of her life.

Virtually all studies of advanced ovarian cancer have shown that the amount of residual disease before the initiation of chemotherapy is the single most important determinant of prognosis. Figure 11.5 shows the highly significant improvement in survival for patients with residual nodules 1 cm or less in the Dutch Cooperative Study.[13] As mentioned earlier, analysis of our own recent experience in Sydney showed that the only patients with a reasonable likelihood of long term survival were those with no macroscopic residual disease (fig. 11.2). We have also shown that patients with peritoneal carcinomas have the same survival as patients with stage III ovarian carcinoma if treated in a similar manner.[14]

Intervention cytoreduction

Recently the European Organisation for Research in Treatment of Cancer (EORTC) have published the results of their prospective study of interval cytoreductive surgery.[47] Patients with residual tumour nodules larger than 1 cm after primary surgery were eligible for the study. After three cycles of cisplatin and cyclophosphamide, patients with no evidence of progressive disease were randomised between intervention cytoreduction and no further surgery, followed by at least three further cycles of cisplatin and cyclophosphamide. Patients with progressive disease on first-line treatment were excluded on the grounds that they had biologically aggressive, chemoresistant tumours for which no currently available treatment was likely to be effective.

Four hundred and twenty five patients were entered into the study and 319 (75%) were randomised. Patients in each arm were comparable. At intervention debulking surgery, tumour resection to lesions less than 1 cm in diameter was possible in 37 of 83 patients.[45] Thirty five percent of patients were found to have residual disease less than 1 cm in size at the time of intervention surgery. The surgery was generally well tolerated without serious morbidity and no mortality.

At the time of analysis three quarters of the patients had been followed up for more than 2.2 years and a quarter for more than 4.3 years. Maximal follow-up was 5.6 years. Progression free and overall survival were significantly longer in the group that had undergone further surgery (p=0.01) with a difference in median survival of six months (fig. 11.6). The risk of progression and the overall risk of death were reduced by one third in patients in the surgical arm. In a multivariate analysis intervention surgery was an independent prognostic factor for both progression-free and overall survival (p=0.012). Overall, after all other prognostic factors had been considered, intervention surgery reduced the risk of death by 33% (95% confidence interval 10% to 50%; p=0.008).

This study does not mean that intervention cytoreduction is preferable to primary cytoreduction, but it does imply that if patients are referred after an inadequate operation at a community hospital, an intervention operation should be offered after two or three cycles of chemotherapy as long as there is no progression during chemotherapy.

It is my practice to do primary cytoreduction for all patients with suspected advanced ovarian cancer, except for patients with gross ascites and large pleural effusions. Such patients can have the diagnosis of adenocarcinoma confirmed by tapping some of the ascites, and chemotherapy can then be initiated. If the tumour is chemosensitive the effusions will dry up after one or two cycles of chemotherapy, and intervention cytoreduction can then be done, without the high risk of pulmonary morbidity which would otherwise be present. If the effusions do not diminish with the chemotherapy the patient should be offered palliative care only, and aggressive surgery avoided.

Second-look laparotomy

During the past 20 years it has been common practice to do second-look laparotomies at the end of primary chemotherapy to obtain a more accurate indication of the state of the disease, and to allow further treatment for patients with persisting tumours. Currently, however, the routine use of second-look laparotomy is being challenged because of the lack of evidence that the operation increases survival[48] and the high recurrence

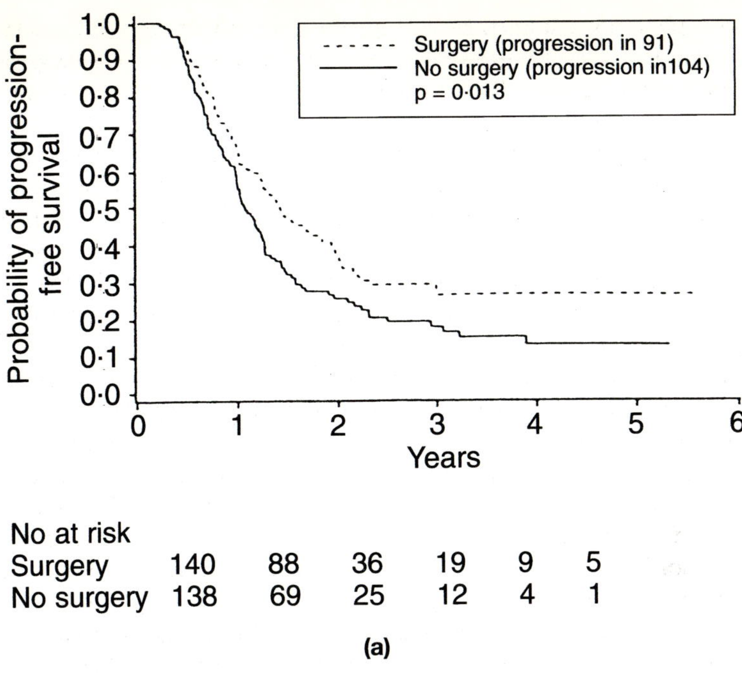

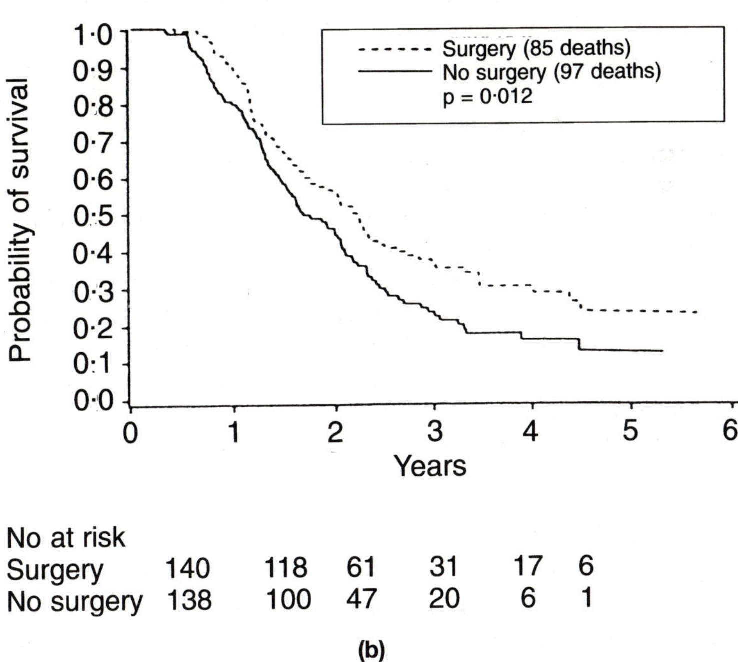

Fig. 11.6 Progression free survival (*a*) and overall survival (*b*) for patients treated with or without cytoreduction.

rate in patients in whom there is no sign of disease at the second-look.[49] In addition this operation produces great anxiety for patients, which cannot be justified if no benefits will accrue.

There are currently no non-invasive investigations which are sufficiently sensitive to assess accurately the degree of pathological response to first-line chemotherapy, so if an experimental protocol is being tested and an accurate measure of the completeness of the pathological response is required a second-look laparotomy will be necessary.

Table 11.3 Outcome of second-look laparotomies

First author	Reference No	Study Design	No of patients	Recurrence (%)
Omura	50	Prospective	226	50
Luesley	51	Prospective	120	43
Omura	52	Prospective	62	59
Berlinson	53	Retrospective	103	45
Rubin	54	Retrospective	83	50
Ho	55	Retrospective	39	52
Podczaski	56	Retrospective	76	50

With the reduction in the number of cisplatin-based chemotherapy cycles from 12 or more to six to eight in most centres, the recurrence rate after second-look laparotomy has shown no evidence of disease increased to about half (table 11.3).[49] This is adequate justification for attempting consolidation treatment in such patients. Patients with microscopic residual disease would also be suitable candidates for experimental approaches such as intraperitoneal treatment or high-dose chemotherapy with autologous bone marrow transplantation. Second-look laparotomy is also justified if trials are being conducted in a subset of patients who have had an excellent response to first-line chemotherapy, but there is no longer any justification for the routine use of second-look laparotomy to see whether first-line treatment may be discontinued.

Secondary cytoreduction

Secondary cytoreduction covers a heterogeneous group of interventions. In general, secondary cytoreduction may be undertaken in patients who are clinically free of disease after primary chemotherapy but have persistent disease at second-look laparotomy or in patients who develop recurrent disease after a disease-free interval. The early study by Berek *et al* reported 32 patients, 12 of whom (38%) had undergone optimal cytoreduction at a median of 12 months (range 6–48) after diagnosis.[57] The median survival of the 12 patients was 20 months, compared with five months for those whose debulking had been suboptimal (p<0.01), but all patients had received alkylating agents as first-line treatment so cisplatin was still available as second-line chemotherapy. This study comprised a heterogeneous group of patients, some having secondary cytoreduction at

160

the time of a second-look laparotomy, and others having it for recurrent disease after a disease-free interval.

Creasman reviewed published reports about second-look laparotomies that had been done for patients without clinical evidence of disease and found a total of 1207 patients in 16 papers.[58] There were 600 patients with macroscopic disease, and survival was significantly influenced by the residual disease after cytoreduction (table 11.4). Debulking to microscopic residual disease did not confer survival comparable to that for patients with microscopic disease who did not undergo debulking (31% compared with 47% survival; p = 0.0086), but survival was double that of patients with any macroscopic disease after secondary cytoreduction (31% compared with 15%; p = 0.0001). Whether or not the improved survival was associated with the surgical debulking or the biology of the tumour cannot be answered from these retrospective data.

Table 11.4 Effect of findings at second-look laparotomy on survival in epithelial ovarian cancer

	No of patients	No (%) that survived
No evidence of disease	414	294 (71)
Microscopic disease	193	91 (47)
Debulked to microscopic disease	118	37 (31)
Debulked to <5 mm	62	14 (22)
Debulked to <2 cm	213	45 (21)
Debulked to >2 cm	221	22 (10)

Hoskins *et al* have also suggested that secondary debulking to microscopic disease at the time of second-look laparotomy may result in improved survival.[59] They reported 67 patients, 17 of whom had microscopic disease at the time of second-look; 28 had disease that was less than 2 cm, and in 22 it was more than 2 cm. After secondary cytoreduction, 33 had microscopic disease, 26 had disease that was less than 2 cm, and seven had disease that was more than 2 cm (one was unknown). Five-year survival was 62% for patients who had microscopic disease at second-look laparotomy, and 51% for patients in whom the disease was rendered microscopic by secondary cytoreduction (p = 0.55). Patients with any macroscopic residual disease had a five-year survival of less than 10%.

Janicke *et al* reported 30 patients who underwent radical debulking procedures for clinically diagnosed relapses after a median recurrence-free interval of 16 months.[60] Complete tumour resection was accomplished in 47% of patients (14/30), and residual nodules of less than 2 cm remained in 12 (40%). Intestinal resections were necessary in 19 patients (63%). There was one postoperative death (within 32 days). Patients with no residual disease survived significantly longer than those with nodules of more than 2 cm (29 months compared with nine months; p = 0.004). Patients with a recurrence-free interval of more than 12 months also had better survival (median 29 months compared with eight

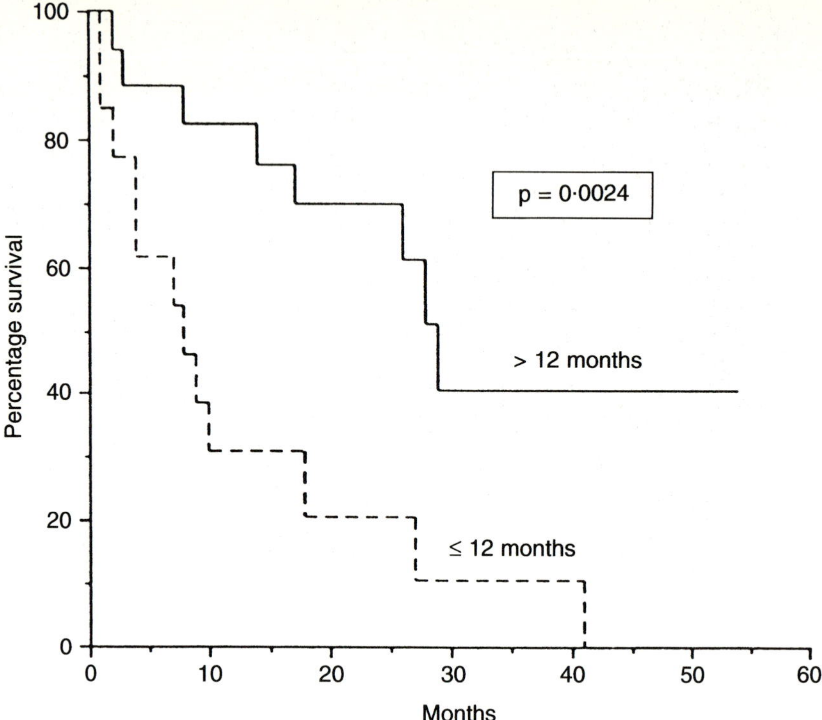

Fig. 11.7 Influence of duration of relapse free interval on survival.

months; p=0.002) (fig. 11.7). Multivariate analysis showed that residual tumour after the second operation was the most important independent variable for predicting survival time, while recurrence-free interval and the provision of any postoperative (second-line) treatment were also significant. FIGO stage, tumour grade, and the patient's age had no predictive value.

The importance of resection to no macroscopic residual disease and the long disease-free interval were also highlighted in a recent paper from Mt Sinai Medical Center in New York.[61] By contrast, a report from the MD Anderson Cancer Center suggested a limited role for secondary cytoreduction, but in this study the 9/35 patients who had all macroscopic tumour resected were not analysed separately.[62] There seems little doubt that this is the only group that is likely to benefit.

The longer the disease-free interval, the more likely it is that patients will present with an isolated recurrence (fig. 11.8), so after two years a rising CA125 concentration should be actively pursued with a CT scan of the pelvis and abdomen, and a chest radiograph. Solitary metastases in the lungs, liver, spleen or other sites should be resected.

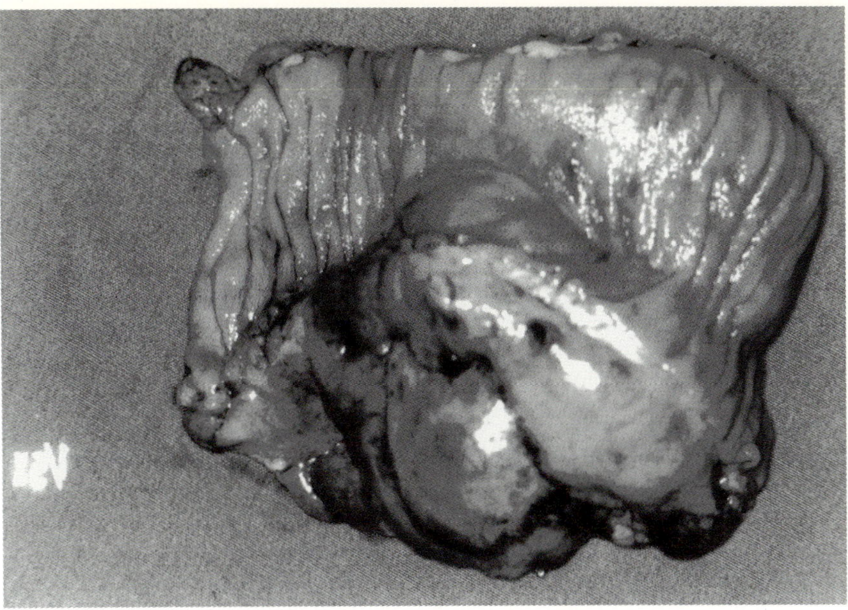

Fig. 11.8 Solitary metastasis in the sigmoid colon which developed seven years after treatment for stage I ovarian cancer.

Bowel obstruction

Because its exfoliated cells implant around the peritoneal cavity, ovarian cancer causes symptoms which are predominantly referable to the gastrointestinal tract. Although bowel obstruction is unusual at initial presentation, it is a common terminal event for patients who are dying of the disease, and almost half of these patients develop small bowel obstruction.[63]

Obstruction is usually the result of a mechanical carcinomatous blockage, although Redman *et al* reported that benign adhesions or radiation-induced strictures were responsible for the obstruction in 23% of 26 patients undergoing laparotomy.[64] Carcinomatous ileus without mechanical obstruction may also be responsible for a functional blockage.

Diagnosis

Diagnosis of bowel obstruction is based on the history, physical examination, and radiological investigations. The typical history is of nausea, vomiting, and increasing abdominal distension. The patient is unable to pass flatus, and there is also progressive constipation. With a high small bowel obstruction the patient vomits shortly after drinking and there is minimal abdominal distension, while with a low obstruction vomiting is delayed, and distension is more pronounced.

Initial radiological investigations should consist of plain supine, erect, and possibly left lateral decubitus films of the abdomen. Patients with bowel obstruction will have dilated, fluid-filled loops of bowel proximal to the obstruction, with air-fluid levels in the erect and left lateral decubitus views. These films provide a baseline evaluation against which progress of the obstruction can be monitored. Contrast studies are usually done, and a barium enema examination is particularly helpful. The large and small bowel are sometimes both obstructed, and it may be difficult to assess the patency of the colon during the operation. As surgical relief of the obstruction will often involve a bypass anastomosis between the small and large bowel it is important to be certain that the colon is patent preoperatively. Upper gastrointestinal contrast studies are less helpful, as the contrast is likely to pass through the colon eventually. The passage of a Cantor tube which will pass on down the small bowel to the point of initial obstruction is more helpful (fig. 11.9). Contrast media may be given down the tube.

Redman *et al* reported that the site of obstruction was small bowel in 42% of cases (often at multiple sites), large bowel in 16%, and both large and small bowel in 42%.[64] Similar experience has been reported by others.[65,66]

Management

Bowel obstruction often causes prolonged hospital stays for patients who have no other symptoms, so relief of obstruction is important because it allows them to function autonomously at home with a reasonable quality of life, even though there may be no remaining therapeutic options for the cancer itself. Medical management should always be tried initially, with surgical management reserved for selected cases.

Medical management

Initial management involves giving fluids intravenously, with careful fluid and electrolyte management, and nothing by mouth. If vomiting continues in appreciable volumes, nasogastric suction or a trial of long tube decompression is indicated.

Helmkamp and Kimmel used long tube decompression to restore small bowel function successfully for at least 30 days in seven of 22 patients (32%) with malignant obstruction, usually from ovarian cancer.[67] The mean duration of tube drainage was 11.3 days. They suggested that surgical intervention should be considered: if the abdominal cramps and nausea persist, or if the patient vomited with the tube clamped; if there was radiological evidence of complete obstruction (failure of the contrast

164

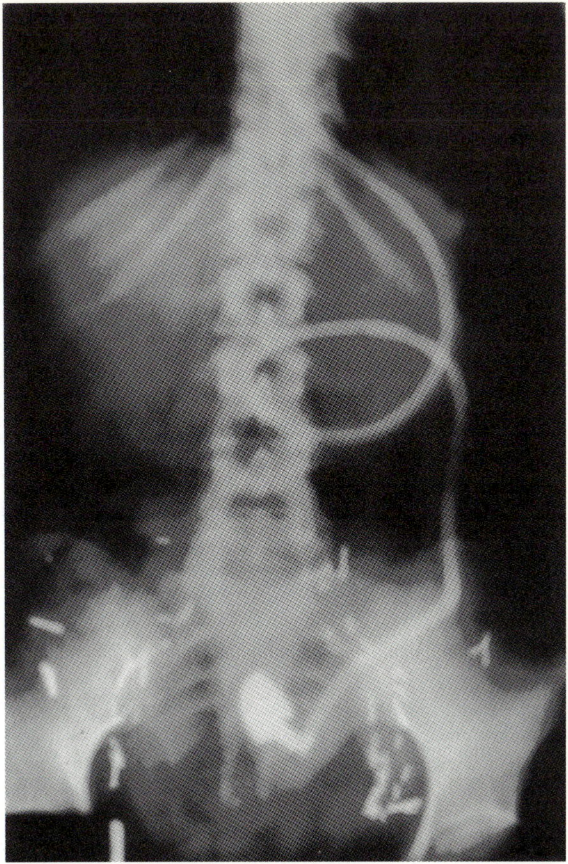

Fig. 11.9 A Cantor tube in the small bowel indicating the site of obstruction.

medium to pass the site of obstruction); or if there was evidence of a strangulating obstruction.

Somatostatin is a peptide that is secreted by the hypothalamus, which has recently been used to control vomiting and relieve small bowel obstruction in selected patients, and a trial of its use may be considered. It has a number of physiological functions including inhibition of the secretion of gastrin and pancreatic enzymes, inhibition of the secretory and motor functions of the intestine, and promotion of the absorption of water and sodium across the mucosa. In a double-blind clinical trial of somatostatin in 54 consecutive patients with mechanical small bowel obstruction, Bastounis *et al* reported that 12 of 27 patients (44%) who did not receive somatostatin required operation, compared with only six of 27 (22%) who received the drug.[68] Although because of the relatively small numbers involved the results did not reach statistical significance, the trend is apparent. In addition, for patients undergoing laparotomy, there was a significantly lower incidence of severe dilatation and necrosis of the intestine proximal to the

165

obstruction among the patients who received somatostatin (16% compared with 83%). The dose of somatostatin was 3 mg/day given intravenously over 12 hours for 48 hours.

A trial of dexamethasone is another consideration for patients whose vomiting is not severe. We have used a dose of 4 mg orally or subcutaneously for three to five days to try and reduce the inflammatory oedema. In our experience, this treatment will sometimes allow the patient to leave hospital on a low residue diet. Contraindications to dexamethasone include diabetes mellitus, active infection (including tuberculosis), active peptic ulceration, and psychosis.

For terminally ill patients, particularly those with a low (ileal) obstruction, the St Christopher's Hospice approach is applicable.[69] No intravenous fluids or nasogastric suction are used, and mild dehydration is encouraged. Antiemetics are used to reduce nausea (for example, haloperidol 2–6 mg/day and hyoscine hydrobromide 0.1 mg every six hours, subcutaneously). Limited oral intake is allowed, and the daily vomiting of a small volume of fluid is accepted. Exquisite mouth care is important to ensure patient comfort.

Surgery

In patients who fail to respond to conservative measures, a decision must be made about the advisability of proceeding to operation. This requires careful clinical judgement and experience. Castaldo *et al* suggested that operation was not advisable unless the patient's life expectancy was at least two months, and this is a reasonable rule of thumb.[70] Using this criterion, Krebs and Goplerud found surgery was most likely to be successful in patients under 65 years who were in a good nutritional state, had no palpable intra-abdominal masses, no clinical ascites, and had not been treated by radiation.[71] In my experience, localised abdominal disease is a common finding and should not be regarded as a contraindication to operation, but diffuse palpable abdominal disease usually indicates a poor prognosis for the surgical relief of obstruction.

The difficulty of patient selection is well illustrated by a report of surgery for intestinal obstruction in advanced ovarian cancer from the Memorial Sloan-Kettering Cancer Center. Rubin *et al* reported unsuccessful relief of obstruction in 20 of 54 patients (37%), although definitive procedures were possible in 43 of the patients (80%).[65]

Because the goal of the operation is palliative, the simplest procedure consistent with relief of obstruction is usually desirable. This will usually be a bypass procedure for small bowel obstruction and a colostomy for large bowel obstruction, but particularly for small bowel obstruction resection and primary anastamosis is a good option if it can be accomplished without difficulty. For example, a matted loop of terminal ileum which can readily be mobilised is best treated by resection, whereas if the ileum is fixed in the pelvis, side-to-side anastomosis between the proximal small bowel and

166

Table 11.5 Operative mortality from bowel obstruction in ovarian cancer

First author	Reference No	No of patients	No (%) that died
Castaldo	70	25	3 (12)
Tunca	72	90	13 (14)
Piver	73	49	9 (18)
Krebs	71	104	12 (12)
Redman	64	26	4 (15)
Rubin	65		54 (7)
		348	45 (13)

ascending colon is desirable. Stapling devices may be used, and care must be taken to avoid sites of malignant serosal implants for the anastomosis.

Surgery for intestinal obstruction from ovarian cancer is associated with appreciable morbidity and mortality. A review of 348 such patients demonstrated an operative mortality of 13% (table 11.5). Morbidity occurs in 30 to 50% of cases, typical complications including sepsis, pulmonary embolism, enterocutaneous fistulas, persistent bowel obstruction, and wound dehiscence.[66] To reduce morbidity, total parenteral nutrition should be considered if weight loss is greater than 10%. Antibiotics given intravenously and calcium heparin are desirable perioperatively as is the use of calf compressors. A large bowel washout should also be undertaken preoperatively if feasible.

In view of the serious morbidity and mortality patient selection is critical, and these patients should be managed in a gynaecological cancer centre where experience is concentrated. Each case must be treated on its merits, and the patient should be actively involved in the decision making process. In general, patients whose obstruction can successfully be relieved have a significantly improved survival. Rubin *et al* reported a mean survival of 6.8 months for those whose operations were successful, compared with 1.8 months for those whose were not.

Conclusion

The current status of surgery for advanced ovarian cancer is reflected in the conclusions of a recent consensus meeting on ovarian cancer held in Copenhagen.[74] The conclusions were as follows:

"There is no doubt that in most cases, the ultimate outcome for the patient is related mainly to the inherent biological properties of the tumour. Equally there is no doubt that all patients should be offered the best available chemotherapy. However, the only randomized prospective study of cytoreductive surgery for patients with epithelial ovarian cancer has revealed a significantly prolonged progression-free and overall survival ($p \leq 0.01$) in patients having intervention surgery and a significant reduction in the risk of disease progression and death. The following reflects the authors' assessment of currently available data on the role of surgery in this disease:

1. All patients with advanced ovarian cancer should have an aggressive attempt at primary cytoreductive surgery, provided their general medical status permits this approach. The objective should be to remove all macroscopic disease.
2. In patients who have had suboptimal primary surgery, an attempt at intervention cytoreduction should be undertaken after two or three cycles of chemotherapy in all patients without progression.
3. At the time of second-look laparotomy, secondary cytoreduction is appropriate if all macroscopic disease can be removed.
4. For disease recurrent after a disease-free interval, secondary cytoreduction is indicated:

   (i)   if all macroscopic disease can be removed
   (ii) if cisplatinum has not previously been used
   (iii) if the disease-free intervals exceeds 12 months.

If more effective second-line therapies become available, it is likely that the role of secondary cytoreduction will expand.

In addition to improving median survival for patients with advanced ovarian cancer, removal of the large tumour masses from the pelvis and upper abdomen almost always improves the patient's quality of life. The latter is an important consideration in any disease with a low likelihood of cure."

References

1 Meigs JV. *Tumors of the pelvic organs*. New York: McMillan, 1934.
2 Munnell EW. The changing prognosis and treatment in cancer of the ovary. *Am J Obstet Gynecol* 1968; **100**: 790–5.
3 Griffiths CT. Surgical resection of tumor bulk in the primary treatment of ovarian carcinoma. *National Cancer Institute Monographs* 1975; **42**: 101–5.
4 Hacker NF, Berek JS, Lagasse LD, *et al*. Primary cytoreductive surgery for epithelial ovarian cancer. *Obstet Gynecol* 1983; **61**: 413–20.
5 Van Lindert ACM, Alsbach GPJ, Barents J, *et al*. The role of the abdominal radical tumor reduction procedure in the treatment of ovarian cancer. In: Heintz APM, Griffiths CT, Trimbos JB, eds. *Surgery in gynecological oncology*. The Hague: Martinus Nijhoff, 1984: 275–87.
6 Hacker NF, Wain GV, Trimbos JP. Management and outcome of stage III epithelial ovarian cancer. In: Sharp F, Mason WP, Creasman W, eds. *Ovarian cancer*. New York: Chapman and Hall Medical, 1992: 351–6.
7 Schwartz PE, Chambers JT, Kohorn EI, *et al*. Tamoxifen in combination with cytotoxic chemotherapy in advanced epithelial ovarian cancer. A prospective randomized trial. *Cancer* 1989; **63**: 1074–9.
8 Griffiths CT. Carcinoma of the ovary: surgical objectives. In: Sharp F, Soutter WP, eds. *Ovarian cancer – the way ahead*. London: Chameleon Press, 1987: 235–44.
9 Heintz APM, Hacker NF, Berek JS, *et al*. Cytoreductive surgery in ovarian carcinoma: feasibility and morbidity. *Obstet Gynecol* 1986; **67**: 783–8.
10 Chen SS, Bochner R. Assessment of morbidity and mortality in primary cytoreductive surgery for advanced carcinoma. *Gynecol Oncol* 1985; **20**: 190–5.

11 Gruppo Interegionale Cooperative Oncologico Ginecologia. Randomized comparison of cisplatin with cyclophosphamide/cisplatin and with cyclophosphamide/doxorubicin/cisplatin in advanced ovarian cancer. *Lancet* 1987; **2**: 353–8.

12 Clinical Oncology Society of Australia. Chemotherapy of advanced ovarian adenocarcinoma: a randomized comparison of combination versus sequential therapy using chlorambucil and cisplatin. *Gynecol Oncol* 1986; **23**: 1–6.

13 Neijt JP, ten Bokkel Huinink WW, van der Burg MEL, *et al*. Randomized trial comparing two combination chemotherapy regimens (Chap-5 v CP) in advanced ovarian carcinoma. *Clin Oncol* 1987; **5**: 1157–62.

14 Hacker NF. Feasibility of primary cytoreductive surgery. In: Sharp F, Soutter WP, eds. *Ovarian cancer – the way ahead*. London: Chameleon Press, 1987: 245–55.

15 Trimbos JP, Hacker NF. The case against aspirating ovarian cysts. *Cancer* 1993; **72**: 828–31.

16 Dordoni D, Zaglio S, Favalli G. The role of sonographically guided aspiration in the clinical management of ovarian cysts. *J Ultrasound Med* 1993; **12**: 27–31.

17 Silberman AW. Surgical debulking of tumors. *Surg Gynecol Obstet* 1982; **155**: 577–82.

18 Hacker NF, Berek JS. Cytoreductive surgery for ovarian cancer. In: Albert DS, Surwit EA, eds. *Ovarian cancer*. The Hague: Martinus Nijhoff, 1985: 53–67.

19 De Vita VT. The relationship between tumor mass and resistance to chemotherapy. Implications for surgical treatment of cancer. *Cancer* 1983; **51**: 1209–13.

20 Gulling PM, Clark SH, Grantham F, *et al*. The interstitial fluid of solid tumours. *Cancer Res* 1964; **24**: 780–6.

21 Tannic I. Cell kinetics and chemotherapy. A critical review. *Cancer Treat Reports* 1978; **62**: 1117–22.

22 Mendelsohn ML. Autoradiographic analysis of cell proliferation in spontaneous breast cancer of C3H mouse. III. The growth fraction. *J Natl Cancer Inst* 1962; **28**: 1015–20.

23 Skipper HE. Thoughts on cancer chemotherapy and combination modality therapy. *JAMA* 1974; **230**: 1033–9.

24 De Vita VT. Cell kinetics and the chemotherapy of cancer. *Cancer Chemotherapy Reports* 1971; **3**: 23–8.

25 Skipper HE, Schabel FM Jr, Wilcox WS. Experimental evaluation of potential anticancer agents XII: On the criteria and kinetics associated with "curability" or experimental leukemia. *Cancer Chemotherapy Reports* 1964; **35**: 1.

26 Morton DL. Changing concepts in cancer surgery; surgery as immunotherapy. *Am J Surg* 1978; **135**: 367–72.

27 Khoo SK, Tillack SV, Mackay EV. Cell-mediated immunity: Effects of female genital tract cancer, pregnancy, and immunosuppressive drugs. *Aust NZ J Obstet Gynaecol* 1975; **15**: 156–61.

28 Pattillo RA, Ruckert ACF, Story MT, *et al*. Immunodiagnosis in ovarian cancer: Blocking factor activity. *Am J Obstet Gynecol* 1979; **133**: 791–7.

29 Goldie JH, Coldman AJ. A mathematic model for relating the drug sensitivity of tumours to their spontaneous mutation rate. *Cancer Treat Reports* 1979; **63**: 1727–32.

30 Venesmaa P, Ylikorkala O. Morbidity and mortality associated with primary and repeat operations for ovarian cancer. *Obstet Gynecol* 1992; **79**: 168–72.

31 Griffiths CT, Park LM, Fuller AF. Role of cytoreductive surgical therapy in the management of advanced ovarian cancer. *Cancer Treat Reports* 1979; **63**: 235–40.

32 Hoskins WJ, Bundy BN, Thigpen JT, *et al*. The influence of initial surgery on recurrent-free interval and survival in small-volume stage III epithelial ovarian cancer. A Gynecologic Oncology Group Study. *Gynecol Oncol* 1992; **47**: 159–66.

33 Heintz APM, van Oosterom AT, Trimbos JBMC, *et al*. The treatment of advanced ovarian carcinoma (1): clinical variables associated with prognosis. *Gynecol Oncol* 1988; **30**: 348–58.

34 Farias-Eisner R, Oliviera M, Teng F, *et al*. The influence of tumour distribution number and size after optimal primary cytoreductive surgery for epithelial ovarian cancer. *Gynecol Oncol* 1992; **46**: 267.

35 Eisenkop S, Nalick R, Teng N. Peritoneal implant excision or ablation during cytoreductive surgery. The impact on survival. *Gynecol Oncol* 1992; **45**: 97.

36 Hunter RW, Alexander NDE, Soutter WP. Management of surgery in advanced ovarian carcinoma: Is maximum cytoreductive surgery an independent determinant of prognosis? *Am J Obstet Gynecol* 1992; **166**: 504–11.

37 Blythe JG, Wahl TP. Debulking surgery. Does it increase the quality of survival? *Gynecol Oncol* 1982; **14**: 396–400.

38 van Houwelingen JC, ten Bokkel Hiunnink WW, van der Burg MEL, *et al.* Predictability of the survival of patients with advanced ovarian cancer. *J Clin Oncol* 1989; **7**: 769–73.

39 Burghardt E, Girardi F, Lahousen M, *et al.* Patterns of pelvic and paraaortic lymph node involvement in ovarian cancer. *Gynecol Oncol* 1991; **40**: 103–6.

40 Burghardt E, Winter R. The effect of chemotherapy on lymph node metastases in ovarian cancer. *Bailliere's Clin Obstet Gynaecol* 1989; **3**: 167–71.

41 Wu PC, Qu JY, Lang JH, *et al.* Lymph node metastases in ovarian cancer: A preliminary survey of 74 cases of lymphadenectomy. *Am J Obstet Gynecol* 1986; **155**: 11103–8.

42 Hudson CN. Surgical treatment of ovarian cancer. *Gynecol Oncol* 1973; **1**: 370–6.

43 Berek JS, Hacker NF, Leuchter RS, *et al.* Urologic operations during cytoreductive surgery for ovarian cancer. *Gynecol Oncol* 1982; **13**: 87–92.

44 Montz FJ, Schlaerth JB, Berek JS. Resection of diaphragmatic peritoneum and muscle: Role in cytoreductive surgery for ovarian cancer. *Gynecol Oncol* 1989; **35**: 338–40.

45 Adelson MD, Baggish MS, Cassell SL, Thompson MA. Cytoreduction of ovarian cancer with the cavitron ultrasonic surgical aspirator. *Obstet Gynecol* 1988; **72**: 140–3.

46 Brand E, Pearlman N. Electrosurgical debulking of ovarian cancer: A new technique using the argon beam coagulator. *Gynecol Oncol* 1990; **39**: 115–18.

47 van der Burg MEL, van Lent M, Buyse M, *et al.* The effect of debulking surgery after induction chemotherapy on the prognosis in advanced epithelial ovarian cancer: An EORTC Gynecological Cancer Cooperative Group Study. *N Engl J Med* 1995; **332**: 629–34.

48 Luesley D, Lawton F, Blackledge G, *et al.* Failure of second-look laparotomy to influence survival in epithelial ovarian cancer. *Lancet* 1988; ii: 599–602.

49 Ferrier AJ, DePetrillo AD. Second-look laparotomy in the routine management of ovarian cancer. In: Sharp F, Mason WP, Creasman W, eds. *Ovarian cancer 2.* New York: Chapman and Hall Medical, 1992: 388.

50 Omura GA, Bundy BN, Berek JS, *et al.* Randomised trial of cyclophosphamide plus cisplatin with or without doxorubicin in ovarian cancer: A gynaecologic oncology group study. *J Clin Oncol* 1989; **7**: 457–65.

51 Luesley DM, Chan KK, Lawton FG, *et al.* Survival after negative second look laparotomy. *Eur J Surg Oncol* 1989; **15**: 205–10.

52 Omura GA, Blessing JA, Ehrlich CE, *et al.* A randomised trial of cyclophosphamide and doxorubicin with or without cisplatin in advanced ovarian cancer. *Cancer* 1986; **57**: 1725–30.

53 Berlinson JL, Lee KR, Jarrell MA, *et al.* Management of epithelial ovarian neoplasms using a platinum-based regimen: a 10-year experience. *Gynecol Oncol* 1990; **37**: 66–73.

54 Rubin SC, Hoskins WJ, Hakes TB, *et al.* Recurrence after negative second-look laparotomy for ovarian cancer: analysis of risk factors. *Am J Obstet Gynecol* 1988; **159**: 1094–8.

55 Ho A, Beller U, Speyer JL, *et al.* A reassessment of the role of second-look laparotomy in advanced ovarian cancer. *J Clin Oncol* 1987; **5**: 1316–21.

56 Podczaski E, Manetta A, Kaminski P, *et al.* Survival of patients with ovarian carcinomas after second look laparotomy. *Gynecol Oncol* 1989; **36**: 43–7.

57 Berek JS, Hacker NF, Lagasse LD, *et al.* Survival of patients following secondary cytoreductive surgery in ovarian cancer. *Obstet Gynecol* 1983; **61**: 189–94.

58 Creasman WT. Evaluation of debulking surgery at second-look laparotomy. In: Sharp F, Mason WP, Creasman W, eds. *Ovarian cancer.* New York: Chapman and Hall Medical, 1992: 375–383.

59 Hoskins WJ, Rubin SC, Dulaney E, *et al.* Influence of secondary cytoreduction at the time of second-look laparotomy on the survival of patients with epithelial ovarian cancer. *Gynecol Oncol* 1989; **34**: 365–71.

60 Janicke F, Holscher M, Kuhn W, *et al.* Radical surgical procedure improves survival time in patients with recurrent ovarian cancer. *Cancer* 1992; **70**: 2129–36.

61 Segna RA, Dottino PR, Mandeli JP, *et al.* Secondary cytoreduction for ovarian cancer following cisplatin therapy. *J Clin Oncol* 1993; **11**: 434–9.

62 Morris M, Gershenson DM, Wharton JT, *et al.* Secondary cytoreductive surgery for recurrent epithelial ovarian cancer. *Gynecol Oncol* 1989; **34**: 334–8.

63 Krebs HB, Goplerud DR. The role of intestinal intubation in obstruction of the small intestine due to carcinoma of the ovary. *Surg Gynecol Obstet* 1984; **158**: 467–72.

64 Redman CWE, Shafi MI, Ambrose S, *et al.* Survival following intestinal obstruction in ovarian cancer. *Eur J Surg Oncol* 1988; **14**: 383–6.

65 Rubin SC, Hoskins WJ, Benjamin I, Lewis JL. Palliative surgery for intestinal obstruction in advanced ovarian cancer. *Gynecol Oncol* 1989; **34**: 16–19.

66 Clarke-Pearson DL, Chin NO, DeLong ER, *et al.* Surgical management of intestinal obstruction in ovarian cancer. I Clinical features, postoperative complications and survival. *Gynecol Oncol* 1987; **26**: 11–18.

67 Helmkamp BF, Kimmel J. Conservative management of small bowel obstruction. *Am J Obstet Gynecol* 1985; **152**: 677–9.

68 Bastounis E, Hadjinikolaou L, Ioannou *et al.* Somatostatin as adjuvant therapy in the management of obstructive ileus. *Hepatogastroenterology* 1989; **36**: 538–9.

69 Baines M, Carter RL, Oliver DJ. Medical management of intestinal obstruction in patients with advanced malignant disease. *Lancet* 1985; **2**: 990–3.

70 Castaldo TW, Petrilli ES, Ballon SC, *et al.* Intestinal operations in patients with ovarian carcinoma. *J Obstet Gynecol* 1981; **139**: 80–5.

71 Krebs H-B, Goplerud DR. Surgical management of bowel obstruction in advanced ovarian carcinoma. *Obstet Gynecol* 1983; **61**: 327–30.

72 Tunca JC, Buchler DA, Mack EA, *et al.* The management of ovarian-cancer-caused bowel obstruction. *Gynecol Oncol* 1981; **12**: 186–91.

73 Piver MS, Barlow JJ, Lele SB, Frank A. Survival after ovarian cancer induced intestinal obstruction. *Gynecol Oncol* 1981; **12**: 219–21.

74 Hacker NF, Van der Burg MEL. Debulking and intervention surgery. *Ann Oncol* 1993; 4 (suppl 4): S17–S22.

12 Assessment of response

A PETER M HEINTZ

Assessment of response to treatment in ovarian cancer is important for a number of reasons. By using internationally agreed response criteria, treatment protocols can be evaluated and compared (table 12.1). Evaluation of response, of course, is also important from an individual point of view in that ineffective treatment can be discontinued or modified, and successful treatment can be stopped in the knowledge that the patient is cured. In addition, there are a number of subjective responses to treatment which, although difficult to quantify, certainly improve patients' wellbeing. These include: improvements in appetite, reduction in intermittent bowel obstruction, resolution of ascites, and relief of pain. In this chapter I will discuss the various methods of assessment of response to treatment.

Table 12.1 Definitions

- **Clinically complete response**
  Complete regression of all clinically detectable tumour for at least four weeks.

- **Pathologically complete response**
  No tumour seen at laparoscopy or laparotomy, and no signs of tumour in biopsy specimens or peritoneal washings.

- **Microscopic disease**
  No tumour seen at laparoscopy or laparotomy, but tumour cells found in peritoneal biopsy specimens or washings, or both.

- **Partial response**
  Reduction by half of total measurable tumour load on two consecutive occasions not less than four weeks apart. In the case of a single lesion reduction in the tumour area by a half or more.
  In the case of multiple lesions reduction in the sum of the products of the perpendicular diameters of the lesions by a half or more.

- **Progressive disease**
  An increase in the size of the lesion by more than a quarter, or (multiple lesions) an increase in the products of the perpendicular diameters by more than a quarter, or development of a new lesion.
  Development of pleural effusion or ascites if confirmed to be malignant by cytological examination.

- **No change**
  Reduction of less than half or increase of less than a quarter in the sum of the products of the largest perpendicular diameters of all measurable lesions.

Response evaluated by both invasive and non-invasive methods.[1]

172

Surgical assessment

All patients with low stage ovarian cancer (stage I–IIa, IIc) and most patients with advanced stage disease (IIb, III, IV) who have had optimal cytoreductive surgery and subsequent cytotoxic treatment will have no evidence of disease after treatment has been stopped. Unfortunately, there is no reliable laboratory test or radiological technique to diagnose subclinical disease. For this reason second-look operations (laparoscopy and laparotomy) have been advocated for formal reassessment.

The term "second-look operation" was first introduced by Wangensteen *et al* in 1949 and applied to patients with colonic cancer whose abdomens were re-explored while they remained clinically free of disease after primary treatment.[2] The idea was that if recurrences were discovered they might be treated in the preclinical stage. Rutledge and Burns introduced the concept into the planned treatment of ovarian carcinoma and used the operation to evaluate the need for continuation of chemotherapy in patients with asymptomatic disease.[3]

The term "second-look operation" should be used only for the first surgical evaluation in patients who are clinically free of disease after adequate first-line treatment. The benefit of this approach for the patient has always been earlier discontinuation of chemotherapy in those with pathologically confirmed complete remission, or a change to a more effective treatment in the asymptomatic phase.

Because it is known that the number of patients with a complete response will not increase after six courses of platinum-based combination chemotherapy, however, and because there is no effective second-line treatment available, second-look operations have ceased to be of importance for patients who have completed their first-line treatment. For this reason, these reassessment operations should be made only in the interest of research when second-line treatments are being studied and after informed consent from the patient. Three types of procedures are advocated as reassessment operations: laparoscopy, laparotomy, and a combination of the two.

Second-look laparoscopy

The laparoscope has been advocated by several authors for the inspection of the peritoneal cavity after treatment. Because second-look laparoscopy is always done for patients with a history of surgery or radiotherapy, or both, it is not surprising that inadequate or unsuccessful procedures are often reported. If adequate visualisation is defined as the ability of the surgeon to visualise the entire peritoneal cavity from the cul-de-sac to the diaphragm including the paracolic gutters, adequate laparoscopy has been reported to be successful in 50%–72% of the procedures.[45]

If peritoneal washings are taken during laparoscopy the incidence of persistent disease is reported to be 20%–50%,[4-8] but, a second-look laparoscopy that fails to show disease does not correlate with complete pathological response, as persistent disease at laparotomy after a laparoscopy has failed to show disease is reported in 30%–50% of patients (table 12.2). These reports clearly indicate that second-look laparoscopy is of limited value in assessing extent of disease in the follow-up of a patient with ovarian carcinoma.

Table 12.2 Incidence of persistent disease at second-look laparotomy after laparoscopy had failed to show evidence of disease in patients with ovarian carcinoma

First author	Reference No	Total No of patients	No (%) with tumour at second-look laparotomy
Mangioni	5	11	6 (54)
Heintz	8	15	8 (53)
Rosenoff	9	4	2 (50)
Smith	10	11	5 (45)

In addition, laparoscopy after one or more previous laparotomies carries a certain morbidity. The reported incidence of serious complications ranges between 1% and 10% with bowel injury being most common.[5 10 11] For this reason, the Palmer test should always be done after insufflation in patients who have had previous laparotomies. In selected cases, open laparoscopy is indicated.

Second-look laparotomy

The diagnostic value of a second-look laparotomy is illustrated by the reported incidences of persistent disease (50%–75%) in patients with advanced disease who are clinically free of disease.[8 13–26] However, the value of a pathologically confirmed complete response is also questionable, as 13%–50% of patients with such a response have been reported to develop recurrences sooner or later.[8 12–16 18–25 27–29 31–37] The technique of the second-look laparotomy is similar to the technique and guidelines used for a staging laparotomy.

Second-look laparoscopy before second-look laparotomy

It is frustrating to find disease during a second-look laparotomy in a patient without clinical evidence of this residuum because this means an unnecessary laparotomy. Because of the inaccuracy of the laparoscope alone, a few authors have mentioned the possibility of combining laparoscopy and laparotomy for second-look procedures, which gives the opportunity of selecting only those patients for laparotomy in whom no tumour is detected by laparoscopy. In our experience this combination prevents half the second-look laparotomies.[8 38]

Conclusion

Second-look procedures are very accurate in evaluating the degree of cancer in patients who are clinically free of disease after an adequate number of courses of chemotherapy. In this respect, the combination of laparoscopy and laparotomy can prevent unnecessary laparotomies, but the benefit of second-look operations for individual patients has never been proved. The results of the reassessment operations have undoubtedly contributed to our understanding of the biological behaviour of the disease, and continue to contribute to our understanding of the value of new treatment strategies. This conclusion also underlines clearly the limited value of the procedure for individual patients who are not treated in a research setting.

Imaging in response assessment

It is essentially impossible to assess intrahepatic or retroperitoneal disease, or metastases within lymph nodes or the bowel mesentery by clinical means. In addition, fibrosis in the pelvis secondary to extensive surgery further inhibits clinical examination. Ultrasound, computed tomography (CT), and magnetic resonance imaging (MRI) may all have a role in assessing response and have potential advantages over surgical assessment in that the examination can be repeated at intervals over a long-term period.

CT has been studied widely in patients with ovarian cancer, and although it can be helpful in making the diagnosis of recurrent disease, its value in detecting clinically undetectable tumour deposits is limited. Shields *et al* showed that it was unable to differentiate between adhesions and tumour deposits, or tumour and benign unrelated lesions.[39] Small peritoneal deposits are also usually missed. When the results of CT evaluation are compared with those of second-look laparotomy the false negative rate of CT ranges from 17%–64%,[40 41] so a negative CT should be interpreted with care as an indicator of complete remission.

MRI may be particularly useful in the pelvis because of the contrasting appearances of various pelvic structures, particularly bowel gas, fat, ascites, and urine, but the available data suggest that the resolution of MRI is insufficient to detect isolated metastases smaller than about 1–2 cm in diameter.[42–44] Consequently, neither MRI nor CT are particularly useful in the assessment of response in patients with a complete clinical response. Both methods can, however, be valuable in following behaviour of assessable tumour deposits during chemotherapy.

Ultrasound is widely used in gynaecological oncology, particularly in the diagnosis of a pelvic mass. In general, the technique is ideal for measuring solid tumours. The method is cheap, easy to use, and gives a clear impression of the response to chemotherapy by simply measuring the volume of the mass. It is the first choice in the diagnosis and investigation of liver metastases, but it is not sensitive enough to diagnose occult disease.

Immunoscinitigraphy is a technique that is based on the principle that radio-labelled tumour specific antibody binds to the antigen in malignant tissue and forms a hot spot which can be detected by a gamma camera, but although the idea and the results in research settings are promising, there are no immunoscintigraphic tests available at present that can localise clinically occult disease.[44]

Tumour markers

For the assessment of response it is important to have a tumour marker with a high sensitivity so that as many patients as possible with residual disease, even those with occult disease, can be identified. There are no markers with a 100% sensitivity. The most suitable marker seems to be CA125 in serum and many authors have reported the association between serum CA125 titres and disease status at second-look.[45-50] In general, an increase in the CA125 titre, particularly in patients who had raised CA125 titre at the beginning of treatment, is likely to reflect residual disease. The titre can, however, be normal in patients with small volume residual disease. Rubin *et al* reported the presence of tumour in 62% of patients who had a normal CA125 titre at the time of operation.[50] Berek *et al* reported that the predictive value of a positive test, that is a CA125 titre of more than 35 U/ml, was 100% in patients who underwent a second-look procedure. The predictive value of the negative test was only 56%, however, and disease was present in 44% of the patients with a normal test who underwent a second-look procedure.[46]

CA125 has proven to be a useful way to monitor the response to treatment during chemotherapy in those patients who initially have a raised titre.[51] In addition, rising titres indicate resistance to treatment and progression of disease and therefore the need to change the treatment.

Conclusions

Second-look operations remain the most accurate way of assessing patients' status after primary surgery and chemotherapy. For patients in complete remission, serum CA125 estimation seems the most appropriate means of continuing assessment, the advantage over imaging techniques being the ability to detect small volume disease (tumour nodules less than 1 cm in diameter).

The results of all assessment procedures, however, invasive or non-invasive, are only of clinical importance if the results initiate effective further treatment.

The interval between a rise in CA125, or a combination of markers, some months before clinically apparent relapse, may provide an advantage in terms of instituting second-line treatment earlier. Whether or not this manoeuvre would improve the duration of second-line response is yet to be tested in large numbers of patients.

References

1 Miller AB, Hoogstraten B, Staquet M, Winkler A. Reporting results of cancer treatment. *Cancer* 1981; **47**: 207–14.
2 Wangenstein OH. Cancer of the colon and rectum with special reference to (1) earlier recognition of alimentary tract malignancy; (2) secondary delayed re-entry of the abdomen in patients exhibiting lymph node involvement; (3) subtotal primary excision of the colon; (4) operation in obstruction. *Wisconsin Medical Journal* 1949; **48**: 591–7.
3 Rutledge F, Burns BC. Chemotherapy of advanced ovarian cancer. *Am J Obstet Gynecol* 1966; **96**: 761–72.
4 Berek JS, Griffith CT, Leventhal JM. Laparoscopy for second-look evaluation in ovarian cancer. *Obstet Gynecol* 1981; **58**: 192–8.
5 Mangioni C, Bolis G, Molteni P, *et al.* Indications, advantages and limits of laparoscopy in ovarian cancer. *Gynecol Oncol* 1979; **7**: 47–51.
6 Quinn MA, Bischop GJ, Campbell JJ, *et al.* Laparoscopic follow-up of patients with ovarian carcinoma. *Br J Obstet Gynaecol* 1980; **87**: 1132–9.
7 Ozols RF, Fisher RI, Anderson T, *et al.* Peritoneoscopy in the management of ovarian cancer. *Am J Obstet Gynecol* 1981; **140**: 611–9.
8 Heintz APM, van Oosteron AT, Trimbos JBMC, *et al.* The treatment of advanced ovarian carcinoma (II): interval reassessment operations during chemotherapy. *Gynecol Oncol* 1988; **30**: 359–71.
9 Rosenoff SG, de Vita VT, Hubbard S, *et al.* Peritoneoscopy in the staging and follow-up of ovarian carcinoma. *Semin Oncol* 1975; **2**: 223–8.
10 Smith GW, Day TG, Smith JP. The use of laparoscopy to determine the results of chemotherapy in ovarian cancer. *J Reprod Med* 1977; **18**: 257–60.
11 Lacey CG. Laparoscopy in gynecologic oncology. In: Morrow CP, Bonnar J, O'Brien TG, Gibbons WE, eds. *Recent clinical developments in gynecologic oncology.* New York: Raven Press, 1983: 181–9.
12 Curry SL, Zembo MM, Nahhas WA, *et al.* Second-look laparotomy for ovarian cancer. *Gynecol Oncol* 1981; **11**: 114–18.
13 Webb MJ, Snijder JA, William TJ, *et al.* Second-look laparotomy in ovarian cancer. *Gynecol Oncol* 1982; **14**: 285–93.
14 Raju KS, McKinna JA, Barker GH, *et al.* Second-look operations in the planned management of advanced ovarian carcinoma. *Am J Obstet Gynecol* 1982; **144**: 650–4.
15 Roberts WS, Hodel K, Rich WM, Di Saia PJ. Second-look laparotomy in the management of gynecologic malignancy. *Gynecol Oncol* 1982; **13**: 345–55.
16 Phibbs GD, Smith JP, Stanhope CR. Analysis of sites of persistent cancer at "second-look" laparotomy in patients with ovarian cancer. *Am J Obstet Gynecol* 1983; **147**: 611–17.
17 Berek JS, Hacker NF, Lagasse LD, *et al.* Second-look laparotomy in stage 3 epithelial ovarian cancer. Clinical variables associated with disease status. *Obstet Gynecol* 1984; **64**: 207–12.
18 Barnhill DR, Hoskins WJ, Heller PB, *et al.* The second-look surgical reassessment for epithelial ovarian carcinoma. *Gynecol Oncol* 1984; **19**: 148–54.
19 Gershenson DM, Copeland LJ, Wharton JT, *et al.* Prognosis of surgically determined complete responsers in advanced ovarian cancer. *Cancer* 1985; **55**: 1129–35.
20 Podratz KC, Malkasian G Jr, Hilton JF, *et al.* Second-look laparotomy in ovarian cancer: evaluation of pathologic variables. *Am J Obstet Gynecol* 1985; **152**: 230–8.
21 Cain J, Saigo P, Pierce V, *et al.* A review of second-look laparotomy for ovarian cancer. *Gynecol Oncol* 1986; **23**: 14–25.
22 Carmichael JA, Shelley WE, Brown LB, *et al.* A predictive index of cure versus no cure in advanced ovarian carcinoma patients—replacement of second-look laparotomy as a diagnostic test. *Gynecol Oncol* 1987; **27**: 269–81.

23 Podczaski ES, Stevens CJ, Manetta A, *et al*. Use of second-look laparotomy in the management of patients with ovarian epithelial malignancies. *Gynecol Oncol* 1987; **28**: 205–14.

24 Ghatage P, Krepart GV, Lotocki R. Factor analysis of false negative second-look laparotomy. *Gynecol Oncol* 1990; **36**: 172–5.

25 Lund B, Williamson P. Prognostic factors for outcome of and survival after second-look laparotomy in patients with advanced ovarian carcinoma. *Obstet Gynecol* 1990; **76**: 617–22.

26 Schwartz PE, Smith JP. Second-look operations in ovarian cancer. *Am J Obstet Gynecol* 1980; **138**: 1124–30.

27 Smirz LR, Stehman FB, Ulbright TM, Sutton GP, Ehrlich CE. Second-look laparotomy after chemotherapy in the management of ovarian maligancy. *Am J Obstet Gynecol* 1985; **152**: 661–8.

28 Copeland LJ, Gershenson DM. Ovarian cancer recurrences in patients with no macroscopic tumour at second-look laparotomy. *Obstet Gynecol* 1986; **86**: 873–4.

29 Miller DS, Ballon SC, Teng NN, Seifer DB, Soriero OM. A critical reassessment of second-look laparotomy in epithelial ovarian carcinoma. *Cancer* 1986; **57**: 530–5.

30 Gallup DG, Talledo OE, Dudzinski MR, Brown KW. Another look at the second assessment procedure for ovarian epithelial carcinoma. *Am J Obstet Gynecol* 1987; **157**: 590–6.

31 McCusker MC, Hoffman JS, Curry SL, Koulos JP, Gondos B. The role of second-look laparotomy in treatment of epithelial ovarian cancer. *Gynecol Oncol* 1987; **28**: 83–8.

32 Lippman SM, Alberts DS, Slijmen DJ, *et al*. Second-look laparotomy in epithelial ovarian carcinoma. Prognostic factors associated with survival duration. *Cancer* 1988; **61**: 2571–7.

33 Podratz KC, Malkasian GJ, Wieand HS, *et al*. Recurrent disease after negative second-look laparotomy in stages III–IV ovarian carcinoma. *Gynecol Oncol* 1988; **29**: 274–82.

34 Rubin SC, Hoskins WJ, Hakes TB, *et al*. Recurrence after second-look laparotomy for ovarian cancer: analysis of risk factors. *Am J Obstet Gynecol* 1988; **159**: 1094–8.

35 Sonnendecker EW. Is routine second-look laparotomy for ovarian cancer justified? *Gynecol Oncol* 1988; **31**: 249–55.

36 Luesley DM, Chan KK, Lawton FG, *et al*. Survival after negative second-look laparotomy. *Eur J Surg Oncol* 1989; **15**: 205–10.

37 Rubin SC, Hoskins WJ, Saigo PE, *et al*. Prognostic factors for recurrence following negative second-look laparotomy in ovarian cancer patients treated with platinum-based chemotherapy. *Gynecol Oncol* 1991; **42**: 137–41.

38 Heintz APM. Ovarian cancer, surgical treatment and prognosis. Thesis. Leiden: University of Leiden, 1985.

39 Shields RA, Peel KR, McDonald HN, *et al*. A prospective trial of computed tomography in the staging of ovarian malignancy. *Br J Obstet Gynaecol* 1985; **92**: 407–12.

40 Stern J, Buscema J, Rosenshein N, Siegelman S. Can computed tomography substitute for second-look operation in ovarian carcinoma. *Gynecol Oncol* 1981; **11**: 82–8.

41 Stehmann FB, Calkins Ar, Wass JL, Smirz LR, Sutton GP, Ehrlich CE. A comparison of findings at second-look laparotomy with pre-operative computed tomography in patients with ovarian cancer. *Gynecol Oncol* 1988; **29**: 37–42.

42 Brenner DE, Shaft MI, Jones HW, *et al*. Abdominopelvic computed tomography: Evaluation in patients undergoing second-look laparotomy for ovarian carcinoma. *Obstet Gynecol* 1985; **65**: 715–9.

43 Buy JN, Moss AA, Ghossian MA, *et al*. Peritoneal implants from ovarian tumours: CT findings. *Radiology* 1988; **169**: 691–4.

44 Dershaw DD, Panicek DM. Radiologic evaluation of ovarian cancer. In: Markham M, Hoskins WJ, eds. *Cancer of the ovary*. New York: Raven Press, 1993: 133–52.

45 Nilof JM, Bast RJ, Schaetzl EM, Knapp RC. Predictive value of CA125 antigen levels in second-look procedures for ovarian cancer. *Am J Obstet Gynecol* 1985; **151**: 981–6.

46 Berek JS, Knapp RC, Malkasian GD, *et al*. CA125 serum levels correlated with second-look operations among ovarian cancer patients. *Obstet Gynecol* 1986; **67**: 685–9.

47 Atack DB, Misker JA, Allen HH, *et al*. CA125 surveillance and second-look laparotomy in ovarian carcinoma. *Am J Obstet Gynecol* 1986; **154**: 287–9.

48 Potter ME, Morade M, To AC, *et al*. Value of serum CA125 levels: does the result preclude second-look? *Gynecol Oncol* 1989; **33**: 201–3.

49 Meier W, Streber P, Eiermann W, *et al*. Serum levels of CA125 and histological findings at second-look laparotomy in ovarian carcinoma. *Gynecol Oncol* 1989; **35**: 44–6.

50 Rubin SC, Hoskins WJ, Hakes TB, *et al*. Serum CA125 levels and surgical findings in patients undergoing secondary operations for epithelial ovarian cancer. *Am J Obstet Gynecol* 1989; **160**: 667–71.
51 Lavin PT, Knapp RC, Malkasian GD, *et al*. CA125 for the monitoring of ovarian carcinoma during primary therapy. *Obstet Gynecol* 1987; **69**: 223–7.

13 Treatment options at relapse

MANFRED LAHOUSEN

Recurrence is defined as the development of a new focus of cancer after a tumour-free interval of at least six months. It is important to make the distinction between resistant and relapsed disease because patients with resistant disease are unlikely to respond to salvage treatment, particularly if there has been only a partial response to primary platinum-based chemotherapy, whilst those with relapsed disease may respond to treatment again, particularly if the interval between diagnosis and relapse has been many months.

Although most patients with ovarian cancer will relapse, there are surprisingly few reports detailing the location and timing of the recurrence. Table 13.1 shows the sites of recurrence disease in 52 patients with stage I, II, or completely resected stage III disease. The sites of relapse in patients with stage I or II disease were evenly distributed between the pelvis, liver, and lymph nodes, while nearly half of the recurrences in the patients with stage III disease were pelvic.

Table 13.1 Site of recurrent disease in patients with stage I, II, and III ovarian cancer with no macroscopic residuum after primary surgery

Site	Stage		
	I	II	III
Pelvis	8	2	15
Lung	0	0	6
Lymph nodes:			
Pelvic	1	0	0
Para-aortic	0	1	1
Supraclavicular	2	1	1
Abdomen	1	0	2
Total	14	5	33

Treatment of early relapse

Early relapse can be difficult to diagnose, particularly if the volume is small and limited to the pelvis or abdomen. The information gained from pelvic examination in a woman who has previously undergone extensive surgery may be limited, and the resolution of imaging techniques reduces their usefulness in the pelvis although small volume, extra-abdominal soft tissue recurrences may be recognised early.[12] A recommended follow-up policy is shown in table 13.2.

It has been shown that CA125 titres may rise three to nine months before recurrent disease becomes clinically detectable.[34] The tumour markers carcinoembryonic antigen, ferritin, tissue polypeptide antigen, and CA125 were measured at two monthly intervals in 170 patients followed up for at least a year and raised titres indicated recurrent disease two to nine months before the recurrence was diagnosed clinically.[56]

Tumour markers can therefore permit the diagnosis of recurrent cancer earlier than clinical examination alone, but does this earlier diagnosis lead to a better chance of response to second-line treatment? We constructed a biomathematical model to circumvent the difficulties of interpreting the course of tumour marker titres.[7] A characteristic value was calculated with the model—a high value indicating recurrent disease—and its usefulness was tested in a prospective study. Twenty four patients without clinical evidence of disease had a high characteristic value and were treated with second-line etoposide when recurrence was clinically evident four to seven months after. At twelve months 92% of these patients had died of recurrent disease. A second group of 13 patients were treated as soon as the characteristic value was high—that is before there was clinical confirmation of recurrence. Nine patients (69%) were still alive at one year, so initiation of treatment on the basis of rising titres of tumour markers alone has the potential to improve the salvage rate at one year by 60%.

Patients with recurrent disease can be divided into three groups. Some patients will have had little or no response to first-line therapy and will have either static or progressive disease, some will have had a partial response initially and the disease will progress at a later date, and the third group of patients will have had a complete response to treatment before relapsing. Patients with recurrent disease are, in general, less well able to tolerate toxic treatment regimens, which in any case are likely to be given with palliative rather than curative intent. The decision to treat patients with recurrent disease may, therefore, largely be based on a desire to palliate symptoms or to assess new treatment regimens.

Some patients may undergo further operations. Exenterative type procedures have been advocated but are not usually appropriate because of the widespread intra-abdominal nature of most cases of recurrent ovarian cancer. When it has been possible to carry out such surgery, however, survival has been increased.[8] The benefit of less aggressive surgery for recurrent disease has been evaluated in only a few studies.[9-11] It is apparent

Table 13.2 Methods and intervals of follow-up of patients with cancer of the ovary

	First and second years						Third year				Fourth and fifth year		Over five years
	2	4	6	8	10	12	3	6	9	12	6	12	12
History	+	+	+	+	+	+	+	+	+	+	+	+	+
Weight	+	−	−	+	−	+	−	+	−	+	+	+	+
Clinical examination	+	+	+	+	+	+	+	+	+	+	+	+	+
Colposcopy/cytology	+	−	−	+	−	+	−	+	−	+	+	+	+
Assay of tumour markers	+	+	+	+	+	+	+	+	+	+	+	+	−
Chest radiograph	−	+	−	−	+	−	−	−	+	−	+	−	+
Ultrasound/CT	−	−	+	−	−	+	−	+	−	+	−	+	−
Laboratory tests	+	+	+	+	+	+	+	+	+	+	+	+	+
Immunoscintigraphy	−	−	−	+	−	−	−	−	+	−	+	−	−

that patients who have a long interval between diagnosis and relapse, sometimes many years, are most likely to benefit from further surgical debulking.[12] No patient in our series of 52 had localised disease which we considered totally resectable.

It is rare to encounter a patient who has not been treated with a first-line regimen containing platinum. Patients who have received alkylating agents as first line chemotherapy, however, can often achieve a durable response to second-line cisplatin or carboplatin. Treatment after failure of first-line platinum-containing regimens is less effective, but paclitaxel induces remissions in 20%–30% of such patients. Again there is evidence of the effect that the quality and duration of first-line chemotherapy have on the response to further cytotoxics; patients who have only a short response to first-line treatment are unlikely to respond a second time.[13] This emphasises the point that any type of salvage treatment, surgical or cytoxic, is likely to be more effective in patients who responded well to first-line treatment, and the importance of distinguishing patients with progressive, resistant and recurrent cancer.

Second-line treatment

Second-line treatment often presents both psychological and ethical problems. After extensive debulking procedures and primary chemotherapy patients may be unwilling to undergo renewed treatment with further toxic agents, particularly as long term survival is rare.

The effect of second-line chemotherapy is influenced by the volume and site of the recurrence and the patient's general condition, which may reflect both symptoms of the recurrence and long term side effects of the primary cytotoxic regimen. As far as the biology of the tumour is concerned, aggressive regimens consisting of combinations of agents with different mechanisms of action seem to offer the highest probability of success. Potentially useful agents include platinum analogues, etoposide, ifosfamide, 5-fluorouracil, and hexamethylmelamine. Tamoxifen is extremely well tolerated and produces responses in the order of 8%–20% in this setting and deserves further study.

Our 52 patients, who received cisplatin, doxorubicin, and cyclophosphamide chemotherapy (PAC) after primary surgery were treated with a number of regimens when they relapsed. The main second-line regimen was etoposide which was used to treat 18 patients who survived for between three and 40 months. Four patients were given further courses of PAC and survived a mean of 21.2 months. Five patients were given cisplatin only and survived a mean of 10.2 months, and one survived for 22 months after treatment with both cisplatin and carboplatin. Mitozantrone was given to five patients who survived a mean of 10 months. Overall only three patients have had a disease free interval of longer than 24 months.

Because there are few convincing data that whole-abdominal radiotherapy has an impact on survival in patients with persistent disease after treatment with platinum,[14] its role for patients with recurrent ovarian cancer will also be of limited value. Limitations include the relative radioresistance of hypoxic cells within recurrent tumour and the inability to deliver high doses of radiation without exceeding normal tissue tolerance. Patients with macroscopic disease are most unlikely to gain any useful response, although it may be used for palliation in patients with recurrent, localised, pelvic disease in whom chemotherapy is considered inappropriate. The mean survival time for the small number of patients in our series who were treated with radiation for pelvic sidewall recurrence was only three months. The combination of chemotherapy and radiotherapy, which has shown promise as primary therapy, may be of use in treating recurrent disease.[15]

Experimental regimens

In research institutions, patients with recurrent ovarian cancer may be included in protocols designed to test the activity of experimental regimens. These may include testing new drugs and agents, using established drugs at higher concentrations, or using new methods of giving the drugs. As for most of its natural history ovarian cancer is confined (at least macroscopically) to the peritoneal cavity, there may be a role for the use of intraperitoneal treatments—either cytotoxics or biological agents. It is apparent, however, that intraperitoneal treatment still produces systemic side effects, that local complications may be troublesome, and that patients with anything more than minimal recurrent disease (nodules 1 mm or less in diameter) are unlikely to benefit.[16] Other innovations such as giving lymphokine activated killer (LAK) cells or antibody-toxin or antibody-radioisotope conjugates directly into the peritoneal cavity are under investigation.[17–19] It is probable that useful activity with these agents will also be limited to those patients with minimal peritoneal disease, implying that before such treatments could be instituted, patients would need to undergo further laparotomy to attempt surgical cytoreduction.

At the present time, the prognosis for patients with recurrent ovarian carcinoma is poor. Nevertheless, advances in our understanding of the biology of the disease and our ability to monitor its course coupled with the development of new treatments could lead to improvements in outcome. Randomised studies with large numbers of patients are required to test the potential impact on prognosis that earlier detection of recurrence with tumour markers might provide.

References

1 Triller J, Goldhirsch A. Computertomographie, primare Laparotomie und second-look-Operation bei Ovarialkarzinomen. *Fortschritte auf dem Gebiete der Routgenstrahlen* 1984; **140**: 294–303.
2 Dooms GC, Hricak H, Crooks LE, Higgins CB. Magnetic resonance imaging of the lymph nodes: comparison with CT. *Radiology* 1984; **153**: 719–24.

3 Bast RC Jr, Siegal FP, Runcowicz C, *et al*. Elevation of serum CA125 prior to diagnosis of an epithelial ovarian carcinoma. *Gynecol Oncol* 1985; **22**: 115–20.

4 Zurawski VR Jr, Orjaster H, Andersen A, Jellum E. Elevated serum CA125 levels prior to diagnosis of ovarian neoplasia: relevance for early detection of ovarian cancer. *Int J Cancer* 1988; **42**: 677–80.

5 Lahousen M, Stettner H, Purstner P. A tumour-marker combination versus second-look surgery in ovarian cancer: 1. clinical experience. *Bailliere's Clin Obstet Gynaecol* 1989; **3**: 201–8.

6 Lahousen M. Rezidiverkennung beim Ovarialkarzinom. In: Meerpohl MG, Pfleiderer A, Profous CHB, eds. *Das Rezidiv in der gynakologischen Onkologie*. Berlin: Springer, 1990: 201–11.

7 Lahousen M, Stettner H, Pickel H, Urdl W, Purstner P. The predictive value of a combination of tumour markers in monitoring patients with ovarian cancer. *Cancer* 1987; **60**: 2228–32.

8 Lawhead RA Jr, Clark DG, Smith DH, Pierce VK, Lewis JL Jr. Pelvic exenteration for recurrent or persistent gynecologic malignancies: a 10-year review of the Memorial Sloan-Kettering Cancer Center experience (1972–1981). *Gynecol Oncol* 1989; **33**: 279–82.

9 Berek JS, Hacker NF, Lagasse LD, Nieberg RK, Elashoff RM. Survival of patients following secondary cytoreductive surgery in ovarian cancer. *Obstet Gynecol* 1983; **61**: 189–93.

10 Morris M, Gershenson DM, Wharton JT, Copeland LJ, Edwards CL, Stringer CA. Secondary cytoreductive surgery for recurrent epithelial ovarian cancer. *Gynecol Oncol* 1989; **34**: 334–9.

11 Teufel G, Nicolai M, Aisslinger U, Simonis H, Meerpohl MG, Pfleiderer A. Operative Therapie progredienter oder rezidivierender maligner Ovarialtumoren. *Geburtshilfe Frauenheilkd* 1991; **51**: 186–93.

12 Wiltshaw E, Shepherd JH, Crowther M. Salvage surgery – results. In: Sharp F, Soutter WP, eds. *Ovarian cancer – the way ahead*. London: Royal College of Obstetricians and Gynaecologists, 1987: 313–20.

13 Blackledge G, Lawton FG, Redman C, Kelly K. Responses of patients in phase II studies of chemotherapy in ovarian cancer: implications for patient treatment and the design of phase II trials. *Br J Cancer* 1989; **59**: 650–3.

14 Peters WA III, Blasko JC, Bagley CM Jr, Rudolph RH, Smith MR, Rivkin SE. Salvage therapy with whole-abdominal irradiation in patients with advanced carcinoma of the ovary previously treated by combination chemotherapy. *Cancer* 1986; **58**: 880–2.

15 Petru E, Pickel H, Heydarfadai M, Lahousen M. Remission induction with carboplatin-epirubicin-prednimustine followed by consolidation radiotherapy in advanced ovarian cancer. *Int J Clin Pharmacol Res* 1992; **12**: 205–7.

16 Markman M. Intraperitoneal antineoplastic agents for tumors principally confined to the peritoneal cavity. *Cancer Treat Rev* 1986; **12**: 219–42.

17 Stewart JA, Belinson JL, Moore AL, *et al*. Phase I trial of intraperitoneal recombinant interleukin2/lymphokine activated killer cells in patients with ovarian cancer. *Cancer Res* 1990; **50**: 6302–10.

18 Stewart JSW, Hird V, Snook D, *et al*. Intraperitoneal yttrium-90-labelled monoclonal antibody in ovarian cancer. *J Clin Oncol* 1990; **8**: 1941–50.

19 Pai LH, Bookman MA, Ozols RF, *et al*. Clinical evaluation of intraperitoneal pseudomonas exotoxin immunoconjugate OVB3-PE in patients with ovarian cancer. *J Clin Oncol* 1991; **9**: 2095–2103.

14 Optimum chemotherapy regimens, dose intensity, and new drugs

JAN P NEIJT

After extensive surgical cytoreduction, chemotherapy is the treatment of choice for patients with advanced ovarian cancer (International Federation of Gynecology and Obstetrics (FIGO) stages IIb, III, IV). For the last decade there has been continuous debate about "what is optimum chemotherapy?" At a recent consensus meeting it was agreed that the use of a combination of agents including cisplatin should be recommended because platinum-based chemotherapy yields superior response rates and progression-free survival, and may be well lead to superior long term survival as well. The consensus was that either cyclophosphamide 750 mg/m^2 plus cisplatin 75 mg/m^2 every three weeks or cyclophosphamide 500 mg/m^2 plus doxorubicin 50 mg/m^2 plus cisplatinum 50 mg/m^2 every three weeks (CAP) was acceptable standard treatment.[1] Carboplatin and more recently paclitaxel (taxol) are two drugs to be considered in "standard treatment" regimens.

A number of new treatments are currently being tested in clinical trials and hold promise to change the future for patients with ovarian cancer. Among them are methods to overcome tumour resistance, the use of dose intensive regimens, and new drugs. The dose of the drugs delivered seems to be important in clinical outcome but new drugs are needed to improve the results in advanced disease. Recently the introduction of paclitaxel has engendered hope that the initial treatment of ovarian cancer may lead to improved survival.

Optimum chemotherapy

Cisplatin-based treatment

The results of several studies have now established that cisplatin-based combination chemotherapy is more effective than single agents or combinations of drugs without cisplatin in ovarian cancer. Clinical trials, meta-analyses, and population based studies have all shown the importance

186

of cisplatin in the treatment of advanced disease. The five and 10 year survival rates have doubled with cisplatin-based combinations compared with the use of alkylating agents either alone or combinations without cisplatin. For these advantages, however, we pay the price of enhanced toxicity.

Clinical trials

Two questions that must be asked in ovarian cancer chemotherapy are whether cisplatin based treatment is better than non-platinum regimens (for example single alkylating agents) and whether cisplatin is better in combination than given alone. A British clinical trial studied the first question. Standard alkylating treatment with chlorambucil was compared with PACe (a combination of cisplatin, doxorubicin, cyclophosphamide). The overall response rate and number of surgical complete remissions favoured the combination.[2] Overall survival was longer in the PACe group, but the number of patients was too small for the difference to be significant. Another clinical study which compared cyclophosphamide and cisplatin-based chemotherapy was ended prematurely because of the large difference in survival. The two year survival rate was 62% for the combination and only 19% for those treated with cyclophosphamide.[3] Two other randomised studies compared combinations with and without cisplatin[4 5]; both confirmed the superiority of the cisplatin combination.

It is important to realise that the results of cisplatin-based chemotherapy may differ considerably in subgroups with different characteristics. For example, there is a significant difference in overall survival in stage III patients between the cisplatin-based combination and the combination without it, but in stage IV this difference does not exist.[6] The same applies for patients with suboptimally debulked tumours. There are no differences in survival between patients with tumours of more than 1 cm treated with a non-cisplatin combination or cisplatin-based chemotherapy, whereas in the other patients with small tumours the cisplatin regimen gives a significant advantage (fig. 14.1). These data explain that studies aimed to detect differences between the efficacy of therapeutic regimens should preferably be done in prognostically favourable patients (stage IIb, III, small residual tumour and a good performance status).

In most studies the overall response achieved with cisplatin-based combinations is between 70% and 80%, and 40%–50% of these responses are clinically complete. In all studies the progression-free survival is longer for patients treated with the platinum regimen (the median for platinum regimens is about 19 months and a quarter are alive progression-free at five years). The overall survival is more difficult to interpret because several trials used a cross-over design. Patients who relapsed on the alkylating drugs received subsequently cisplatin as second-line treatment influenced survival and diminished the difference between the treatments involved.

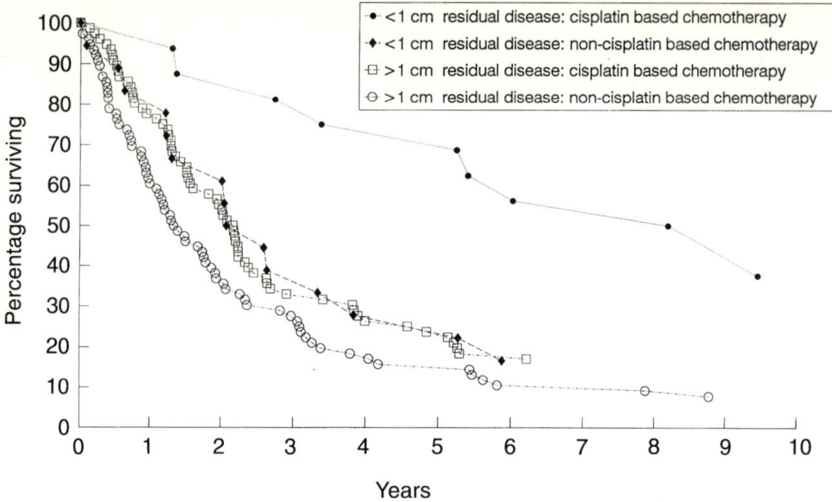

Fig. 14.1 Survival in relation to tumour diameter and treatment in 106 patients with advanced ovarian cancer.

Meta-analysis and population based studies

On the basis of a meta-analysis of randomised trials the Advanced Ovarian Cancer Trialists Group concluded that immediate platinum-based treatment was better than non-platinum regimens in terms of survival and that cisplatin was better in combination than as a single agent.[7]

Clinical studies are not alone in indicating that the introduction of cisplatin offers survival advantage. In a population based study in The Netherlands patients treated from 1981–1985 had more than 10% improved survival compared with those treated in the period between 1975–1980 before cisplatin was introduced.[8] A retrospective analysis of cancer registry data for 2669 women with newly diagnosed ovarian carcinoma from 1983–1988 showed that the use of extensive surgery and platinum-based chemotherapy improved survival for stage III patients. Platinum-based combinations were given to 76% of 221 patients with pathological stage III disease. For patients with no residual disease the five year survival was 50% for those treated with platinum compared with 37% for those who had received non-platinum therapy. Corresponding figures for patients with residual disease were 20% and 5%, respectively.

Long term survival

Several investigators have now published the long term survival results of their randomised studies. The Gruppo Interregionale Cooperativo Oncologico Gynaecologia in Italy compared three different regimens in

Table 14.1 Survival rates of subgroups of patients treated with cisplatin combination chemotherapy.[6]

Characteristics of patients	Survival rates at five years	Median survival (years)
All patients	30	2.1
Response:		
Complete remission	64	7.8
Microscopic disease	35	4.2
Partial remission	12	1.6
Progression	0	0.5
FIGO stage:		
III	34	2.6
IV	14	1.4
Residual tumour before chemotherapy*		
Microscopic	61	6.6
<1 cm	41	3.8
1–2 cm	25	1.8
2–5 cm	25	1.6
≥5 cm	18	1.5
Histological grade (Broder's):		
1	50	5.1
2	26	2.6
3	27	2.2
4	39	1.3
Karnofsky index:		
100	39	3.3
90	34	1.9
80	25	1.8
70	9	1.0

* Largest cross-sectional diameter.

529 cases. After treatment with a combination of cisplatin, cyclophosphamide, and adriamycin 22% of the patients survived for seven years; with the combination of cisplatin and cyclophosphamide 17% survived; and with cisplatin alone 12% survived.[9]

The Netherlands Joint Study Group for Ovarian Cancer updated their experience with cisplatin-based treatment. These studies were initiated in 1979 and 1981 and accrued 377 patients with advanced ovarian cancer. At 10 years 9% of the patients initially treated with a non-cisplatin combination and 21% of the patients assigned to a regimen that included cisplatin were alive. Among the 10 year survivors treated without cisplatin half had developed progressive disease but were alive as a result of retreatment with cisplatin. The survival cures showed that about 60% of the patients who had achieved complete remission according to second-look laparotomy were alive at five years, and 40% at 10 years. These data showed that relapses occurred even after five years of follow up, and in the most favourable group of patients with an incidence of about 5% each year. Patients with microscopic disease at second-look had a less favourable outlook: 35% survived five years.[6] The survival rates of other subgroups of patients treated with a cisplatin combination are shown in table 14.1.

Side effects and complications of treatment

Cisplatin-based treatment is associated with considerable toxicity; the main side-effects are nausea, vomiting, nephrotoxicity, electrolyte disturbances, allergic reactions, ototoxicity, and neurotoxicity.

Vomiting starts within hours after the infusion in many patients, and nausea and vomiting may last as long as a week. One of the regimens to prevent it is high dose metoclopramide combined with diphenydramine and dexamethasone.[10] Newer treatment consists of 5-hydroxytryptamine (5HT$_3$) receptor antagonists such as ondansetron.[11] An Italian group compared the metoclopramide regimen with the more modern ondansetron treatment plus dexamethasone, and in a double blind multicentre study of 287 patients treated with high dose cisplatin chemotherapy the ondansetron plus dexamethasone combination was significantly more efficacious and better tolerated than the metoclopramide regimen. More importantly its efficacy (at least for vomiting) was maintained during subsequent cycles of chemotherapy. Patients were given dexamethasone 20 mg in 15 ml of saline as an intravenous infusion over 15 minutes, 45 minutes before they were given cisplatin. Immediately thereafter they were given 0.15 mg/kg of ondansetron as an intravenous infusion over 15 minutes. Two further doses of ondansetron were given at two hourly intervals. With this regimen 77% of the patients did not vomit during the first cycle of chemotherapy and 74% did not vomit during the third cycle.[12]

Cisplatin-induced nephrotoxicity caused by reduction of the glomerular filtration rate is encountered in half to three quarters of patients. The incidence of severe nephrotoxicity can be reduced by hydration with saline so that the urinary output is maintained at more than 100 ml/hour. Supplementation of magnesium or calcium may be necessary because hypomagnesaemia and hypocalcaemia caused by urinary losses may lead to life threatening tetany.

Peripheral neuropathy after prolonged treatment (above 400 mg/m^2 cumulative dose) is another serious side-effect and results in sensory deficits with loss of proprioception. Patients complain of dysaesthesia and paresthaesiae in hands and feet, and there is loss of deep tendon reflexes, and reduced sensitivity to vibration, touch, and pain sensation.[13] There is no evidence that any form of neuroprotective agent is clinically useful to prevent this. After chemotherapy has been stopped neuropathy may worsen for up to four months, thereafter it improves gradually. Ototoxicity, which is usually bilateral, may lead to hearing loss especially in the high frequency range and is correlated with the total dose of cisplatin.

After combination chemotherapy including an alkylating agent and cisplatin some patients may develop secondary myelodysplasia or acute leukaemia. The risk of leukaemia is greatest four to five years after chemotherapy and is increased for at least eight years after it has finished.[14] As well as cyclophosphamide and cisplatin, carboplatin is also leukaemogenic.[15]

190

The role of other drugs in combination with cisplatin

Apart from alkylating agents and the platinum compounds, several other drugs are effective in ovarian cancer. These include anthracyclines (doxorubicin, epirubicin, and mitozantrone), etoposide, and hexamethylmelamine. Recently paclitazel (taxol) has been added to the armentarium. Most of these drugs have been tested as single agents in phase II studies, and the results obtained vary widely (table 14.2). Several prognostic factors other than the efficacy of these drugs influence the final response.[16 17] Patients with a long interval since chemotherapy, a small disease load, a serous cell type, and a small number of disease sites, have a better chance of responding to a drug that has been tested in phase II studies or retreatment with cisplatin. On the other hand, patients who have had progressive disease during initial platinum therapy (short interval) and who have a large tumour mass are less likely to respond to retreatment with cisplatin or one of the other drugs.

Table 14.2 Variable response rates of drugs in phase II trials in ovarian cancer

First author	Reference No	Drug	Response rate
Hilgers	75	Mitozantrone	0–28
Thigpen	76	Cisplatin	15–20
Eisenhauer	77	Carboplatin	8–32
Muggia	71	Teniposide	0–40
Thigpen	78	Ifosfamide	12–22
		Hexamethylmelamine	10–20
Rowsinky	79	Paclitaxel	20–50

It is questionable whether adding a further drug (table 14.2) to the cisplatin/cyclophosphamide combination increases its efficacy. One problem is that the toxicity of the less active agent added to the combination may prevent the use of optimal doses of the more effective drugs, such as cisplatin, carboplatin, and cyclophosphamide.

Doxorubicin In older studies doxorubicin produced a response rate of 43% when used as a single drug in patients previously treated for ovarian cancer.[18] For this reason the Gynecologic Oncology Group conducted a randomised study in patients with FIGO stage III disease, all with less than 1 cm residual disease, to find out whether there was a difference between cylophosphamide and platinum, and a combination of cyclophosphamide, doxorubicin and cisplatin (CAP). In 349 evaluable patients the group found no significant differences in progression-free interval, number of second-look laparotomies that showed no sign of disease, or survival between the two groups. They concluded that using dosage regimens with equal haematological toxicity the addition of doxorubicin gave no significant advantage.[19] An Italian study also showed no statistically significant improvement of survival attributable to the addition of doxorubicin to cyclophosphamide and platinum.[20] In a study of the Danish Ovarian Cancer Group 267 patients were randomised to cyclophosphamide 500 mg/m^2 plus cisplatin 60 mg/m^2 with or without doxorubicin 40 mg/m^2 every four weeks, and there were no significant differences in response rate or progression-free survival. Overall follow-up at the time of the update ranged from 8–11 years. Median survival was 26 months for the triple regimen and 21 months for the double. At five years 23% of the patients who had received the triple regimen were alive compared with 19% of the other group, but none of these differences was significant.[21]

Although randomised trials could not prove any benefit from the addition of doxorubicin to a cisplatin-based regimen, an overview analysis (or meta-analysis) detected a small survival advantage for regimens using doxorubicin.[22 23] The advantage of this type of analysis is that the increased power may detect moderate differences in effects that are not shown by individual trials, but this type of analysis also raises concern.[24] It remains unclear whether the benefit found is an effect of doxorubicin or a result of the greater dose intensity achieved by adding the drug. It also remains unclear whether the small benefit is clinically relevant. In dosage regimens with equal haematological toxicity the addition of doxorubicin probably gave no significant advantage.[25] A problem with combinations including doxorubicin is that its myelotoxicity prevents the optimal dosing of the more effective drugs such as cisplatin, carboplatin, and paclitaxel.

Hexamethylmelamine Hexamethylmelamine has also been used in initial treatment programmes including cisplatin. It has structural similarities to the classical alkylating agents, but the mechanism of action is different. One study compared cyclophosphamide and cisplatin with hexamethylmelamine, cyclophosphamide, doxorubicin, and cisplatin (H-CAP) and initially showed no survival differences,[26] but after longer follow up an improved progression-free survival was found in the patients who had small volume disease and who were treated with four drug combinations. These long term results stimulated a group from Vanderbilt University Medical Center in Nashville, USA, to compare their experience with cyclophosphamide and cisplatin and H-CAP. They claimed that the addition of hexamethylmelamine to CAP prolonged median survival in patients with limited residual disease.[27] The Netherlands Joint Study Group for Ovarian Cancer who compared the double regimen with CHAP-5 (cyclophosphamide, hexamethylmelamine, doxorubicin, and cisplatin) could not confirm these results, and randomised trials that positively prove the value of hexamethylmelamine in first line treatment regimens are lacking. For this reason in the United States the Food and Drug Administration decided to withhold approval for primary treatment based on hexamethylmelamine.[25]

Ifosfamide Ifosfamide is an analogue of cyclophosphamide. Data about its use in previously treated patients have been conflicting. Two studies have failed to show that it has any effect in pretreated ovarian cancer,[28 29] but a Gynecologic Oncology Group study of 41 evaluable patients showed

7% complete responses and 14% partial responses. Response to ifosfamide was not correlated with previous response to cisplatin-based treatment and the Group concluded that further study of this drug was warranted.[30]

An interesting study was done by the London Gynaecological Oncology Group in which 152 patients were randomised to receive either sequential therapy with three cycles of ifosfamide followed by three cycles of carboplatin, or six cycles of single agent carboplatin. After three cycles of treatment only two patients in the ifosfamide arm achieved complete remission and 12 partial remission (overall response rate 29%), whereas in the carboplatin arm 10 patients achieved complete remission and 23 partial remission (overall response rate 63%). Although both treatments were generally well tolerated, 47% of the patients in the ifosfamide-carboplatin arm developed alopecia sufficient to require a wig compared with only 2% in the carboplatin arm. The group concluded that ifosfamide is clearly less effective and more toxic than carboplatin. Because those patients who did not respond to ifosfamide were subsequently treated successfully with carboplatin, no significant survival differences were found.[13]

Carboplatin-based treatment

Carboplatin (CBDCA, or JM8) is a cisplatin analogue, less nephrotoxic and less emetogenic than its parent cisplatin, and causes little or no neurotoxicity, ototoxicity, or nephrotoxicity. Unfortunately the drug is myelosuppressive, and thrombocytopenia is a major side effect. Three randomised studies of single agent carboplatin compared with cisplatin in previously untreated patients with ovarian cancer showed that a dose of $400\,mg/m^2$ of carboplatin monthly is equivalent to $100\,mg/m^2$ of cisplatin monthly.[32 33 34] One of the conclusions common to these three studies was that carboplatin was less toxic than cisplatin in all respects except bone marrow suppression.

As carboplatin is excreted by the kidney, reduced renal function can lead to reduced clearance of the drug, and can be correlated with the efficacy and toxicity of the drug. Considerable interpatient variability in clearance can be expected, and so a pharmacologically based method of calculating the dose of carboplatin seems better than a dose based on body surface area.[35] Calvert *et al* proposed a simple formula to calculate the dose of carboplatin for a given patient using the area under the curve (AUC) of a plot of drug concentration against time after the drug was given. The dose is calculated as: dose (mg) = AUC (mg/ml × minute) × (glomerular filtration rate (ml/minute) + 25). Using this formula a better prediction of drug toxicity is possible.[36]

The AUC of carboplatin probably also correlates with the chance of responding to treatment. An increase in the AUC from 2 to 5 increased the likelihood of response from 20% to more than 40%. Increasing the AUC above 5 did not further improve the response rate,[37] so it seems likely that an adequate dose of carboplatin is an AUC of 5–6 mg/ml × minute.

The optimal AUC for carboplatin in combination with other drugs is uncertain, and the influence of using this method of selecting the dose of carboplatin on progression-free survival and survival has not been investigated.

Clinical trials

Several studies have been done to explore the role of carboplatin as part of combination chemotherapy. A study at the Mayo Clinic compared cyclophosphamide plus carboplatin with cyclophosphamide and cisplatin and showed better progression-free survival with the carboplatin combination.[38] Unfortunately the dose of carboplatin used in this study was very low: 150 mg/m^2. This is probably the reason that such a large difference was found between the two treatment arms. Using higher doses of carboplatin in combination chemotherapy the myelotoxicity of the drug becomes an important side effect. Adjusting the dose of carboplatin or delaying treatment cycles to permit recovery of leucocytes and platelets may reduce the effect.

Using higher doses of carboplatin (300 mg/m^2 every four weeks) the Southwest Oncology Group reported equal efficacy for the combinations carboplatin plus cyclophosphamide and cisplatin 100 mg/m^2 plus cyclophosphamide. Patients treated with carboplatin combination had a clinical response rate of 61% and those who received cisplatin 52%. The number of pathologically complete responses and median survival were similar in both arms.[39] Unexpectedly there were similar percentages of granulocytopenia in both arms, which may be because postponement of treatment was allowed in the protocol. Twenty to fifty percent of the patients did not receive cisplatin in the third to the sixth courses. It was not reported how many patients in the cisplatin group were subsequently treated with carboplatin.

The National Cancer Institute of Canada completed a similar study in patients with macroscopic residual disease. Of those patients who underwent repeated surgery, 35% of cisplatin-treated and 26% of carboplatin-treated patients achieved a histologically complete response. Overall response and survival were similar for the cisplatin and carboplatin arms, but it must be kept in mind (as shown in fig. 14.1) that is unlikely that in these patients who had suboptimal debulking there would be a detectable difference between the two treatment arms.[40]

Another large study that used carboplatin in combination was done by the European Organisation for Research in Treatment of Cancer (EORTC) Gynecological Group. In this study 341 patients were randomised and received combination chemotherapy with cyclophosphamide, hexamethylmelamine, adriamycin, and platinum for five days (CHAP-5), or a regimen to CHAP-5, replacing cisplatin for five days with carboplatin on one day (CHAC-1). The response rate achieved with the cisplatin regimen was 63% and with the carboplatinum combination 48%.[41] In patients with less than 1 cm residual disease both overall survival and progression-free survival were better in the cisplatin arm. In this group with a better prognosis the median overall survival and median progression-free survival were 59.2 and 39.5 months with CHAP-5 and 42.5 and 17.9 months with CHAC-1 (95% confidence intervals (CI) were overlapping on both occasions).[42]

A French cooperative group have reported results that contrast with those of most other trials. This group compared the CAP-regimen (cyclophosphamide, doxorubicin and cisplatin $75\,mg/m^2$) with the same regimen but with cisplatin replaced by carboplatin ($300\,mg/m^2$). The report described the results of 144 eligible patients. The pathological and overall response rates were significantly higher for the cisplatin combination than for the carboplatin combination being 33% and 15%, and 73% and 47%, respectively. At the median follow-up of 27 months, the median survival was significantly higher with the CAP than with the carboplatin regimen (27.9 months compared with 20.6 months).[43] One reason for these disappointing results with the carboplatin combination is probably that a cross-over from carboplatin to cisplatin (at relapse or earlier) was not allowed.

Meta-analysis

The question of whether carboplatin is as effective as cisplatin was also addressed by the British Advanced Ovarian Cancer Trialists Group using meta-analysis. The survival curves computed for patients treated with cisplatin or carboplatin alone or in combination showed better survival after three years for those treated with cisplatin. The authors concluded that "at this time, platinum combinations should be accepted as the optimal standard therapy" and that "the comparison of cisplatin and carboplatin is potentially flawed by cross-overs."[44] We can conclude that at present the position of carboplatin as initial treatment and combined with other myelosuppressive agents is questionable.

Conclusion

Data from studies of carboplatin have to be analysed with special attention to cross-over, the dose delivered, postponement of treatment, and follow-up. Although most trials suggest equivalent results between carboplatin and cisplatin regimens, on the presently available data it is recommended that carboplatin should not routinely replace cisplatin in the management of patients with potentially curable low volume disease.[42] During an international workshop it was accepted as a consensus statement that carboplatin based treatment is an acceptable choice only in patients with suboptimal stage III and IV ovarian cancer. Although a growing enthusiasm was recognised for calculating the dose of carboplatin according to the AUC, all clinical data comparing carboplatin-based and cisplatin-based treatment to date have calculated the dose according to body surface area, so the recommendation refers to a regimen in which the dose is calculated according to the more traditional method of body surface. The optimal recommended dose is therefore carboplatin 300 mg/m^2 and cyclophosphamide 600 mg/m^2 every four weeks.[1]

Paclitaxel (taxol) based treatment

Among many drugs tested in the last few years, paclitaxel is the most promising. It is a natural product from the bark of the Pacific yew *Taxus brevifolia* which increases the polymerisation of tubulin to stabilise the cellular microtubule network. These microtubules are an important element in cells for forming the mitotic spindle.

Side effects and doses

The dose-limiting toxicity of paclitaxel is myelosuppression, primarily neutropenia. Other adverse effects included hypersensitivity reactions, myalgias, arthralgias, total alopecia, diarrhoea, nausea, vomiting, mucositis, cardiac arrhythmias, and peripheral neuropathy.[45] It is not nephrotoxic and can be delivered even if renal function is compromised. Less than 10% of paclitaxel is recoverable in the urine and most of it is probably excreted in bile.

In most studies paclitaxel has been given as a 24-hour infusion. To assess its safety as a three-hour infusion, 24 European and Canadian institutions used it in 407 patients with ovarian cancer who had been pretreated with platinum. Patients were treated with either 135 or 175 mg/m^2 of paclitaxel as a three or 24 hour infusion. To avoid the hypersensitivity effects seen in early trials, all patients received premedication with a steroid, an antihistamine, and a H$_2$-receptor antagonist. A total of 2331 cycles were given. Pharmacokinetic studies conducted in a subset of patients showed non-linear pharmacokinetic variables for the patients treated by the three-hour infusion, which resulted in greater drug exposure to the drug for this regimen compared with the 24 hour one.[46] Neutropenia and stomatitis were encountered less often in the three-hour infusion arm. Neuropathy seemed to be dose dependent; in the high dose arm 51% of the patients had some grade of neuropathy.

One of the conclusions that can be drawn from this study is that the sensory neuropathy, neutropenia, and probably also the myalgia are dose dependent. Patients who received the 24-hour infusion had a much higher incidence of grade IV neutropenia compared with those who received the three-hour infusion. Hypersensitivity symptoms were not affected by the way the drug was given or the total dose. Although minor hypersensitivity reactions occurred in about 40% of the patients, severe hypersensitivity reactions (angio-oedema, respiratory distress and hypertension, or generalised urticaria) occurred in only 1%–2% of the patients. The

Table 14.3 Premedication schedule

Premedication regimen	Time before paclitaxel
Oral steroid	6 and 12 hours
Antihistamine iv	30 minutes
H_2-receptor antagonist iv	30 minutes

hypersensitivity reactions are probably caused by polyethoxylated castor oil, the pharmaceutical vehicle for paclitaxel, which induces the direct release of histamines from the circulating cells. For patients who experience major hypersensitivity reactions while receiving paclitaxel the drug should be discontinued, but there are reports that retreatment of patients the next day may be successful.[47]

In the final analysis response rates of 15% and 20% were found in the groups treated with 135 mg/m^2 and 175 mg/m^2 respectively (p = 0.2). In patients whose disease had progressed while receiving a platinum analogue, the response rate achieved was only 12%.[48] Progression-free survival was 14 and 19 weeks, respectively. There were no differences in response rate or progression-free survival between the three and 24-hour infusion rates. It was concluded that the three-hour infusion of 175 mg/m^2 of paclitaxel offers the best therapeutic index and is the preferred dosage regimen in patients with previously treated ovarian cancer.[49]

Using the three-hour infusion of 175 mg/m^2 and premedication according to the guidelines in table 14.3 the side-effects to be expected are: hypersensitivity reactions (see above), arthralgia/myalgia of any grade in 66% of the patients, neuropathy in about 50% of the patients, stomatitis in 27%, hair loss in almost 90%, nausea and vomiting in 7%, and dysrhythmias in less than 1% of patients. Neutropenia grade IV is seen in half, with febrile neutropenia in 7%. Thrombocytopenia grade IV is encountered in 4% (Ten Bokkel Huinink WW. European randomised trial of paclitaxel in refractory ovarian cancer. Paper presented at symposium "Taxol: a novel advance in chemotherapy." Amsterdam, 1993.)

Clinical trials

Paclitaxel produces responses in patients with ovarian carcinoma that is clinically resistant to the platinum compounds, an observation which suggests that taxanes are at least partially non-cross-resistant (table 14.4).[50 51]

After reports had been published about the activity of paclitaxel in refractory ovarian cancer, a large study was initiated in the United States in September 1991. Only patients with epithelial ovarian cancer who had received three of more previous regimens of chemotherapy were eligible. All patients had platinum resistant tumours that had progressed or persisted despite platinum-based treatment or had recurrent tumours within three months of completing a platinum-based regimen. Patients were treated with paclitaxel 135 mg/m^2 given by 24 hour continuous infusion every three

Table 14.4 Single agent paclitaxel: responses in cisplatin-resistant ovarian cancer (tumour progression while on platinum-containing treatment)[80]

Group undertaking study	Reference No	Resistant	Relapse <3 months	Relapse 3–6 months
National Cancer Institute	81	8/20	5/10	3/7
Gynecologic Oncology Group	82	4/19	1/4	1/2
National Cancer Institute	81	4/6	0/1	1/1
Johns Hopkins	83	3/22	3/5	1/5
Albert Einstein	80	2/16	0/1	2/2
Total		21/83	9/21	8/17
Complete and partial remission rate		25%	43%	47%

weeks. Again, leucopenia was the most common toxic effect, with 78% of the patients experiencing grade III or IV toxicity. Fifteen treatment-related details were reported (1.5%) among the first 1000 patients registered. The objective response rate was 22% (95% CI for overall response 19% to 25%). The median time to progression from initiation of treatment was 7.1 months in patients who responded and 4.5 months overall. The median duration of survival was 8.8 months. The treatment related deaths were secondary to neutropenia and most occurred during the first cycle.[52] The results confirm that paclitaxel is able to induce remissions in patients who have progressive disease during platinum treatment. Because the duration of survival and median time to progression were similar as in other phase II studies, the impact on survival of this second-line treatment is probably absent or minimal.

The observation that treatment with paclitaxel resulted in remissions in platinum resistant disease encouraged the use of this drug as a first line treatment. The Gynecologic Oncology Group performed a study in a total of 388 eligible patients, all with suboptimally debulked stage III (more than 1 cm residual disease) or stage IV disease. In the cyclophosphamide/cisplatin combination, cyclophosphamide was replaced by paclitaxel 135 mg/m^2 over 24 hours and compared with cyclophosphamide and cisplatin for six courses. The paclitaxel combination resulted in a superior response rate with more clinically complete responses (54% of the cisplatin/paclitaxel patients as opposed to a third of the cisplatin/cyclophosphamide patients), and more pathologically complete responses. More importantly there was a significant improvement in progression-free interval (median 18.1 compared with 13.6 months). (Williams SD. Phase III trial comparing cisplatin/ cyclophosphamide with cisplatin/paclitaxel in advanced ovarian cancer. Paper presented at symposium "Taxol: a novel advance in chemotherapy." Amsterdam, 1993.) At the time of the report median survival was 23.3 months for the cylophosphamide/cisplatin regimen with the median survival yet to be reached for the paclitaxel combination. Based on these data that were obtained in a poor prognostic subgroup with a large volume of residual disease, the combination of paclitaxel and cisplatin may be the most effective treatment available today in patients with a poor prognosis.

Although this study has engendered hope that treatment may be improving, it is not yet sure that this regimen will replace the standard treatment with cisplatin and cyclophosphamide. The integration of paclitaxel into the standard treatment has been criticised because of the inconvenience of the 24 hour regimen, the side-effects of paclitaxel plus cisplatin, and the expense. New studies are under way with plans to confirm the findings of this important study with a modified regimen ($175 \, \text{mg/m}^2$ of paclitaxel in three hours) in both optimal and suboptimal patients.

Dose intensity

The importance of dose intensity (defined as $\text{mg/m}^2/\text{time period}$) in relation to clinical outcome in ovarian cancer has been reviewed by Levin and Hryniuk.[53] The average relative dose intensity of cisplatin (relative dose intensity was calculated as a fraction of a dose of a drug in the standard regimen of cyclophosphamide, hexamethylmelamine, doxorubicin, and platinum, CHAP) correlated significantly with clinical response and with the median survival time. Since their first report, numerous additional studies have been published. For this reason the analysis was updated,[54] and again an association was found between the outcome in dose intensity for platinum alone or in multiagent regimens. For other drugs like cyclophosphamide and doxorubicin this association was of borderline significance. As might be expected, multiagent regimens containing platinum produced better response rates than platinum alone for any fixed relative dose intensity. Whether the dose of the active drugs delivered in a certain time period is crucial for clinical outcome has yet to be confirmed by randomised clinical studies. A further question is whether an increase above a standard dose will further improve results whilst maintaining acceptable toxicity.

Clinical trials of dose intense regimens

Most clinical studies of dose intense regimens have been small and not randomised. So far few randomised studies of cisplatin dose intensification have been published.[55-7] A key study about the question of cisplatin dose was performed by the Scottish Ovarian Cancer Study Group. Patients were given treatment with $50 \, \text{mg/m}^2$ (low dose) or $200 \, \text{mg/m}^2$ (high dose) cisplatin plus $750 \, \text{mg/m}^2$ cyclophosphamide for a maximum of six cycles. Overall median survival was 17 months in the low dose group and 28 months in the high dose group. Because of unacceptable side effects or patients' refusal, six in the low dose and 25 in the high dose group stopped treatment. The price paid for the improvement of survival consisted of neurotoxicity, hearing loss, alopecia, vomiting, and anaemia.[58]

199

In another randomised study of the Gynecologic Oncology Group increased toxicity was the only difference between the dose intense regimen and the standard treatment.[59] In this study the total dose of cisplatin received in the two groups was, however, kept the same: 400 mg/m². It seems likely that both dose intensity and total dose influence the outcome of treatment. Moreover in this study, only patients with tumours of more than 1 cm in diameter were eligible. This is a group of patients in which differences between treatments are expected to be smaller than in patients with smaller tumours to start with.

A well designed trial was done at the University of Milan.[60] The cumulative dose of cisplatin was similar in both treatment arms and the variable that differed was the dose intensity. Patients were randomised to receive either standard cisplatin (75 mg/m² every three weeks for six courses), or a more dose intense regimen consisting of cisplatin 50 mg/m² weekly for eight courses. The latter group received a double dose intensity (50 mg/m² compared with 25 mg/m²) a week with a similar total dose of 450 mg/m². Second-line treatment was standardised. Two hundred and ninety six patients were randomised with a median follow-up of 41 months; the median progression-free survival was 21 months compared with 18 months, and overall survival was 46 months compared with 43 months for the high and low dose arms, respectively. No differences were observed in complete remission rate or overall response rate. Although the total dose was kept the same in both treatment arms, a point of concern is that the patients randomised to the 50 mg/m² weekly for eight courses received treatment for only two months. So far the study has been published only in abstract and the final publication has to be awaited before final conclusions can be drawn.

Escalating the dose of cisplatin beyond 100 mg/m² is not clinically relevant because of the severe toxicity, but carboplatin may be completely different. Carboplatin lacks neurotoxicity and this makes it suitable for high dose chemotherapy. At present none of the studies using carboplatin in a dose intensive schedule have been completed, and results are awaited.

Another approach to increasing the platinum dose-intensity is to combine carboplatin and cisplatin. The Belgian Study Group for Ovarian Carcinoma performed a phase I-II trial of escalating doses of cisplatin (50–100 mg/m²) plus carboplatin (300–400 mg/m²). They concluded that both drugs could be safely combined at reasonably high doses (cisplatin 100 mg/m² and carboplatin 300 mg/m²) fairly well over a six month period. The authors correctly concluded that randomised studies are needed before final conclusions can be drawn.[61]

The above studies nevertheless leave the crucial clinical question unanswered: will an increase above the conventional dose range have an equal effect on survival while maintaining acceptable toxicity? If there is a benefit for survival, is it a trend towards better palliation (improved median survival and response rate without reducing the quality of life) or an improvement in long term survival?

Single high dose chemotherapy

High dose chemotherapy with autologous bone-marrow transplantation and peripheral blood stem cell transfusions are now feasible in patients with ovarian cancer.[62] A number of drugs with proved efficacy in ovarian cancer are suitable for use in dose escalation regimens or at ultrahigh doses. Among them are melphalan, cyclophosphamide, ifosfamide, VP16, mitozantrone, thiotepa, and carboplatin. The dose of carboplatin can be escalated to 2000 mg/m^2 before hepatotoxicity and renal toxicity become dose-limiting. Although the first studies have been done in patients with refractory ovarian cancer, high dose chemotherapy is most likely to be of benefit to patients who have small volume disease that is not yet drug resistant. A number of studies have now been done in such patients but none of them was conclusive. In bulky disease most responses were of short duration and the toxicity of the treatment remains substantial. There is no evidence that high dose chemotherapy with or without haematological support is curative in patients with refractory disease. Although high dose studies have reported encouraging results, final conclusions about the efficacy of this type of treatment could not be drawn.[63 64] Once the optimum standard regimen for ovarian cancer has been established, prospective randomised trials are needed that will compare high dose treatment with the standard.

New drugs

Docetaxel (taxotère)

Docetaxel is a semisynthetic compound structually related to taxol and is derived from the needles of the European yew tree (*Taxus baccata*) which has undergone phase I and phase II clinical trials in the United States and Europe. Its toxicity is in many ways similar to that of paclitaxel, but more pronounced skin toxicity has been reported and in addition many patients developed oedema after prolonged treatment.[65] In three phase II trials a total of 87 evaluable patients have been treated with an overall response rate of 32% (95% CI 23% to 43%).[66] The Early Clinical Trial Group of EORTC reported their experience in 75 patients, and even in the group of patients whose disease was progressing while they were taking cisplatin or with progressive disease within four months of stopping a cisplatin-containing regimen, a response rate of 25% was found.[67] These results are encouraging, but the skin toxicity is serious and more data have to be awaited before the drug can be used in previously untreated patients.

Gemcitabine

Gemcitabine (2′,2′-difluorodeoxicytidine) is a primary antimetabolite with a close resemblance to cytosine-arabinoside. A multicentre phase II

Table 14.5 Variables associated with eight patients who responded to gemcitabine

Variable	No of patients
<6 Months since last chemotherapy	8
Progression while on treatment with platinum	7
Bulk of disease >5 cm	7
FIGO stage IV	5
Poorly or undifferentiated tumour	6
Best previous response complete remission	1
Best previous response partial remission	3

study was done by Lund *et al* from Denmark with the purpose of assessing its activity in ovarian carcinoma and to find out how toxic it was.[68] Patients were admitted with advanced epithelial ovarian cancer who had received a maximum of two previous treatment regimens. Retreatment with the same regimen and substitution of cisplatin with carboplatin or vice versa because of toxicity were considered as one treatment regimen. Patients received gemcitabine $800 \, mg/m^2$, given as a 30 minute intravenous infusion weekly for three consecutive weeks, followed by a fourth week of rest.

A total of 51 patients entered the study and 50 patients were eligible. Most of the patients had bulky disease and all patients had received previous platinum-containing combination chemotherapy. A total of 42 patients was evaluable for response. Eight patients (19%; 95% CI 9% to 34%) achieved a partial response. Median duration of response was 8.1 months (range 4.4–12.5 months). Table 14.5 summarises the prognostic characteristics of the patients who responded to gemcitabine. It is noticeable that all respondents had bad prognostic features and little chance to respond to gemcitabine as a second-line drug.

Treatment with gemcitabine was well tolerated. Grade III nausea and vomiting occurred in only six patients and no hair loss was observed. Specific treatment-related, non-haematological side-effects consisted of transient proteinuria, haematuria, and increase in transaminase activities; 14 patients developed a flu-like syndrome a few hours after the injection of gemcitabine which lasted up to 24 hours. Leucopenia and thrombocytopenia were the main reasons for omitting doses (27% and 14%, respectively) and for reducing doses (37% and 21%, respectively). The activity of gemcitabine was also observed in a small American phase II study in seven previously treated patients, two of whom responded. Assessment of response was based only on the reduction in the titre of the tumour marker CA125. Unfortunately there was no information about the previous treatment in the two responders.[69]

These data indicate that gemcitabine may not be cross-resistant to platinum. It is an interesting new drug in ovarian cancer, well tolerated, and with activity in platinum resistant disease.

Compared with the data about paclitaxel, results of the phase II study of Lund *et al* are encouraging. The response rate achieved with paclitaxel in the Canadian/European study was only 11.9% in patients who had progressed while receiving a platinum analogue.[48] Although the results with gemcitabine are encouraging, more studies are needed to confirm the results obtained and to assess the activity of gemcitabine in untreated ovarian cancer.

Etoposide (VP16)

Although etoposide is not a new drug it deserves attention because data from continuous infusion studies has shed new light on its activity. With etoposide (VP16) response rates up to 31% have been reported.[70 71] The drug is specific to phases of the cell cycle and extending the duration of treatment can probably optimise the benefits of treatment. Markman *et al* therefore hypothesised that long term oral etoposide may result in more killing of tumour cells than if the drug is given a few days each month. Preliminary results of a phase II trial of long term low dose oral etoposide in platinum-resistant ovarian cancer were encouraging. Of nine patients evaluable for response there was one partial remission; a second patient with rapidly rising CA125 titres before etoposide underwent stabilisation of the CA125.[72]

Thirty four patients were entered in another study. All had increasing measurable disease that was resistant to platinum analogue therapy. VP16 was given in a total dose of 100 mg orally for 14 days every 21 days. The response rate was 26% (95% CI 11% to 41%). Twenty eight patients who had cancer that progressed while they were receiving a platinum analogue had a response rate of 21% (95% CI 6% to 36%).

Durations of response were short but the study clearly showed that prolonged treatment with VP16 has activity against platinum-resistant ovarian cancer.[48] These findings have been confirmed by a British group who found a 24% response rate among 41 assessable patients (95% CI 12% to 41%). Oral etoposide has activity in platinum-resistant ovarian cancer and is probably a useful palliative treatment.[73]

Optimum chemotherapy: conclusions

There is no uniform agreement worldwide of what is the optimum chemotherapy for patients with advanced ovarian cancer. During a recent consensus meeting certain features of an optimal regimen were defined. The inclusion of a platinum compound in first-line treatment is essential and that combination chemotherapy should also be used. Most members of the consensus group preferred cyclophosphamide 750 mg/m^2 plus cisplatin 75 mg/m^2 every three weeks as optimum standard treatment. A combination of cyclophosphamide 500 mg/m^2 plus doxorubicin 50 mg/m^2 and cisplatin 50 mg/m^2 every three weeks was also accepted as reasonable.

The choice of the platinum compound is subject to debate. It was agreed that carboplatin based treatment is acceptable in patients with suboptimal stage III and stage IV ovarian cancer, but not in those who have little residual tumour and a good prognosis. The data from more recent clinical studies strongly suggest that paclitaxel 135 mg/m^2 over 24 hours followed by cisplatin 75 mg/m^2 every three weeks will be at least as good as the cyclophosphamide-cisplatin treatment. It was thought, however, that it was too early to recommend the paclitaxel-cisplatin regimen as the new standard treatment in the absence of a confirmatory trial.

Questions remain about the importance of dose intensity of the currently available agents. So far there is no evidence that cisplatin doses of more than 25 mg/m^2 a week in multi-agent regimens are useful. Furthermore there is no evidence that using cytokines in neutropenia is necessary and superior to reducing drug doses in the face of severe myelosuppression. There is also no conclusive evidence that high dose therapy with autologous bone marrow transplantation or peripheral blood stem cell support is beneficial in any subset of patients with epithelial ovarian cancer.

Though most patients with a clinically complete remission will relapse, there is no firm evidence that prolonged treatment with any agent or technique prevents or delays recurrences. With regard to second-line treatment it is important to keep in mind that patients who receive retreatment are not curable, so the toxic effects of treatment must be balanced against the probable benefit of any intervention. For those patients who relapse after completing their first line platinum treatment, the probability that they will respond again to a platinum compound increases with the increasing length of the intertreatment interval. These patients should all receive a second trial with the platinum compound. Patients who have progressive disease while receiving cisplatin are unlikely to benefit from any of the other agents currently available. Low response rates are to be expected from anthracyclines, taxanes, and etoposide. Enrollment of such patients into phase II trials of investigational approaches is an option.

During chemotherapy CA125 titres should be measured routinely at regular intervals. If the titre rises other appropriate clinical assessments such as pelvic examination, radiography, and ultrasound should be made. At present there is no consensus that CA125 can be used as the sole criterion for diagnosing progressive disease and making management decisions. There is no consensus about the routine use of CA125 titres for follow-up of ovarian cancer.[74] Once there is any suspicion of relapse based on increased CA125 titre, however, it is recommended that additional clinical assessments be used. Although not proved, it is likely that early recurrent disease detected early is easier to treat effectively than large tumours detected later.

Although the paclitaxel-cisplatin combination has engendered hope that treatment can be improved further, confirmatory trials are needed. The question arises whether cisplatin in combination with taxol can be replaced by carboplatin. The use of carboplatin in combination with paclitaxel would allow the treatment to be given in outpatients with no admission to hospital necessary, and a combination may be better tolerated. It is likely that in the next decade a new standard treatment will emerge but today the cisplatin-cyclophosphamide combination remains the "gold standard".

References

1 Neijt JP, Wiltshaw E, Lund B. Advanced epithelial ovarian cancer: 1993 consensus statement. *Ann Oncol* 1993; **4** (suppl 4): 83–9.

2 Williams CJ, Mead GM, Macbeth GR, *et al*. Cisplatin combination chemotherapy versus chlorambucil in advanced ovarian carcinoma: Mature results of a randomized trial. *J Clin Oncol* 1985; **3**: 1455–62.

3 Decker DG, Fleming JR, Malkasian GD, Webb MJ, Jefferies JA, Edmonson JH. Cyclophosphamide plus cisplatinum in combination: A treatment program for stage III or IV ovarian carcinoma. *Obstet Gynecol* 1982; **60**: 481–7.

4 Omura G, Blessing JA, Erlich CE, *et al*. A randomized trial of cyclophosphamide and doxorubicin with or without cisplatin in advanced ovarian carcinoma. Gynecologic Oncology Group Study. *Cancer* 1986; **57**: 1725–30.

5 Neijt JP, ten Bokkel Huinink WW, van der Burg MEL, *et al*. Randomized trial comparing two combination chemotherapy regimens (HEXA-CAF vs. CHAP-5) in advanced ovarian carcinoma. *Lancet* 1984; **ii**: 594–600.

6 Neijt JP, ten Bokkel Huinink WW, van der Burg MEL, *et al*. Longterm survival in ovarian cancer. Matured data from The Netherlands Joint Study Group for Ovarian Cancer. *Eur J Cancer* 1991; **27**: 1367–72.

7 Advanced Ovarian Cancer Trialists Group. Chemotherapy in advanced ovarian cancer; an overview of randomized trials. *BMJ* 1991; **303**: 884–93.

8 Balvert-Locht HR, Coeberg JW, Hop WC. Improved prognosis of ovarian cancer in The Netherlands during the period 1975–1985, a registry-based study. *Gynecol Oncol* 1991; **42**: 3–8.

9 Gruppo Interregionale Cooperativo Oncologico Ginecologia. Long-term results of a randomized trial comparing cisplatin with cisplatin and cyclophosphamide with cisplatin, cyclophosphamide, and adriamycin in advanced ovarian cancer. *Gynecol Oncol* 1992; **45**: 115–17.

10 Kris MG, Gralla RJ, Tyson LB, *et al*. Improved control of cisplatin-induced emesis with high-dose metoclopramide and with combinations of metoclopramide, dexamethasone and diphenhydramine. *Cancer* 1985; **55**: 527–34.

11 Kirchner V, Aapro M, Alberto E. Ondansetron for the prevention of cancer-treatment induced nausea and vomiting in multiple side of treatment. *Eur J Cancer* 1991; **27** (suppl 2): 297.

12 Italian Group for Anti-emetic Research. Difference in persistence of efficacy of two anti-emetic regimens on acute emesis during cisplatin chemotherapy. *J Clin Oncol* 1993; **11**: 2396–2404.

13 Gerritsen v.d. Hoop R, Van Houwelingen JC, van der Burg MEL, ten Bokkel Huinink WW, Neijt JP. The incidence of neuropathy in 395 patients with ovarian cancer treated with or without cisplatin. *Cancer* 1990; **66**: 1697–1702.

14 Kaldor JM, Day NE, Pettersson EF, *et al*. Leukemia following chemotherapy for ovarian cancer. *N Engl J Med* 1990; **322**: 1–6.

15 Colon-Otero G, Malkasian GD, Edmonson JH. Secondary myelodysplasia and acute leukemia following carboplatin containing combination chemotherapy for ovarian cancer. *J Natl Cancer Inst* 1993; **85**: 1858–60.

16 Blackledge G, Lawton F, Redman C, Kelly K. Response of patients in phase II studies of chemotherapy in ovarian cancer: implications for patient treatment and the design of phase II trials. *Br J Cancer* 1989; **59**: 650–3.

17 Markman M, Rothman R, Hakes T, *et al*. Second line platinum therapy in patients with ovarian cancer previously treated with cisplatinum. *J Clin Oncol* 1991; **9**: 389–93.

18 Thigpen T. Single agent chemotherapy in the management of ovarian carcinoma. In: Alberts D, Surwit E, eds. *Ovarian cancer* Boston: Martinus Nijhoff, 1985: 115–46.

19 Omura GA, Bundy BN, Berek JS, Curry S, Delgado G, Mortel R. A randomized trial of cyclophosphamide plus cisplatin with or without doxorubicin in ovarian carcinoma: A Gynecologic Oncology Group study. *J Clin Oncol* 1989; **7**: 457–65.

20 Conte PF, Bruzzone M, Chiari S, Sertoli MR, Daga MG, Rubagotti A. A randomized trial comparing cisplatin plus cyclophosphamide versus cisplatin doxorubicin and cyclophosphamide in advanced ovarian cancer. *J Clin Oncol* 1986; **4**: 965–971.

21 Thigpen JT, Bertelsen K, Eisenhauer EA, Hacker NF, Lund B, Sessa C. Long-term follow-up of patients with advanced ovarian carcinoma treated with chemotherapy. *Ann Oncol* 1993; **4** (suppl 4): 35–40.

22 Fanning F, Bennett TZ, Hilgers RD. Meta-analysis of cisplatin, doxorubicin and cyclophosphamide versus cisplatin and cyclophosphamide chemotherapy of ovarian carcinoma. *Obstet Gynecol* 1992; **80**: 954–60.

23 The Ovarian Cancer Meta-analysis Project. Cyclophosphamide plus cisplatin versus cyclophosphamide, doxorubicin and cisplatinum chemotherapy of ovarian carcinoma: a meta-analysis. *J Clin Oncol* 1991; **9**: 1668–74.

24 Omura GA, Brady MF. Meta-analysis of cisplatin, doxorubicin and cyclophosphamide versus cisplatin and cyclophosphamide chemotherapy of ovarian carcinoma. *Obstet Gynecol* 1993; **81**: 641–2.

25 McGuire WP. Primary treatment of epithelial malignancies. *Cancer* 1993; **71** (suppl): 1541–50.

26 Edmonson JH, McCormack GW, Wieand HS for the Northern Central Cancer Treatment Group and Mayo Clinic: Late emerging survival differences in a comparative study of HCAP vs. CP in stage III–IV ovarian carcinoma. In: Salmon SE ed. *Adjuvant therapy of cancer*. Vol VI. Philadelphia: WB Saunders, 1990: 512–21.

27 Hainsworth JD, Jones HW, Burnett LS, *et al.* The role of hexamethylmelamine in the combination chemotherapy of advanced ovarian cancer: A comparison of hexamethylmelamine, cyclophosphamide, doxorubicin and cisplatin (H-CAP) versus cyclophosphamide, doxorubicin and cisplatin (CAP). *Am J Clin Oncol* 1990; **13**: 410–15.

28 Jungi WF, Sessa C, Engeler V, *et al.* Phase II trial with high-dose ifosfamide (IFO) + Mesna in advanced pretreated ovarian cancer. *Proceedings of the American Society of Clinical Oncology* 1985; **4**: 115.

29 Willemse PHB, Van der Burg MEL, Gaast A, *et al.* Ifosfamide given as a 24-h infusion with mesna in patients with recurrent ovarian cancer, preliminary results. *Cancer Chemother Pharmacol* 1990; **26**: 51–4.

30 Sutton GP, Blessing JA, Homesley HD, Berman ML, Malfetano J. Phase II trial of ifosfamide and mesna in advanced ovarian carcinoma, a Gynecologic Oncology Group Study. *J Clin Oncol* 1989; **7**: 1672–6.

31 Perren TJ, Wiltshaw E, Harper P, *et al.* A randomized study of carboplatin versus sequential ifosfamide/carboplatin for patients with FIGO stage III epithelial ovarian carcinoma. *Br J Cancer* 1993; **68**: 1190–4.

32 Wiltshaw E, Evans B, Harland S. Phase III randomized trial cisplatin versus JM8 (carboplatin) in 112 ovarian cancer patients, stages III and IV. *Proceedings of the American Society of Clinical Oncology* 1985; **4**: 121.

33 Adams M, Kerby IJ, Rocker I, Evans A, Johansen K, Franks CR. A comparison of the toxicity and efficacy of cisplatin and carboplatin in advanced ovarian cancer. The Swon Gynaecological Cancer Group. *Acta Oncol* 1989; **28**: 57–60.

34 Mangioni C, Bolis G, Pecorelli S, *et al.* Randomized trial in advanced ovarian cancer comparing cisplatin and carboplatin. *J Natl Cancer Inst* 1989; **81**: 464–71.

35 Hande KR. Pharmacologic-based dosing of carboplatin: a better method. *J Clin Oncol* 1993; **11**: 2295–6.

36 Calvert AH, Newell DR, Gumbrel LA, *et al.* Carboplatin dosage: prospective evaluation of a simple formula based on renal function. *J Clin Oncol* 1989; **7**: 1748–6.

37 Jodrell DI, Egorin MJ, Canetta RM, *et al.* Relationships between carboplatin exposure and tumor response and toxicity in patients with ovarian cancer. *J Clin Oncol* 1992; **10**: 520–8.

38 Edmonson JH, McCormack GM, Wieand HS, *et al.* Cyclophosphamide-cisplatin versus cyclophosphamide-carboplatin in stage III–IV ovarian carcinoma, a comparison of equally myelosuppressive regimens. *J Natl Cancer Inst* 1989; **81**: 1500–4.

39 Alberts DS, Green S, Hannigan EV, *et al.* Improved therapeutic index of carboplatin plus cyclophosphamide versus cisplatin plus cyclophosphamide, final report by the Southwest Oncology Group of a phase III randomized trial in stages III and IV ovarian cancer. *J Clin Oncol* 1992; **10**: 706–17.

40 Swenerton K, Jeffery J, Stuart G, *et al.* Cisplatin-cyclophosphamide versus carboplatin-cyclophosphamide in advanced ovarian cancer: a randomized phase III study of the

National Cancer Institute of Canada. Clinical Trialists Group. *J Clin Oncol* 1992; **10**: 718–26.

41 Ten Bokkel Huinink WW, van der Burg MEL, *et al.* Carboplatin in combination therapy for ovarian cancer. *Cancer Treat Rev* 1988; **15**: 9–15.

42 Vermorken JB, ten Bokkel Huinink WW, Eisenhauer EA, *et al.* Carboplatin versus cisplatin. *Ann Oncol* 1993; **4** (suppl 4): 41–8.

43 Belpomme D, Bugat R, Rives M, *et al.* Carboplatin versus cisplatin in association with cyclophosphamide and doxorubicin as first line therapy in stage III and IV ovarian carcinoma, results of an artac phase III trial. *Proceedings of the American Society of Clinical Oncology* 1992; **11**: 227.

44 Williams CJ, Stewart L, Parmar M, Guthrie D. Meta-analysis of the role of platinum compounds in advanced ovarian carcinoma. *Semin Oncol* 1992; **19**: 120–8.

45 Runowicz CD, Wiernik PH, Einzig AI, Goldberg GL, Horwitz SB. Taxol in Ovarian Cancer. *Cancer* 1993; **71**: 1591–6.

46 Huizing MT, Keunig ACF, Rosing H, *et al.* Pharmaco-kinetics of paclitaxel and metabolites in a randomized comparative study in platinum pretreated ovarian cancer patients. *J Clin Oncol* 1993; **11**: 2127–35.

47 Peereboom DM, Donehower RC, Eisenhauer EA, *et al.* Successful retreatment with Taxol after major hypersensitivity reactions. *J Clin Oncol* 1993; **11**: 885–90.

48 Hoskins PJ, Swenerton KD. Oral etoposide is active against platinum-resistant ovarian cancer. *J Clin Oncol* 1994; **12**: 60–3.

49 Eisenhauer EA, Ten Bokkel Huinink WW, Swenerton KD, *et al.* European-Canadian randomized trial of paclitaxel in relapsed ovarian cancer: High vs low dose and long vs short infusion. *J Clin Oncol* 1994; **12**: 2654–66.

50 Einzig AI, Wiernik PH, Sasloff J, Runowicz CD, Goldberg GL. Phase II study and long-term follow-up of patients treated with taxol for advanced ovarian adenocarcinoma. *J Clin Oncol* 1992; **10**: 1748–53.

51 Thigpen T, Blessing J, Ball H, Hummel S, Barret R, Buffalo J. Phase II trial of Taxol as a second-line therapy for ovarian carcinoma: A Gynecologic Oncology Group Study. *Proceedings of the American Society of Clinical Oncology* 1990; **9**: 604.

52 Trimble EL, Adams JD, Vena D. Paclitaxel for platinum refractory ovarian cancer: results from the first 1000 patients registered to the National Cancer Institute Treatment Referral Center, 9103. *J Clin Oncol* 1993; **11**: 2405–10.

53 Levin L, Hryniuk WM. Dose intensity analysis of chemotherapy regimens in ovarian carcinoma. *J Clin Oncol* 1987; **5**: 756–67.

54 Levin L, Simon R, Hyrniuk WM. Importance of multi-agents chemotherapy regimens in ovarian carcinoma: dose intensity analysis. *J Natl Cancer Inst* 1993; **85**: 1732–42.

55 Ozols F, Thigpen JT, Dauplat J, Colombo N, Piccart MJ, Bertelsen K, *et al.* Dose intensity. *Ann Oncol* 1993; **4** (suppl 4): 49–56.

56 Murphy D, Crowther D, Rennison J, *et al.* A randomised dose-intensity study in ovarian carcinoma comparing chemotherapy given at four week intervals for six cycles with half dose chemotherapy given for twelve cycles. *Ann Oncol* 1993; **4**: 377–84.

57 Hong Kong Ovarian Carcinoma Study Group. A randomized study of high-dose versus low-dose cisplatinum combined with cyclophosphamide in the treatment of advanced ovarian cancer. *Chemotherapy* 1989; **35**: 221–7.

58 Kaye SB, Lewis CR, Paul J, *et al.* Randomized study of two doses of cisplatin with cyclophosphamide in epithelial ovarian cancer. *Lancet* 1992; **340**: 329–33.

59 McGuire WP, Hoskins WJ, Brady MF, Homesly HD, Clarke-Pearson DL. A phase III trial of dose intense versus standard dose cisplatin and cytoxan in advanced ovarian cancer. *Proceedings of the American Society of Clinical Oncology* 1992; **11**: 226.

60 Colombo N, Pittelli MR, Parma G, Marzola M, Torri W, Mangioni C. Cisplatin (P) dose intensity in advanced ovarian cancer (AOC): a randomized study of conventional dose (DC) vs dose-intense (DI) cisplatin chemotherapy. *Proceedings of the American Society of Clinical Oncology* 1993; **12**: 255.

61 Piccart MJ, Nogaret JM, Marcelis L, *et al.* Cisplatin combined with carboplatin, a new way of intensification of platinum dose in the treatment of advanced ovarian cancer. *J Natl Cancer Inst* 1990; **82**: 703–7.

62 de Vries EGE, Hamilton TC, Lind M, Dauplat J, Neijt JP, Ozols RF. Drug resistance, supportive care and dose intensity in ovarian carcinoma. *Ann Oncol* 1993; **4**: 57–62.

63 Viens P, Marninchi D, Legros M, *et al.* High-dose melphalan and autologous marrow rescue in advanced epithelial ovarian carcinomas: A retrospective analysis of 35 patients treated in France. *Bone Marrow Transplant* 1990; **5**: 227–33.

64 Dauplat J, Legros M, Condat P, Ferriere JP, Ben-Ahmed S, Plagne R. High dose melphalan and autologous bone marrow support for treatment of ovarian carcinoma with positive second-look operation. *Gynecol Oncol* 1989; **34**: 294–8.

65 Hurris H, Irvin R, Kuhn J, *et al*. Phase I clinical trial of Taxotère administered as either a two hour or a six hour intravenous infusion. *J Clin Oncol* 1993; **11**: 950–8.

66 Hansen HH, Eisenhauer EA, Hansen M, *et al*. New cytotoxic drugs in ovarian cancer. *Ann Oncol* 1993; **4** (suppl 4): 63–70.

67 Piccart MJ, Gore M, ten Bokkel Huinink WW, *et al*. Taxotère (RP56976, NSC628503): An active new drug for the treatment of advanced ovarian cancer. (OVCA). *Proceedings of the American Society of Clinical Oncology* 1993; **12**: 820.

68 Lund B, Hansen OP, Theilade K, Hansen M, Neijt JP. Phase II study of Gemcitabine (2′,2′,-defluorodoxycytidine) in previously treated ovarian cancer. *JNCI* 1994; **86**: 1530–3.

69 Morgan-Ihrig C, Lembersky B, Christopherson W, Tarassoff P. A phase II elevation of difluorodeoxy-cytidine (dFdC) in advanced stage refractory ovarian cancer. *Proceedings of the American Society of Clinical Oncology* 1991; **10**: 196.

70 Kuhnle H, Meerpohl HG, Lenaz L, *et al*. Etoposide in cisplatin-refractory ovarian cancer. *Proceedings of the American Society of Clinical Oncology* 1988; 7: 137.

71 Muggia FM, Russell CA. New chemotherapies for ovarian cancer. Systemic and intraperitoneal podophyllotoxins. *Cancer* 1991; **67**: 225–30.

72 Markman M, Hakes T. Reichman B, *et al*. Exploring the use of chronic low-dose oral etoposide in ovarian cancer: Is there a role for this "New Drug" in the management of platinum-refractory disease? *Semin Oncol* 1992; **19** (suppl 14): 25–27.

73 Seymour MT, Mansi JL, Gallagher CJ, *et al*. Protracted oral etoposide in epithelial ovarian cancer: a phase II study in patients with relapsed or platinum-resistant disease. *Br J Cancer* 1994; **69**: 191–5.

74 Rustin GJS, van der Burg MEL, Berek JS. Tumour markers. *Ann Oncol* 1993; **4** (suppl 4): 71–7.

75 Hilgers RD, Rivkin SE, Von Hoff DD, Alberts DS. Mitoxantrone in epithelial carcinoma of the ovary. A Southwest Oncology Group study. *Am J Clin Oncol* 1984; 7: 499–501.

76 Thigpen T, Blessing JA. Current therapy of ovarian carcinoma: an overview. *Semin Oncol* 1985; **12**: (suppl 4): 47–52.

77 Eisenhauer EA, Swenerton KD, Sturgeon JFG, Fine S, O'Reilly SEO, Canetta R. Carboplatin therapy for recurrent ovarian carcinoma: National Cancer Institute of Canada experience and a review of the literature. In: Bunn PA, Canetta R, Ozds RF, Rozencweig M, eds. *Carboplatin (JM-8). Current perspectives and future directions.* Philadelphia: WB Saunders, 1990: 133–40.

78 Thigpen JT, Vance RB, Khansur T. Second-line chemotherapy for recurrent carcinoma of the ovary. *Cancer* 1993; **71** (suppl 4): 1559–64.

79 Rowinsky EK, Onetto N, Canetta RM, Arbuck SG. Taxol: the first of the taxanes. An important new class of antitumor agents. *Semin Oncol* 1992; **19**: 646–62.

80 Carter SK. Future directions in Taxol research. *International Cancer Communiqué* 1993; 1: 7–8.

81 Sarosy G, Kohn E, Adamo D, *et al*. Taxol dose intensification (D.I.) in patients with recurrent ovarian cancer. *Proceedings of the American Society of Clinical Oncology* 1992; **11**: 226.

82 Sarosy G, Kohn E, Stone DA, Rothenberg M, Jacob J, Orvis Adama D, *et al*. Phase I study of Taxol and granulocyte colony-stimulating factor in patients with refractory ovarian cancer. *J Clin Oncol* 1992; **10**: 1165–70.

83 McGuire WP, Rowinsky EK, Rosenshein NB, *et al*. Taxol: A unique antineoplastic agent with significant activity in advanced ovarian epithelial neoplasms. *Ann Intern Med* 1989; **111**: 273–9.

15 Role of radiotherapy in ovarian cancer

NINA EINHORN

The role of radiotherapy in the treatment of ovarian cancer can be represented by a sinus curve. During the 1950s and 1960s it was often used. When alkylating agents were reported to have an impact on epithelial tumours of the ovaries the use of radiotherapy was questioned and a controversy developed during the 1970s. At the end of the 1970s after the publication of results from the Princess Margaret Hospital in Toronto interest in radiotherapy increased, and there have been several publications during the 1980s and in recent years.

My intention is to describe the role of radiotherapy up to the present time. To make it easier for the reader and to make the issue more practical from the clinical point of view, I have divided the subject according to the timing of the treatment in the multidisciplinary approach.

Historical development

In 1912, Eymer reported long term remission in eight patients with advanced ovarian carcinoma treated with roentgen rays.[1] The problem associated with ortho voltage radiotherapy was the poor penetration of the beam and the limitation of the dose by the skin reaction, which did not permit a sufficient dose to reach the deep tumours. Nevertheless, even with ortho voltage therapy it was possible to irradiate the pelvis, and to achieve 23% survival in serous and 60% survival in endometrial and mesonephroid tumours limited to the pelvis.[2]

With the development of super voltage radiotherapy the feasibility of treating ovarian cancer increased. The problem with super voltage therapy was not the penetration to deep tumours, but the limitations of the doses which could be given to the upper abdomen where potential damage to the kidneys and liver did not permit doses high enough to sterilise tumours. It became obvious that because of the radiosensitivity of the kidneys and liver the doses that it was possible to give to the lower abdomen were much too high for the upper abdomen. Even for the lower abdomen doses exceeding 50 Gy were more than the small bowel could tolerate. This consideration has a major role in radiotherapy for advanced ovarian cancer;

even if there is sufficient proof that epithelial ovarian cancer is a radiosensitive tumour, the site of the tumour and the tendency to dissemination of tumour cells to the whole abdominal cavity are considerable limitations.

Because of the lack of early symptoms in ovarian cancer, most patients have disseminated tumour cells throughout the abdominal cavity at the time of diagnosis.[3]

In spite of many years of investigation, there is still no real consensus about the role of radiotherapy in ovarian cancer. The discussion can be divided into three main parts: the role of adjuvant radiotherapy in early stages of ovarian cancer; the role of radiotherapy in advanced stages primarily after operation; and the role of radiotherapy after primary surgery, induction chemotherapy, and second-look surgery.

Experimental approaches such as intraoperative irradiation and the use of radiolabelled monoclonal antibodies are under investigation.

Early stages of ovarian cancer (Ia, Ib, Ic and IIa)

One of the first studies of early stage disease was done at the MD Anderson Hospital in which pelvic irradiation was compared with melphalan in a randomised trial.[4] The study included all stages of ovarian cancer and presented an analysis for each of the stages. In 108 evaluable patients treated from April 1968 to 1974 there was an advantage in favour of irradiation in stage I but no differences in five year actuarial survival for any stage between the randomised groups. One of the groups received 12 cycles of melphalan 0.2 mg/kg/day for five days and the other group whole abdominal irradiation (2600–2800 cGy in 21 fractions over two weeks) using a ^{60}Co moving strip technique with liver shielding and later a 2000 cGy pelvic boost. Unfortunately, prognostic factors such as grade and substages of stage I were not stratified. The conclusion was that melphalan was the treatment of choice, causing less toxicity and costing less. The study, published in 1975, had an impact on further development and a switch towards chemotherapy followed.

The next randomised study was done by the Gynecologic Oncology Group between 1971 and 1978 and published in 1980.[5] The questions asked were whether postoperative treatment with melphalan or pelvic irradiation reduced the incidence of relapse compared with surgery alone, and if so which of the postoperative treatments was superior to the other.

The results showed 17% recurrence in the observation group, 30% in the pelvic irradiation group, and 6% in melphalan group. There was no significant difference between the melphalan and control groups or between the pelvic irradiation and control groups. Unfortunately, 49% of the patients randomised had to be withdrawn from the study for various reasons, which laid the study open to criticism. It was also questionable how the patients were distributed between the three groups with regard to prognostic variables.

210

At the same time a study on stage Ia ovarian cancer was done at the Princess Margaret Hospital.[6] During the period 1971–77, 54 patients with stage Ia ovarian cancer were studied. They were stratified by prognostic factors and the treatment was randomised between observation and pelvic irradiation with 4500 cGy in 20 fractions. Nine relapses distributed between the two study groups indicated no curative benefit for pelvic irradiation. The difference between the two groups was in the distribution of relapses; in the pelvic irradiation arm the ratio of relapse in abdomen to pelvis was 4:1 whereas in the observation group it was 1:3. The conclusion was that control of pelvic disease was inadequate regardless of the treatment and that treatment should be directed to the whole peritoneal cavity.

At the Radiumhemmet in Stockholm between March 1979 and February 1981 we did a randomised study of the early stages of seropapillary ovarian cancer (Einhorn N, *et al.* Treatment in early stages of ovarian carcinoma—a randomized study. Presented at Nordisk Förening för Obstetrik och Gynekologi, Gothenburg, 1980.) For stages Ia, Ib and IIa patients were randomised to receive lower abdominal irradiation, or melphalan, or a combination of the two. For stage Ic or ruptured tumours the patients were randomised between chemotherapy or a combination of chemotherapy and lower abdominal irradiation. There was a significant difference in patients with stage Ic tumours for whom the combination therapy was superior to chemotherapy alone. Of particular interest was the observation that patients with well differentiated tumours all survived regardless of the treatment given.

In 1971, the Gynecologic Oncology Group started two new randomised studies in early ovarian cancer.[7] In the first group, patients with stage Ia or Ib well and moderately differentiated cancer were randomised to observation only or treatment with melphalan. After six years of observation there were no differences with regard to tumour-free survival or total survival, the rates being 91% compared with 98%, and 94% compared with 98%, respectively. In the second study in stage Ia (poorly differentiated tumours), patients were randomly assigned to treatment with melphalan or with intraperitoneal ^{32}P. No difference was found between the two groups.

An Italian cooperative group did two randomised studies (Pecorelli S *et al.* Adjuvant therapy in early ovarian cancer: results of two randomised trials. Presented at the Society of Gynecologic Oncologists, Orlando, 1994), in the first of which patients with stages Ia and Ib tumours were randomised between cisplatin treatment and observation alone. The results showed the same survival rate for both groups—85% for cisplatin and 86% for the observation group. In the second study patients with stage Ic disease were randomly assigned to be treated by either cisplatin or ^{32}P. The survival times were similar—for cisplatin 80% and for ^{32}P 78%.

In another study the Danish cooperative group randomised 124 patients with stages Ib, Ic, and II tumours between whole abdominal irradiation and pelvic irradiation plus cyclophosphamide. There was no difference between the two groups with respect to recurrence-free survival or four-year overall survival, which were 55% and 63%, respectively.[8]

A study was performed at the Princess Margaret Hospital during the period 1971–75 which included patients with stages Ib, II and III ("asymptomatic") disease.[9] The patients were stratified by age, stage, pathological type, grade and amount of residual tumour. Pelvic irradiation was given to all patients. The objective was to find out whether survival could be improved by adding either chlorambucil 6 mg/day for two years or irradiation of the upper abdomen. The results showed that for patients treated with whole abdominal irradiation survival was significantly better than for those given treatment to the pelvis with or without chlorambucil. The difference was seen only in patients who had had complete excision of macroscopic tumour. If the resection was not complete, neither of the treatments was superior. The weakness of the study was that the stages were not separated, probably because the patients had been operated on at other hospitals and there was no exact information about the primary surgery.

The strength of the study, on further analysis, was that it resulted in a useful prognostic classification dividing ovarian cancer into low, intermediate, and high risk groups. The conclusion was that stage I grade 1 was a low risk group that had no need of adjuvant therapy. All other stage I, as well as stage II were in an intermediate group for whom adjuvant radiotherapy was important.[10]

In another Canadian randomised study done from 1975 to 1984 stage III was also included with the early stages.[11] Patients were randomised after they had received pelvic irradiation to receive further treatment with whole abdominal irradiation, melphalan or ^{32}P. The last treatment was subsequently shown to be too toxic and it was abandoned in 1980. The results showed no significant difference among the three groups.

Another Italian cooperative group recently published a study on early stage disease. Patients were randomised to receive cisplatinum and cyclophosphamide, or whole abdominal irradiation. Because of low accrual and poor compliance with the protocol the study was stopped prematurely when only approximately half the projected number of patients had been entered. There was no significant difference between the two treatment groups but the trend was in favour of chemotherapy; the toxicity was described as less severe in the chemotherapy group.[12]

Conclusion—early stage disease

From a thorough review of all the trials of early stage disease, we can conclude that the biological prognostic factors in the early stages of ovarian cancer are more important than the treatment given. The important observation from previous studies of early stage disease was that well differentiated tumours do not require any postoperative adjuvant treatment. For patients with poor prognostic factors, randomised trials have been started in both Europe and in the United States in which one of the groups of patients receives no further treatment.

Advanced ovarian cancer—stages IIb, III, and IV

During the late 1970s, attention started to be given to the accuracy of staging procedures in ovarian cancer through the work of the Ovarian Cancer Study Group under the direction of the National Cancer Institute.[13] It became clear that even with an accurate staging procedure about 30% of the patients were upstaged. This led to the identification of "real" stages I and II, resulting in better adjuvant policy in early stages, and the upstaging placed some patients in a more actively treated group.

The group of patients with microscopic and minimal disease in the advanced stages is the one which, judging from several publications during the last few years, gains the most from radiotherapy. Recent studies analysing the outcome of radiotherapy have clearly shown that the size of the largest residual tumour mass after primary surgery is the most important factor in predicting survival after either chemotherapy or irradiation. The probability of sterilising the tumour by radiotherapy is highly dependent on the sensitivity of the surrounding tissue to the tumouricidal dose that is necessary. Taking into consideration the pronounced tendency for ovarian cancer to disseminate to the whole abdominal cavity it becomes obvious that the radiosensitivity of some critical organs will limit doses to the upper abdomen. The tolerance dose of kidneys and liver is about 2500 cGy.[14] With that in mind it is necessary to look at the radiation doses in relation to the amount required to sterilise different volumes of tumour. Correlation of tumour size with tumouricidal doses in ovarian carcinoma is estimated for tumours >2 cm to be 4000 to 6000 cGy, for tumours 0.5–2.0 cm 4500–5000 cGy, and for microscopic tumours 2500–3000 cGy.[15] The combination of these two factors—the sensitivity of critical organs in the upper abdomen and the tumouricidal doses needed to sterilise different sizes of tumour in ovarian carcinoma—makes it clear that irradiation of the upper abdomen must not exceed the limits tolerated by critical organs and this makes it impossible to treat large tumour masses in the upper abdomen.

The lower abdomen, on the other hand, is much more accessible as it can cope with larger tumour doses of up to 5000 cGy. In consequence, radiotherapy can sterilise tumour masses of larger volume in the pelvis, although to limit irradiation to the lower abdomen or pelvis only will in advanced stages (with few exceptions) be unsatisfactory.

Techniques

Intraperitoneal instillation of ^{32}P (colloidal chromic phosphate)

Radioactive colloids have been given intraperitoneally to treat intra-abdominal disease for the last four decades. At first radioactive gold (^{198}Au)

213

was the isotope used most, especially for the treatment of ascites. Over the last 30 years, however, colloidal radioactive phosphorus (^{32}P) has been used, particularly for adjuvant treatment of early stage I, II, or stage III disease with microscopic residuum. ^{32}P is a β energy source with an effective range from 4–6 mm in tissue and a half life of 14.2 days. The absence of γ radiation makes it easy to handle in terms of exposure of staff and complications. Experience with intra-abdominal ^{32}P has shown that the risk of complications is minimal. Usually an abdominal radionuclide scan is carried out before injection of ^{32}P. Technetium sulphur colloid $7.4 - 11.1 \times 10^7$ mBq is applied through a peritoneal catheter and the whole abdomen is scanned to assess the potential distribution of the colloid.

The precise distribution of ^{32}P given intraperitoneally is not known exactly. It seems that there is both abdominal and systemic distribution during the instillation. The patient's position has to be changed frequently to encourage dispersion.

External beam radiation

Generally, external beam radiation can be divided into two types—lower abdominal or pelvic and whole abdominal, and there are some differences between techniques.

Whole abdominal irradiation

Two different techniques have been developed with the aim of treating the whole abdominal cavity. The moving strip technique takes in a field 10 cm wide which is moved in increments of 2.5 cm (strips) so that each of the strips receives eight or 10 fractions. This technique was developed at the MD Anderson Hospital. The aim is to give 2600–2800 cGy delivered over 8–12 days. The liver is shielded routinely on the anterior posterior strip and the kidney on the posterior strip. The liver and kidney receive roughly 1800 cGy. The pelvis is then later boosted with 15×15 cm portals and a 2000–3000 cGy dose.[4]

At the Princess Margaret Hospital in Toronto a comparative study was done comparing the moving strip technique with the open field technique which includes the whole abdomen and diaphragm. The results showed that the two techniques are comparable but that the open field technique was associated with fewer complications.[16] The open field technique includes the whole abdomen with the inferior margin extending below the obturator foramen and the lateral border extending beyond the peritoneum; the upper field includes the diaphragm. The standard method at the Princess Margaret Hospital was to give 2250 cGy in mid-plane in 22 fractions with renal shielding after 1500 cGy but without liver shielding.

A modification of whole abdominal irradiation is the technique introduced by Martinez.[17] The whole peritoneal cavity, after treatment with 1000 cGy, is shielded for kidneys posteriorly and at 1500 cGy the liver is shielded to

50%. In total 3000 cGy are given to the upper abdomen with 2250 cGy to the liver, 2000 to the kidneys, and 3000 to the pelvis. In the second phase of the treatment the para-aortic nodes, the diaphragm, and the true pelvis are treated with a total dose of up to 4200 cGy. The third phase consists only of a booster to the pelvis bringing the dosage up to 5100 cGy.

In Stockholm a six-field technique has been used for whole abdominal irradiation given with 2000 cGy to the whole abdominal cavity without shielding. A 2000–2500 cGy booster is given to the lower abdominal field with an upper limit of L4 and two lateral fields to the upper abdomen given anteriorly from the kidneys with a further 2000 cGy.[18]

Lower abdominal irradiation

The booster to the pelvis and lower abdomen can be given either by using a pelvic field of 15×15 cm or by using a whole abdominal lower field including the peritoneal cavity and with an upper field to the junction between L4 and L5. While the total pelvic dose can be as high as 5000 cGy, the risk of bowel complications will increase with doses to the lower abdominal cavity that exceed 4500 cGy.

The results of treatment in advanced stages

Five studies have presented long term survival or relapse free rates among patients known to have macroscopic residual disease. With a residual tumour less than 2 cm in diameter Dembo achieved a 43% 10 year recurrence-free survival in stage III patients[19]; Fuller *et al* presented a 62% 10 year relapse-free survival[20]; Goldberg and Peschel a 41% six-year survival[23]; Martinez *et al* a 50% 15 year failure-free survival[17]; and Weiser *et al* a 42% 10-year survival.[22] All these patients were treated after primary resection with no adjuvant method other than radiotherapy. These data can be interpreted as strong evidence that radiotherapy has curative potential as an adjuvant to resection with known macrocopic residual disease.

There are, however, some critical points which must be mentioned. These are all retrospective analyses, even if the study at the Princess Margaret Hospital was prospectively planned. In one of the studies by Fuller *et al* many of the patients treated with total abdominal pelvic irradiation also received chemotherapy with a single alkylating agent.

In the study by Martinez *et al* the survival for stage III is less than 20%, which reflects the fact that the residual disease of less than 2 cm applied mainly to stage II disease, and in the paper by Weiser *et al* stages II and III are considered together with regard to residual disease.

Irradiation combined with chemotherapy in advanced disease

Fuks *et al*[23] and Coltart *et al*[24] were the first to postulate that sequential surgery, chemotherapy, second-look laparotomy, and radiotherapy in advanced ovarian cancer could be a model for treatment. Since then, several (mostly retrospective) reports have been published, although a few randomised studies have been carried out. Considerable controversy has been aroused by these reports. There are almost as many negative as there are positive conclusions, and even some of the negative conclusions show equivocal results. There can be no doubt that the attitude of the investigators influences the presentation of the results and that they are highly heterogeneous with respect to the assessment of residual disease, techniques used, and selection of patients. Among the three randomised studies, one suggested that after second-look laparotomy at which residual tumour is found, irradiation was inferior to continued chemotherapy.[25] The other two indicated that abdominal pelvic irradiation and further chemotherapy were equivalent.[26 27] A conclusion can be drawn that radiotherapy is to be of benefit to patients with advanced ovarian carcinoma there must be little or no residual disease.

The reasons why the results of irradiation in advanced stages of ovarian cancer are so controversial can be explained by the varying quality of the studies that have been reported during the last 20 years. Few of them were randomised although some of the material is of good quality even if presented in a retrospective manner. There are, however, several weak points. There is a considerable heterogeneity with respect to such prognostic factors as the grade, histology, age, stage and patient performance status. Furthermore in many cases there were too few patients, the surgical staging was inaccurate, the pathological classification was inaccurate, and there was no clear definition of how the patients were selected for the different treatments.

Experimental studies

Intraoperative irradiation

Three studies have been reported during the last few years in which intraoperative irradiation was used as palliation for small groups of patients. The doses varied between 12.5 and 25 Gy in one dose and the toxicity was acceptable. According to the authors the results deserve further investigation in patients with ovarian cancer.[28–30]

Treatment with radio-labelled monoclonal antibodies

During recent years three groups have presented results of treatment with radio-labelled monoclonal antibodies. Paganelli *et al* used HMOV 18 ^{111}In labelled monoclonal antibodies. They concentrated mainly on pharmacokinetic distribution in relation to the site of the tumour. No results of treatment have been published.[31] An English group led by Epenetos has published several studies with monoclonal antibodies HMFG 1, HMFG 2, AUA 1, HI 712 radio-labelled with either ^{131}I or [90]Y. Two

216

conclusions can be drawn from their work: the toxicity is acceptable and there is an indication that the treatment can be effective.[32-36]

The Boston group led by Knapp has published two studies in which CA125 monoclonal antibodies were radio-labelled with [131]I. Toxicity was acceptable. This study also suggested that treatment with monoclonal antibodies should be used only in patients with little or no residual or microscopic tumour.[37 38]

Conclusions—advanced stages

What is clear from the reports is that the amount of residual tumour is of great importance and it becomes clear that patients with gross residual tumour should not be candidates for radiotherapy in the first place. It seems that radiotherapy can be curative as an adjuvant after primary resection in a certain subset of patients. In addition, radiotherapy followed by chemotherapy can have an influence on survival as consolidation treatment. There is, however, still a lack of sufficient data from large studies comparing different regimens and which also consider the biological prognostic factors that have an extremely important role in the behaviour of ovarian carcinoma.

References

1 Eymer H. Beeinflussung ovn proliferierenden Ovarialtumoren durch Roentgenstrahlen. *Strahlen* 1912; **1**: 358–61.
2 Kottmeier H-L. Ovarian cancer with special regard to radiotherapy. *AJR* 1971; **CXI**: 417–21.
3 Einhorn N, Nilsson B, Sjövall K. Factors influencing survival in carcinoma of the ovary: Study from a well-defined Swedish population. *Cancer* 1985; **55**: 2019–25.
4 Smith J, Rutledge FN, Delclos L. Results of chemotherapy as an adjunct to surgery in patients with localized ovarian cancer. *Semin Oncol* 1975; **2**: 277–81.
5 Hreshchyshyn MM, Park RC, Blessing JA, *et al*. The role of adjuvant therapy in stage I ovarian cancer. *Am J Obstet Gynecol* 1980; **138**: 139–45.
6 Dembo AJ, Bush RS, Beale FA, Bean HA, Pringle JF, Sturgeon JF. The Princess Margaret Hospital study of ovarian cancer: stage I, II and asymptomatic III presentations. *Cancer Treat Reports* 1979; **63**: 249–54.
7 Young RC, Walton LA, Ellenberg SS, *et al*. Adjuvant therapy in stage I and stage II epithelial ovarian cancer. *N Engl J Med* 1990; **322**: 1021–7.
8 Sell A, Bertelsen K, Andersen JE, Ströyer I, Panduros J. Randomized study of whole-abdomen irradiation versus pelvic irradiation plus cyclophosphamide in treatment of early ovarian cancer. *Gynecol Oncol* 1990; **37**: 367–73.
9 Dembo AJ, Bush RS, Beale FA, *et al*. Ovarian carcinoma: Improved survival following abdominopelvic irradiation in patients with a completed pelvic operation. *Am J Obstet Gynecol* 1979; **134**: 793–800.
10 Dembo J. Radiation therapy in the management of ovarian cancer. *Clin Obstet Gynecol* 1983; **10**: 261–78.
11 Klassen D, Shelley W, Starreveld A, *et al*. Early stage ovarian cancer: A randomized clinical trial comparing whole abdominal radiotherapy, melphalan, and intraperitoneal chromic phosphate: A National Cancer Institute of Canada Clinical Trials Group Report. *J Clin Oncol* 1988; **6**: 1254–63.

12 Chiara S, Conte PF, Franzone P, *et al*. High-risk early stage ovarian cancer. Randomized clinical trial comparing cisplatin plus cyclophosphamide versus whole abdominal radiotherapy. *Am J Clin Oncol* 1994; **17**: 72–6.

13 Young RC, Decker DG, Wharton JT, *et al*. Staging laparotomy in early ovarian cancer. *JAMA* 1983; **250**: 3072–6.

14 Rubin P, Cooper R, Phillip TL. Radiation biology and radiation pathology syllabus *(Set RT I: Radiation Oncology)*. Chicago: American College of Radiology, 1974.

15 Mychalczak BR, Fuks Z. The current role of radiotherapy in the management of ovarian cancer. *Hematol Oncol Clin North Am* 1992; **6**: 895–913.

16 Dembo AJ, Bush RS, Beale FA, *et al*. A randomized clinical trial of moving strip versus open field whole abdominal irradiation in patients with invasive epithelial cancer of ovary. *Int J Radiat Oncol Biol Phys* 1983; **9**: 97–101.

17 Martinez A, Schray MF, Howes AE, Bagshaw MA. Postoperative radiation therapy for epithelial ovarian cancer: The curative role based on a 24-year experience. *J Clin Oncol* 1985; **3**: 901–11.

18 Einhorn N, Von Hamos K, Hindmarsh T, *et al*. Radiation therapy of ovarian carcinoma: Presentation of a six-field technique. *Radiother Oncol* 1986; 7: 125–31.

19 Dembo AJ. Abdominopelvic radiotherapy in ovarian cancer. *Cancer* 1985; **55**: 2285–90.

20 Fuller DB, Sause WT, Plenk HP, Menlove RL. Analysis of postoperative radiation therapy in stage I through III epithelial ovarian carcinoma. *J Clin Oncol* 1987; **5**: 897–905.

21 Goldberg N, Peschel RE. Postoperative abdominopelvic radiation therapy for ovarian cancer. *Int J Radiat Oncol Biol Phys* 1988; **14**: 425–9.

22 Weiser EB, Burke TW, Heller PB, Woodward J, Hoskins WJ, Park RC. Determinants of survival of patients with epithelial ovarian carcinoma following whole abdomen irradiation (WAR). *Gynecol Oncol* 1988; **30**: 201–8.

23 Fuks Z, Rizel S, Anteby SO, *et al*. The multimodal approach to the treatment of stage III ovarian carcinoma. *Int J Radiat Oncol Biol Phys* 1982; **8**: 903–8.

24 Coltart RS, Nethersell BW, Brown CH. A pilot study of high dose abdominopelvic radiotherapy following surgery and chemotherapy for stage III epithelial carcinoma of the ovary. *Gynecol Oncol* 1986; **23**: 105–10.

25 Bruzzone M, Repetto L, Chiara S, *et al*. Chemotherapy versus radiotherapy in the management of ovarian cancer patients with pathological complete response or minimal residual disease at second look. *Gynecol Oncol* 1990; **38**: 392–5.

26 Lawton F, Luesley D, Blackledge G, *et al*. A randomized trial comparing whole abdominal radiotherapy with chemotherapy following cisplatinum cytoreduction in epithelial ovarian cancer. West Midlands Ovarian Cancer Group Trial II. *Clin Oncol* 1990; **2**: 4–9.

27 Lambert HE, Rustin GJS, Gregory WM, Nelstrop AE. A randomized trial comparing single-agent carboplatin with carboplatin followed by radiotherapy for advanced ovarian cancer: A North Thames Ovary Group Study. *J Clin Oncol* 1993; **11**: 440–8.

28 Calkins A, Lester S, Stehman F, Cline H, Calvo F. Intraoperative radiotherapy in advanced or recurrent gynecologic cancer. *Ann Radiol* 1989; **32**: 502–3.

29 Konski AA, Neisler J, Phibbs G, Bronn DG, Dobelbower RR. A pilot study investigating intraoperative electron beam irradiation in the treatment of ovarian malignancies. *Gynecol Oncol* 1990; **38**: 121–4.

30 Garton GR, Gunderson LL, Webb MJ, *et al*. Intraoperative radiation therapy in gynecologic cancer: The Mayo Clinic experience. *Gynecol Oncol* 1993; **48**: 328–32.

31 Paganelli G, Belloni C, Magnani P, *et al*. Two-step tumour targeting in ovarian cancer patients using biotinylated monoclonal antibodies and radioactive streptavidin. *Eur J Nucl Med* 1992; **19**: 322–9.

32 Stewart JSW, Hird V, Sullivan M, Snook D, Epenetos AA. Intraperitoneal radioimmunotherapy for ovarian cancer. *Br J Obstet Gynecol* 1989; **96**: 529–36.

33 Hird V, Stewart JSW, Snook D, *et al*. Intraperitoneally administered 90Y-labelled monoclonal antibodies as a third line of treatment in ovarian cancer. A phase 1-2 trial: problems encountered and possible solutions. *Br J Cancer* 1990; **62** (suppl X): 48–51.

34 Stewart JSW, Hird V, Snook D, *et al*. Intraperitoneal Yttrium-90-labelled monoclonal antibody in ovarian cancer. *J Clin Oncol* 1990; **8**: 1941–50.

35 Hird V, Maraveyas A, Snook D, *et al*. Adjuvant therapy of ovarian cancer with radioactive monoclonal antibody. *Br J Cancer* 1993; **68**: 403–6.

36 Maraveyas A, Snook D, Hird V, *et al*. Pharmacokinetics and toxicity of an yttrium-90-CITC-DTPA-HMFG1 radioimmunoconjugate for intraperitoneal radioimmunotherapy of ovarian cancer. *Cancer* 1994; **73**: 1067–75.

37 Finkler NJ, Muto MG, Kassis AI, *et al*. Intraperitoneal radiolabeled OC 125 in patients with advanced ovarian cancer. *Gynecol Oncol* 1989; **34**: 339–44.

38 Muto MG, Finkler NJ, Kassis AI, *et al*. Intraperitoneal radioimmunotherapy of refractory ovarian carcinoma utilizing iodine-131-labeled monoclonal antibody OC125. *Gynecol Oncol* 1992; **45**: 265–72.

16 Immunology and immunotherapy of ovarian cancer

JONATHAN S BEREK, OTONIEL MARTINEZ-MAZA

The immune system plays an essential part in host defence, and can respond to host cells that have undergone transformation to become neoplastic cells. These responses, whether natural or induced, can in some cases lead to tumour regression. As more is learnt about the regulation of immune responses, new opportunities for novel immunotherapeutic approaches develop. Immunodiagnostic procedures using antitumour marker antibodies also show great promise as diagnostic or prognostic tools, or both. In this chapter, we will give a brief introduction to the human immune system, followed by a summary of the immune effector mechanisms that are involved in antitumour responses, and end with a summary of experimental immunotherapeutic approaches to ovarian cancer.

Immunological mechanisms involved in anti-tumour responses

The human immune system has the potential to respond to tumour cells in various ways. Some of these immune responses occur in an innate or antigen non-specific manner, while others are adaptive, or antigen-specific. Adaptive responses are specific not only for a given antigen, but also establish a memory, allowing a more rapid and vigorous response to the same antigen in future encounters.[1,2] Various innate and adaptive immune mechanisms are involved in responses to tumours including cytotoxicity directed to tumour cells mediated by cytotoxic T cells, natural killer (NK) cells, macrophages, and antibody-dependent cytotoxicity mediated by complementation activation.[3]

Adaptive or specific immune responses are made up of humoral and cellular responses. Humoral responses refer to the production of antibodies which are antigen-reactive, soluble, bifunctional molecules composed of specific antigen-binding sites that react with foreign antigens. They are associated with a constant region that directs the biological activities of the antibody such as the binding of antibody molecules to cells including

220

phagocytic cells, or the activation of complement. Cellular immune responses are antigen-specific immune responses mediated directly by activated immune cells rather than by the production of antibodies. The distinction between humoral and cellular responses is historical, and originates from the experimental observation that humoral immune function can be transferred by serum, while cellular immune function requires the transfer of cells. Most immune responses include both humoral and cellular components.

Several types of cells including cells from both the myeloid and lymphoid lineages make up the immune system. Specific humoral or cellular (or both) immune responses to foreign antigens involve the coordinated action of populations of lymphocytes operating in concert with each other and with phagocytic cells (macrophages). These cellular interactions include both direct cognate interactions involving cell-cell contact, and cellular interactions involving the secretion of and response to cytokines or lymphokines. Lymphoid cells are found in lymphoid tissues such as lymph nodes or spleen or in the peripheral circulation. The cells that make up the immune system originate from stem cells in the bone marrow.

B cells, humoral immunity, and monoclonal antibodies

The cells that synthesise and secrete antibodies are B lymphocytes.[12] Mature, antigen-responsive B cells develop from pre-B cells (committed B cell progenitors) and differentiate to become plasma cells, which are cells that produce large quantities of antibodies. Pre-B cells originate from bone marrow stem cells in adults, after the rearrangement of immunoglobulin genes from their germ-cell configuration to that seen in B cells. Mature B cells express cell surface immunoglobulin molecules which these cells use as their receptors for antigen.

On interaction with antigen, and in the presence of appropriate cell-cell stimulatory signals and cytokines, mature B cells respond to become antibody-producing cells. Although the generation of antibodies directed to tumour cells is generally not thought to have a central role in antitumour immune responses, the production of monoclonal antibodies to tumour cell antigens has shown great potential for immunotherapy and tumour detection, or both.

Kohler and Milstein developed monoclonal antibody technology more than 10 years ago, and in recent years there has been considerable interest in the use of monoclonal antibodies for detection and monitoring of tumours and for treatment.[4] Monoclonal antibodies that react with tumour-associated antigens may provide new therapeutic agents for ovarian cancer. Immunotoxin-conjugated monoclonal antibodies directed to human ovarian adenocarcinoma antigens can induce tumour cell killing, and can prolong survival in mice implanted with a human ovarian cancer cell line.[5] However, many obstacles limit the clinical use of monoclonal antibodies including tumour cell antigenic heterogeneity, modulation of tumour-associated antigens, and cross-reactivity of normal host and tumour-associated antigens. No unique tumour specific antigens have been identified; all tumour antigens that have been identified to date are tumour-related antigens, which have been seen to be expressed to some extent in non-malignant tissues. Because most monoclonal antibodies are murine, the host's immune system can also recognise and respond to these foreign mouse proteins, but the use of genetically engineered monoclonal antibodies composed of human constant regions with specific antigen-reactive murine variable regions should result in reduced antigenicity to the host. This might help eliminate many of the problems associated with using murine monoclonal antibodies.

T lymphocytes and cellular immunity

T lymphocytes have a central role in the generation of immune responses by acting as helper cells in both humoral and cellular immune responses, and by acting as effector cells in cellular responses.[12] T cell precursors originate in bone marrow and home to the thymus where they mature into functional T cells. During their thymic maturation T cells learn to recognise antigen in the context of the major histocompatibility complex (MHC) type of the individual person. It also seems that T cells with the capability of responding to self are removed during development in the thymus.

T cells can be distinguished from other types of lymphocytes by their cell surface phenotype (the pattern of expression of various molecules on the cell surface), as well as by differences in their biological functions. All mature T cells express certain cell surface molecules such as the cluster determinant (CD)3 molecular complex, and the T cell antigen receptor, which is found in close association with the CD3 complex. The expression of cell surface molecules can be quantified using monoclonal antibodies specific for these molecules. The availability of monoclonal antibody reagents specific for such markers has led to great progress in understanding the organisation of the immune system in recent years. Certainly such monoclonal antibodies are of great value in monitoring the effects on the human immune system of experimental treatment with biological response modifiers or cytokines.

T cells recognise antigen through the cell surface T cell antigen receptor. The structure and molecular organisation of this molecule are similar to those of antibody molecules, which are the B cell receptor for antigen. The T cell receptor gene undergoes similar gene arrangements during T cell development to those seen in B cells, but there are important differences between the antigen receptors on B cells and T cells. The T cell receptor is not secreted, and its structure is somewhat different from that of antibody molecules. The way in which the B cell and T cell receptors interact with antigens is also quite different. T cells can respond to antigens only when these antigens are presented in association with MHC molecules on antigen-presenting cells. Effective antigen presentation involves the processing of

222

antigen into small fragments of peptide within the antigen-presenting cell, and the subsequent presentation of these fragments of antigen in association with MHC molecules expressed on the surface of the antigen-presenting cell. T cells can respond to antigen only when presented in this manner, unlike B cells, which can bind antigen directly, without processing and presentation by antigen-presenting cells.

There are two major subsets of mature T cells which are phenotypically and functionally distinct: T helper/inducer cells, which express the CD4 cell surface marker, and T suppressor/cytotoxic cells, which express the CD8 marker. The expression of these markers is acquired during the passage of T cells through the thymus. CD4 T cells can provide help to B cells (resulting in the production of antibodies by B cells), and interact with antigen presented by antigen-presenting cells in association with MHC class II molecules. CD4 T cells can also act as helper cells for other T cells. CD8 T cells include cells that are cytotoxic (cells that can kill target cells bearing appropriate antigens), and they interact with antigen presented on target cells in association with MHC class I molecules. The CD8 T cell subset also contains suppressor T cells. Suppressor T cells are cells that can inhibit the biological functions of B cells or other T cells.

Although the primary biological role of cytotoxic T cells seems to be lysis of virus-infected autologous cells, cytotoxic immune T cells can mediate the lysis of tumour cells directly. Presumably cytotoxic T cells recognise antigens associated with MHC class I molecules on tumour cells through their antigen-specific T cell receptor, and then a series of events occurs that ultimately results in the lysis of the target cell.

Monocytes and macrophages

Monocyte/macrophages, which are myeloid cells, have important roles in both innate and adaptive immune responses; macrophages play a key part in the generation of immune receptors. T cells do not respond to foreign antigens unless those antigens are processed and presented by antigen-presenting cells. Macrophages (and B cells) express MHC class II molecules and are effective antigen-presenting cells for CD4 T cells.[12] Helper/inducer (CD4) T cells that bear a T cell receptor of appropriate antigen and self specificity are activated by this antigen-presenting cell to provide help (various factors—lymphokines—that induce the activation of other lymphocytes).

In addition to their role as antigen-presenting cells, macrophages play an important part in innate responses by ingesting and killing microorganisms. Activated macrophages, in addition to their many other functional capabilities, can act as cytotoxic, antitumour killer cells.

Natural killer (NK) cells

A third major population of lymphocytes includes NK cells.[12] These do not consistently bear cell surface markers that are characteristic of T or B cells although they can share certain cell surface molecules with other types of lymphocytes. NK cells characteristically have a large granular lymphocyte morphology.[6]

NK cells are effector cells in an innate type of immune response: the non-specific killing of tumour cells or virus-infected cells, or both. NK activity therefore represents an innate form of immunity that does not require an adaptive, memory response for optimal biological function, but the anti-tumour activity can be increased by exposure to several agents, particularly cytokines such as interleukin-2 (IL-2).

While NK cells can express certain cell surface receptors, particularly a receptor for the crystallisable fragment (Fc) portion of antibodies and other NK-associated markers, it seems that cells with NK function are phenotypically heterogeneous, at least when compared with T or B cells. The cells that can carry out antibody-dependent cellular cytotoxicity, or antibody-targeted cytotoxicity, seem to be NK-like cells. Antibody-dependent cellular cytotoxicity by NK-like cells has been shown to result in the lysis of tumour cells *in vitro*, but the mechanisms of this tumour cell killing are not clearly understood although close cellular contact between the effector cell and the target cell seems to be required.

Biological response modifiers

Most immunotherapeutic agents used in the treatment of cancer have been non-specific agents which when introduced into the human system elicit a generalised inflammatory reaction and immune response, probably mediated by the secretion of a range of cytokines by many different types of cells. These agents have diverse and broad biological effects, and are often referred to as immunomodulators or biological response modifiers.

The response of a given patient to treatment with biological response modifiers depends on the ability to react to treatment with a generalised immune response. It is possible that some elements of the immune response elicited by immunotherapeutic agents or biological response modifiers may be counterproductive, possibly causing immune suppression, inducing the production of cytokines that enhance tumour growth, or inducing an unfavourable or inappropriate immune response.

BCG has been widely used in many tumour systems, either systemically, by injection into the lesion, or by scarification.[7] Occasionally it has been mixed with whole irradiated tumour cells and injected into the patient as a vaccine. In a large series intracutaneous injection of melanoma lesions with BCG resulted in some tumour regression in patients with cutaneous recurrence,[8] but visceral or parenchymal metastatic disease is resistant to this treatment. While there have been some preliminary observations about the use of BCG as an adjuvant in children with acute lymphocytic leukaemia and with stage II melanoma, randomised studies have not shown any appreciable responses.

224

Table 16.1 Sources, target cells, and biological activities of cytokines involved in immune responses

Cytokine	Cellular source	Target cells	Biological effects
IL-1	Monocytes and macrophages Tumour cells	T cells, B cells Neurons Endothelial cells	Co-stimulator Pyrogen activation
IL-2	T cells (TH1)	T cells B cells NK cells	Growth Activation and antibody production Activation and growth
IL-3	T cells	Immature haemopoietic stem cells	Growth and differentiation
IL-4	T cells (TH2)	B cells T cells	Activation and growth; isotype switch to IgE; increased MHC II expression Growth
IL-6	Monocytes and macrophages T cells, B cells Ovarian cancer cells Other tumours	B cells T cells Hepatocytes Stem cells Tumour cells	Differentiation, antibody production Co-stimulator Induction of acute-phase response Growth and differentiation Autocrine/paracrine growth and viability-enhancing factor
IL-10	T cells (TH2) Monocytes and macrophages	T cells (TH1) Monocytes and macrophages B cells	Inhibition of cytokine synthesis Inhibition of Ag presentation and cytokine production Activation
IL-12	Monocytes	T cells (TH1) NK cells	Induction
IFN gamma	T cells (TH1) NK cells	Monocytes/ macrophages NK cells T cells B cells	Activation Activation Activation Enhances responses
TNFα	Monocytes and macrophages T cells	Monocytes/ macrophages T cells, B cells Neurons (hypothalamus) Endothelial cells Muscle and fat cells	Monokine production Co-stimulator Pyrogen Activation, inflammation Catabolism/cachexia

Cytokines, lymphokines, and immune mediators

Many events in the generation of immune responses (as well as during the effector phase of immune responses) require or are enhanced by cytokines which are soluble mediator molecules (table 16.1).[12] They are pleiotropic in that they have multiple biological functions that depend on the type of target cell or its maturational state. Cytokines are also heterogeneous: while some cytokines seem to be related in the sense that they share structural features and seem to have evolved from a common ancestral precursor, most cytokines share little structural or amino acid homology. Cytokines (also called monokines if they are derived from monocytes, lymphokines if they are derived from lymphocytes, interleukins if they exert their actions on leucocytes, or interferons if they have anti-viral effects) are produced by a wide variety of cell types, and seem to have important roles in many biological responses outside the immune response, such as haematopoiesis. They may also be involved in the pathophysiology of a wide range of diseases, and show great potential as therapeutic agents in immunotherapy to cancer.

Though cytokines are a heterogeneous group of proteins they share some characteristics. For instance, most cytokines are low-intermediate molecular weight (10–60 kDa) glycosylated secreted proteins. They are also involved in immunity and inflammation, are produced transiently and locally (they do not generally act in an endocrine manner), are extremely potent in small concentrations, and interact with high affinity cellular receptors that are specific for each cytokine. The cell surface binding of cytokines by specific receptors results in signal transduction followed by changes in gene expression, and ultimately by changes in cellular proliferation or altered cell behaviour, or both. Their biological actions overlap, and exposure of responsive cells to multiple cytokines can result in synergistic or antagonistic biological effects.

Many different cytokines are involved in immune responses, particularly the interleukins (IL-1, IL-2, IL-3, IL-4, IL-6, IL-7, IL-10, IL-11, IL-12) (table 16.1). The interleukins are heterogeneous and do not constitute a family of growth-related molecules. Interleukin-1 (IL-1) has a wide range of biological activities including direct effects on several cells involved in immune responses.[9] It is involved in fever and inflammatory responses, and may be involved in the pathogenesis of several diseases such as rheumatoid arthritis. There are two defined forms of IL-1, IL-1α and IL-1β, which have similar biological activities. IL-1 can be released as a soluble form, or can be found as a cell-associated molecule on the cell surface of macrophages. The primary sources of IL-1 are macrophages, the phagocytic cells of the liver and spleen, some B cells, epithelial cells, certain brain cells, and the cells lining the synovial spaces. IL-1 has a broad range of target cells and biological activities, as do most lymphokines.

Interleukin-2 (IL-2) is a lymphokine that was originally called T cell growth factor, which indicates one of the major biological activities of this molecule. Failure of T cells to produce IL-2 results in the absence of a T cell immune response, and a diminution of the antibody response. Natural human IL-2 is a 15 kDa glycoprotein, and is produced primarily by activated T cells. For IL-2 to exert its proliferation-inducing effects it has to interact with a specific receptor for IL-2 on the surface of the target cell. The high affinity receptor for IL-2 consists of two polypeptides, the α (75 kDa) and β (55 kDa) chains. After activation T cells express greatly increased numbers of this high-affinity receptor for IL-2.

A principal role of IL-1 is in the initiation of early events in immune responses. It functions by inducing antigen-responsive T cells to express the gene for IL-2; these T cells will then express the IL-2 receptor and will respond to IL-2 with increased proliferation. Stimulation of resting T cells with antigen presented in the context of self (antigen associated with an MHC molecule on the surface of an antigen-presenting cell) and with IL-1 therefore induces synthesis and secretion of IL-2. During this activation process, responding T cells undergo a change or alteration in their cell surface receptors including the expression of cell surface receptors for IL-2. Continuing exposure to IL-2 then leads to the proliferation of T cells bearing the IL-2 receptor, thereby acting as an activation and response-amplification stage in the generation of immune responses. Activated T cells not only respond to IL-2, but also produce IL-2. IL-2 can therefore act in an autocrine manner (meaning that the cells producing the lymphokine then respond to it) or in a paracrine fashion (meaning that the IL-2 produced by a T cell is taken up and responded to by neighbouring cells).

IL-1 has other direct effects on cells of the immune system. For instance, it can act as a B cell activation-inducing factor, and can also induce the production of other lymphokines that are involved in immune responses such as IL-6. Since its original description as a T cell growth hormone, IL-2 has been shown to have various other immune functions including the promotion of B cell activation and maturation, and activation of monocytes and NK cells. IL-2 can also lead directly or indirectly to the stimulation of the production of interferon and other cytokines.

B lymphocyte activation and differentiation to immunoglobulin-secreting plasma cells is increased by cytokines produced by helper T lymphocytes or monocytes (table 16.1).[12] Several cytokines originally described as B cell stimulating factors (IL-4, IL-5, and IL-6) were seen to have additional biological activities. For instance, IL-6 (a factor that can induce B lymphocyte differentiation to immunoglobulin-secreting cells), is a pleiotropic cytokine with biological activities that include the induction of cytotoxic T lymphocyte differentiation, the induction of acute phase reactant production by hepatocytes, and activity as a colony-stimulating factor for haematopoietic stem cell.[10] IL-6 is produced primarily by activated monocyte/macrophages and T lymphocytes. Interestingly, several types of tumour cells produce IL-6, and it has been proposed as an autocrine/paracrine growth factor for different types of neoplasms.[11–16] It may prove to be an effective antitumour agent by virtue of its ability to enhance antitumour T cell-mediated immune responsiveness.[16 17]

IL-10, a 35–40 kDa cytokine, also called cytokine synthesis inhibitory factor because of its activity as an inhibitor of cytokine production, is produced by a subset of CD4 T cells: "type 2" (TH2) cells, and inhibits the cytokine production by another CD4 T cell subset, "type 1" (TH1) cells.[18] TH1 and TH2 are two T helper cell subpopulations that control the nature of an immune response by secreting characteristic and mutually antagonistic sets of cytokines[19]: TH1 clones produce IL-2 and IFN-2 and IFN gamma, while TH2 clones produce IL-4, IL-5, IL-6, and IL-10. A similar dichotomy between TH1 and TH2 type responses has been reported in humans.[20 21] Human IL-10 inhibits the production of IFN gamma and other cytokines by human peripheral blood mononuclear cells,[22] as well as suppressing the release of cytokines (IL-1, IL-6, IL-8 and tumour necrosis factor (TNF) α by activated monocytes.[23–25] IL-10 also downregulates class II MHC expression on monocytes resulting in a strong reduction in antigen-presenting capacity of these cells.[25] Together, these observations support the concept that IL-10 has an important role as an immune-inhibitory cytokine.

Because epithelial cancers of the ovary usually remain confined to the peritoneal cavity, even in the advanced stages of the disease, it has been suggested that the growth of ovarian cancer intraperitoneally could be related to a local deficiency of antitumour immune effector mechanisms.[26] Recent studies have shown that ascitic fluid from patients with ovarian cancer contained increased concentrations of IL-10.[27] Various other cytokines are also seen in ascites obtained from women with ovarian cancer; concentrations of IL-6, IL-10, TNFα, granulocyte colony stimulating factor (G-CSF) and granulocyte macrophage colony stimulating factor (GM-CSF) were appreciably raised.[28] A similar pattern was seen in serum samples from women with ovarian cancer, with IL-6 and IL-10 often detected. Preliminary results have indicated that ovarian cancer cells do not produce IL-10, so high concentrations of IL-10 could certainly result in a peritoneal environment characterised by immune unresponsiveness and promotion of tumour growth.

IL-3, a factor that can increase the early differentiation of haematopoietic cells,[29] may find a role in immunotherapy because of its ability to induce haematopoietic differentiation in people undergoing aggressive chemotherapeutic treatment or bone marrow transplantation. TNFα is a cytokine that can be directly cytotoxic for tumour cells, can increase immune cell-mediated cellular cytotoxicity, and can activate macrophages and induce secretion of monokines. Other biological activities of TNFα include the induction of cachexia, inflammation, and fever; it is an important mediator of endotoxic shock.

There are three types of interferons: IFN alfa, IFN beta, and IFN gamma.[1 2 30] They can interfere with viral production in infected cells, and have various effects on the immune system as well as direct antitumour effects. For instance, IFN gamma, (a cytokine produced by T lymphocytes) can affect immune function by increasing the induction of MHC molecules expression, increasing the activity of antigen-presenting cells, and thereby increasing T lymphocyte activation.

As research has provided new information on the biological activities of cytokines, these factors have appeared to be extraordinarily pleiotropic with a bewildering array of biological activities including some outside the immune system.[1 2 10 16] Because some cytokines have direct or indirect

228

antitumour or immune-enhancing effects, or both, several of these factors have been used in the experimental treatment of cancer.

The precise roles of cytokines in antitumour responses have not been completely described. As cytokines are pleiotropic they could exert antitumour effects by many different direct or indirect activities. It is possible that a single cytokine could increase tumour growth directly by acting as a growth factor, while at the same time increasing immune responses directed to the tumour. The potential of cytokines to increase antitumour immune responses has been tested in experimental adoptive immunotherapy by exposing the patient's peripheral blood cells or tumour infiltrating lymphocytes to cytokines such as IL-2 *in vitro*, and so generating activated cells with antitumour effects that can be given back to the patient.[31 32] Some cytokines can also exert direct anti-tumour effects: TNF can induce cell death in sensitive tumour cells.

The effects of cytokines in patients with cancer might be modulated by soluble receptors or blocking factors. For instance, blocking factors for TNF and for lymphotoxin were found in ascites from patients with ovarian cancer.[33] Such factors could inhibit the cytolytic effects of TNF or lymphotoxin and should be taken into account in the design of clinical trials of intraperitoneal infusion of these cytokines.

Cytokines have growth-increasing effects on tumour cells in addition to inducing antitumour effects: they can act as autocrine or paracrine growth factors, or both, for human tumour cells including those of non-lymphoid origin. For instance, IL-6 (which is produced by various types of human tumour cells) can act as a growth factor for human myeloma, Kaposi's sarcoma, renal carcinoma, and epithelial ovarian cancer cells.[11–16]

Clearly cytokines are of great potential value in the treatment of cancer, but, because of their multiple, even conflicting, biological effects a thorough understanding of cytokine biology will be essential for their successful use in the treatment of cancer.

Adoptive immunotherapy

Recently the *ex vivo* enhancement of antitumour immune cell responses including the generation of lymphokine-activated killer (LAK) cells or the activation of tumour-infiltrating lymphocytes has provided new immune system-based approaches for antitumour responses. In particular, adoptive immunotherapy with IL-2 has been studied extensively and can produce regression of tumour in various animal and human tumours such as melanoma and renal cell carcinoma, in conjunction with the adoptive transfer of autologous LAK cells.[31 32]

Exposure of peripheral blood monoclonal cells to cytokines *in vitro* (particularly IL-2) leads to the generation of cytotoxic effector cells called LAK cells.[32] These cells are effective cytotoxic cells for various tumour cells including those that are resistant to NK cell or T cell-mediated lysis. The treatment of patients with such *ex vivo* activated autologous cells together with IL-2 simultaneously forms the basis of adoptive immunotherapy, a form of experimental antitumour immunotherapy that has been the subject of great interest.

Experimental treatment of human subjects with autologous *ex vivo* generated LAK cells and IL-2 has yielded tumour regression in some cases.[31-34] This sort of treatment has resulted in some complete responses,[33] with a combined response rate of 27% in 146 patients with cancer who were treated in two separate studies.[34,35] The overall response rate to LAK treatment is low, however, and this type of adoptive immunotherapy causes high morbidity.[36] It is also costly, impractical in most medical settings.

Much current experimental work is aimed at developing more efficient and practical applications of adoptive immunotherapy. One approach involves the *ex vivo* generation of immune effector cells from tumour-infiltrating lymphocytes, which are lymphocytes that are isolated from tumours and activated and expanded *in vitro* by exposure to IL-2. They are then given concurrently with IL-2.[37,38] This approach is hampered by the need to expand a limited number of tumour-infiltrating lymphocytes *in vitro* to generate enough effector cells for treatment. Much attention has also been directed towards the development of new methods for the generation of LAK cells or tumour-infiltrating lymphocytes, including methods that use cytokines other than IL-2 to stimulate these cells.

Another promising approach that has been explored in animal studies involves the targeting of activated T lymphocytes with a bifunctional monoclonal antibody that binds to both the CD3/T cell receptor complex (on the activated effector T cell) and a tumour-associated antigen (on the target tumour cell).[39] This approach has the potential advantage of allowing a large fraction of the activated lymphocytes to target their effects directly on tumour cells thereby reducing the need to amplify a large number of effector cells from tumour-infiltrating lymphocytes. It also has the potential to reduce some of the side effects associated with LAK treatment which is a more non-specific form of adoptive immunotherapy. Adoptive immunotherapy is an active area of study, and may eventually lead to effective forms of antitumour treatment.

Immunotherapy in ovarian cancer

In recent years increased clinical experience has been gained in immunotherapy for ovarian cancer, and various experimental immunotherapeutic approaches have been examined. Patients with small volume, residual peritoneal disease remain an attractive target for immunotherapy, particularly regional peritoneal immunotherapy.[40] The rapid progress in molecular biology, biotechnology (monoclonal antibody production and conjugation), immunology, adoptive immunotherapy, and cytokine biology have also resulted in many reagents becoming available for clinical evaluation in ovarian cancer.

Assessment of immune status in patients with gynaecological cancers

The relative immune status of patients with gynaecological cancers has been examined in many studies, and has taken the form of quantification of mitogen-induced *in vitro* proliferation of mononuclear cells, the use of skin tests to assess *in vivo* responses to microbial antigens or contact allergens, the enumeration of lymphocyte subsets, and the measurement of immunoglobulin concentrations.[3 41] Because of the complex and relatively insensitive nature of the tests used, however, as well as the lack of appropriate controls, the results of these studies have been difficult to assess. The relative pre-treatment immune status of patients may also not be relevant to their subsequent response to immune enhancing antitumour treatment. Perhaps future study of immune function and immune mechanisms in gynaecological cancers will benefit from new technology and from more stringent experimental design and analysis.

Monoclonal antibodies and immunotherapy

Monoclonal antibodies have the potential to kill tumour cells by complement activation and lysis of tumour cells, direct antiproliferative effects, increasing the activity of non-specific phagocytic cells that recognise the murine immunoglobulin on the surface of the tumour cell, or by antibody-dependent cellular cytotoxicity.[42] Most monoclonal antibodies (murine) are not directly cytotoxic, however, and fail to activate human immune effector systems. Studies that have evaluated the efficacy of monoclonal antibody-directed radiotherapy in gynaecological cancer are limited. Prospective studies using new monoclonal antibodies and different energy sources are ongoing. Another approach has been to link monoclonal antibodies to toxins.[42 43] The efficacy of the clinical use of monoclonal antibodies and immunotoxins in humans requires further evaluation.

OC-125, a monoclonal antibody that reacts with a molecule produced by human epithelial ovarian carcinoma is widely used to monitor the blood CA125 antigen concentrations in women with ovarian cancer.[44] Monoclonal antibodies have also been used in gynaecological oncology for radio-immunodetection. Monoclonal antibodies that recognise tumour-associated antigens on epithelial ovarian cancer cells labelled with radioactive tracers have been used to detect primary and recurrent lesions.[42]

Recent work has indicated that the HER2/*neu* oncogene may have an important role in the pathogenesis of ovarian cancer: increased amounts of HER2/*neu* proto-oncogene expression were seen in about a third of ovarian cancers.[45] This alteration in HER2/*neu* expression was associated with disease behaviour. The expression of HER2/neu in ovarian cancer cells also correlated with resistance to killing.[46] As HER2/*neu* is overexpressed in some cancer cells it has been suggested that the HER2/*neu* antigen, a transmembrane protein tyrosine kinase that is homologous to the human epidermal growth factor receptor, might be used as the target for immunotherapy.[47] Recently a monoclonal antibody directed to HER2/*neu* has shown promise as a potential immunotherapeutic agent[48]; it increased human tumour cell susceptibility to TNF and to cisplatin in a nude mouse model system, suggesting that this anti-HER2/*neu* monoclonal antibody could modulate the effects of HER2/*neu in vivo*. Anti-HER2/*neu* monoclonal antibody reagents may therefore be of value in the immunotherapy of HER2/*neu*-expressing tumours including ovarian cancer.

Immunotherapy with biological response modifiers

There has been great interest in examining the potential role of immunotherapy in ovarian cancer. Because long term survival with ovarian epithelial malignancies is poor, and most patients present with metastatic disease, there is a great need to develop useful biological treatments. Furthermore, patients with advanced disease are considerably immunocompromised,[49] suggesting a role for immune-enhancing treatments.

Most studies in metastatic ovarian cancer have used non-specific immunotherapies. Such trials have involved inoculation with biological response modifiers, including inactivated bacteria such as *Corynebacterium parvum* (a heat-killed, Gram negative anaerobic bacillus), bacillus *Calmette-Guerin* (BCG), which is a live, attenuated strain of *Mycobacterium bovis*, or Freund's complete adjuvant. Modifications of these agents can be produced by extracting fractions of these organisms using biochemical techniques. For instance, fractionation (typically by acid or phenol extraction) can lead to the isolation of active components (glycolipids or carbohydrates) that might more selectively elicit desirable immunological responses while sparing less desirable immune reactions or side effects.[50] For example, MER is a methanol-extracted residue of BCG that retains significant immunomodulatory activity, while potentially avoiding some of the problems associated with viable BCG.

Exposure to *C parvum* results in a variety of immune responses including an acute inflammatory response, predominantly the induction and infiltration of neutrophils, and macrophage attraction, activation, and cytotoxicity.[51] NK cytotoxicity is increased as is T lymphocyte activation on exposure to *C parvum*.[52] In animal systems, *C parvum* has been shown to be active,[53] with tumour rejection being temporally associated with a cellular immune response. Studies of a murine teratocarcinoma model examining biochemically fractionated *C parvum* showed that tumour rejection in animals treated with intraperitoneal injection of the residue of pyridine-extracted *C parvum*, (a fraction that contains the bacterial cell walls[54]) is comparable to the rejection observed after intraperitoneal injection of whole unmodified *C parvum*. Giving these *C parvum* fractions sequentially in combination with bacterial endotoxins resulted in an even greater antitumour effect.[40] Interestingly, bacterial endotoxin is a potent

232

inducer of the production of various cytokines that could mediate antitumour responses, particularly IL-1, IL-6, and TNFα.[14]

C parvum and BCG have been used in the largest single retrospective series to date. Phase I studies of *C parvum* have shown that toxicity generally includes systemic chills, fever, malaise, nausea, and vomiting in most patients.[55–59] Serious toxicity, including hypotension, prolonged raised temperature, and chest pain, is not common. In early studies in patients with ovarian cancer, *C parvum* given subcutaneously was combined with escalating doses of cyclophosphamide, doxorubicin, and 5-fluorouracil (CAF) given monthly.[59] Pretreatment immune variables were within normal limits in patients who responded to treatment compared with reduced values in those who did not. Immune function as measured in these studies, however, was not improved by treatment. In a randomised trial of CAF chemoimmunotherapy, with or without *C parvum* given intravenously, there was no difference in response rates, disease progression-free intervals, or survival.[60]

Creasman *et al* reported a retrospective series of patients treated with either melphalan or melphalan plus *C parvum*.[61] They evaluated 108 patients with untreated stage III ovarian epithelial malignancies, and the melphalan plus *C parvum* combination group had a 53% total response rate compared with 29% in the group treated with melphalan alone. A prospective randomised study that attempted to confirm these findings, however, showed no significant differences between melphalan 7 mg/m^2/day for five days orally given every four weeks and the same regimen plus *C parvum* given intravenously on day 7 after chemotherapy.[62]

In a randomised prospective study by Alberts *et al*, 66 patients with stage III and IV epithelial ovarian carcinomas were treated with either a combination of cyclophosphamide and doxorubicin, or these two agents plus intravenous BCG.[55] Doxorubicin was given in a dose of 40 mg/m^2 on day 1, cyclophosphamide 200 mg/m^2 on days 3–6, and BCG on day 8 and day 15. This cycle was repeated every four weeks. Of the 32 patients who were treated with a combination of chemotherapy and immunotherapy the total response rate was 56%, with two of 32 evaluable patients having a complete response. The median duration of response was 45 weeks, the median survival 93 weeks. This compared with the 34 patients treated with doxorubicin and cyclophosphamide alone, of whom 11 patients (32%) had a partial response. The median duration of response in this group was 26 weeks and the median survival was 59 weeks so the immunotherapy combined with chemotherapy did not improve the survival of patients with ovarian carcinoma. It is important to note, however, that in most experimental animal systems in which immunotherapy and chemotherapy are combined, tumour rejection is augmented most often when the giving of the immunostimulant precedes that of cytotoxic agents by a sufficient interval to permit some positive immunomodulation.[63] We are awaiting the results of a prospective randomised Gynecologic Oncology Group study which compared doxorubicin, cyclophosphamide, and cisplatin with or without BCG given by scarification in patients with suboptimal stage III disease.

Only anecdotal evidence exists for responses using tumour vaccines in ovarian cancer. Graham and Graham treated 232 patients with gynaecological malignancies, 48 of whom had ovarian cancer, with Freund's complete adjuvant.[64] It was mixed either with DNA-protein extract of the tumour or with viable tumour cells. Systemic reactions to these agents were low in most patients whose tumours progressed, but the vaccine did not control tumour growth. BCG combined with an allogenic tumour cell vaccine was given to 10 patients with stage III or IV ovarian cancer. Alkylating agents were given as well as the BCG and vaccine of 10^7 irradiated allogenic tumour cells and this resulted in prolonged survival, compared retrospectively with historical controls.[65] Though these reports suggested some improvement in survival, they were all uncontrolled and retrospective studies involving small numbers of patients. Patients also have been treated with irradiated tumour cells injected intralymphatically, a technique referred to as active specific intralymphatic immunotherapy.[66] A complete response, with the patient free of clinical evidence of disease at 13 months, was reported in seven patients with epithelial ovarian carcinoma treated in this way. Unfortunately, most patients with ovarian cancer have bulky, persistent tumours located in the peritoneal cavity, and do not respond to this type of treatment.

Human leucocyte interferon has some antiproliferative effects on ovarian cancer cells *in vitro*.[67] Purified or recombinant interferon alfa or beta have been given systematically in five studies to patients with ovarian cancer, most of whom had persistent or recurrent metastatic epithelial cancers. The combined response rate of these phase I clinical trials was only 10%.[68–70]

IL-2 has been used for experimental immunotherapy of ovarian cancer: recombinant human IL-2 was given intravenously by various regimens to 23 patients with progressive melanoma, renal, colon, or ovarian cancer.[71] IL-2 treatment induced lymphocytosis, as well as causing significant increases in the number of cells expressing the IL-2 receptor, detectable circulating LAK cells, and increased NK cytotoxicity.

Intraperitoneal immunotherapy

Regional immunotherapy is an appealing concept for the treatment of patients whose tumours are confined and have not yet metastasised beyond a definable body cavity. Malignancies that are isolated in the peritoneal cavity, such as residual ovarian cancer, have been treated in many clinical trials with intraperitoneal drugs, most often with cytotoxic chemotherapy such as cisplatin alone or in combination with other drugs.[72 73] The rationale is based on the premise that the residual tumour cells can be exposed to a higher concentration of the drug if brought into direct contact with it and that this might provoke a response in malignancies that would otherwise be resistant. This approach has proved to be promising as a "salvage" treatment for minimal residual ovarian cancer with complete responses of 20%–30% when a small amount of disease persists after systematic induction chemotherapy with a cisplatin combination.[72 73]

Biological response modifiers or immunotherapy have been given intraperitoneally for similar reasons, and also because it has been postulated that one might be able to activate regional effector mechanisms in the peritoneal cavity.[35 56 74–77] Regional immunotherapy could therefore provide a means by which a biological treatment might be rendered effective, even when it has been found to be ineffective when given intravenously. This might be particularly true for treatment with cytokines, or for adoptive immunotherapies, as activated immune effector cells may require direct contact with the malignant target cells to kill them.[34 35 56 74–76 78–82] In patients with minimal residual epithelial ovarian carcinoma after treatment with combination cytotoxic chemotherapy, Bast *et al* and Berek *et al* reported a total of 21 patients treated with intraperitoneal immunotherapy.[56 75] Of the 19 evaluable patients six responded, two completely. All the responding patients had macroscopic disease less than 5 mm in maximum diameter at the start of treatment. Antibody dependent cell-mediated cytotoxicity is significantly increased during the course of treatment[56] as is NK cytotoxicity.[74 83] The increase of cytotoxic effectors in the peritoneal cavity correlates well with the response to the agent *C parvum* given every two weeks in increasing doses starting at 0.25 mg/m^2 and rising to 4 mg/m^2.

Intraperitoneal *C parvum* has been noted by Montovani *et al* to be useful for the palliation of ascites in women with advanced ovarian cancer.[58] In eight patients *C parvum* 7 to 14 mg given intraperitoneally on days 0, 7, and 28 resulted in complete disappearance of ascites in three patients, and a marked reduction of the effusion in two others. The palliative effect was sustained for 6–13 months. Hernandez *et al* reported the intraperitoneal treatment of nine patients with advanced ovarian epithelial malignancies with sterile, pyrogen-free, rabbit-derived human ovarian antitumour serum. Although follow up was short the clinical response rate was 80%, with a one year survival of 87%. A trial of passive serotherapy with rabbit heteroantiserum is being undertaken prospectively.[84] Patients are being treated with intraperitoneal ^{32}P, total abdominal irradiation, and melphalan, with or without 150 to 200 ml of serum. After a two year follow-up in 13 patients there is no difference in survival between the two groups.

Ohkawa *et al* studied the use of intraperitoneal semisynthesised acid polysaccharides, BCG and OK432 (Picibanil, which is a *streptococcal* preparation). These agents were given for four consecutive days with weekly intraperitoneal injections of doxorubicin, 5-fluorouracil, cyclophosphamide, bleomycin, and mitomycin C.[85] Though this was an uncontrolled study, the 60 evaluable patients treated between 1970 and 1977 had a five year survival of 40%. These results further suggest that locoregional immunotherapy combined with chemotherapy may have a role in the control of ovarian cancer confined to the peritoneal cavity.

Treatment with *C parvum* as expected induced a rather profound local reaction and its toxicity precluded more widespread testing. It produced appreciable peritoneal fibrosis, presumably because it induces cells that promote the deposition of collagen. The attraction of these cells, however, and also the release of vasoactive molecules might be responsible for the rejection of tumour cells by non-specific killing—that is, the malignant cells are overwhelmed by the intense and massive outpouring of chemoattracted white cells and their products into the body cavity. The mechanism of tumour cell killing in these circumstances is not known, although regional effector mechanisms such as antibody-dependent cell-mediated cytotoxicity are increased after intraperitoneal *C parvum*.[76]

It has become possible by using recombinant DNA technology to obtain large quantities of defined cytokines. Several of these agents have been examined in phase 1 and 2 clinical trials, including recombinant interferon (IFN) alfa, recombinant IFN gamma, TNF, and IL-2.

Intraperitoneal treatment with interferon alfa

Treatment with rIFN alfa is well-tolerated locally but has significant systemic side effects. Interferon is a cytokine capable of generating cytotoxicity when autologous peripheral blood lymphocytes are incubated with human ovarian carcinoma cells.[35 77] With intraperitoneal rIFNα, NK cytotoxicity is increased and this phenomenon is associated with tumour rejection.[74–76] However, stimulation of NK is not invariably associated with clinical response. *In vitro* data have suggested that the dominant mechanism responsible for killing tumour cells in the peritoneal cavity involves direct effects of IFN on the cancer cells, similar to the cytotoxic chemotherapeutic agents.[74] Exposure of tumour cells to rIFN alfa may make them more vulnerable to the subsequent effects of cytotoxic drugs such as cisplatin.

In Phase II trials, IFN alfa given systematically has produced responses in up to 18% of patients with advanced ovarian cancer.[69 87] Recombinant IFN alfa given intraperitoneally has also produced results in patients with small volume (minimal) residual disease (microscopic disease or tumour nodules less than 5 mm); that is, persistent ovarian cancer after first line chemotherapy.[82 88] The toxicity of intraperitoneal IFN alfa as a single agent has also been described in these studies[82]; 25–50 million units three times a week was not tolerated because of persistent general malaise, fever, and gastrointestinal toxicity, but those patients treated with the same dose once a week tolerated it for 8–16 consecutive weeks. Notably, there was no appreciable neurotoxicity and renal toxicity. While most of the side effects of single agent IFN alfa seem to be different from those of cisplatin, the general malaise and gastrointestinal toxicity produced by each one could potentially be additive when the agents are combined. Willemse *et al* reported similar results in another trial of intraperitoneal IFN alfa in 20 patients with ovarian cancer; of 17 who had a second look laparotomy, five (29%) had complete responses and 4 (24%) had partial responses.[88]

236

Responses in both studies were confined to patients whose residual disease was minimal (less than 5 mm). The toxicity encountered in this trial was similar to that seen in the phase I Gynecologic Oncology Group trial.[82] Overall, 28 surgically evaluated patients have been treated in these two trials, and 14 (50%) responded, with nine (32%) complete responses. All the responding patients had microscopic or small volume residual disease. The combined surgically defined complete response rate in patients with minimal residual disease was therefore 50% (9 of 18 patients). These results suggest that the use of high-dose intraperitoneal rIFN alfa given frequently can produce regional control of very small volume disease confined to the peritoneal cavity. Survival data are not available on these patients, however, so it is not clear if this approach can produce prolonged disease-progression-free intervals.

There is ample evidence to suggest that there is synergy between various interferons and standard cytotoxic agents.[89-95] Interferon has been shown *in vitro* to act synergistically with cisplatin to kill ovarian cancer cells.[96 97] In several reports, the synergy is seen only when the cells are exposed to the interferon before the cytotoxic agent,[88-90 92] presumably because the cytokine stimulates the proliferation of the cells making them potentially more susceptible to the cytotoxic effects of the drug. *In vivo* studies have shown that interferon potentiates the cytotoxicity of cyclophosphamide and cisplatin in non-small-cell lung cancer xenografts.[94] Preclinical results suggest that the cytotoxic effect can be increased when the cancer cells are exposed to the antiproliferative effects of interferon before the cytotoxic agent is given.[90 92 95] Bezwoda *et al* confirmed these observations of the *in vitro* synergy between cisplatin and interferon,[96] and the antitumour effect of interferon is probably related to the direct inhibitory effect of the molecule on tumour growth and not to its effects on modulating immune responses.

Because of the considerable *in vitro* synergy between cisplatin and other agents, a search for clinically tolerable and effective combinations has been undertaken. Nardi *et al* reported 14 evaluable patients who were treated with weekly doses that alternated with 50 million units of intraperitoneal rIFN alfa and 90 mg/m^2 of cisplatin.[97] In this trial, the surgically-documented complete response rate was 50% (7 of 14 patients), and all of these responses were confined to those patients who started their treatment with minimal residual disease. The toxicity was similar to that seen in the phase I-II trials of IFN alfa alone, although the authors reported somewhat less general malaise and gastrointestinal toxicity. The combination of intraperitoneal cisplatin and interferon therefore seemed to be tolerated in these patients, and resulted in an appreciable response rate. The survival time of those patients who had a complete response was generally longer than those patients who did not, but, as response rates were similar to those reported in other series of single-agent intraperitoneal cisplatin, it was not clear whether the addition of the interferon to the cisplatin had any additive influence on the response rate. Another clinical phase I study conducted at two hospitals showed that combined intraperitoneal treatment with cisplatin and rIFN alfa can safely be given to patients with residual ovarian carcinoma after systemic chemotherapy.[98] The maximum tolerated doses of the combination of rIFN alfa and cisplatin as dictated by the phase I trial are rIFN alfa 25×10^6 IU and cisplatin 60 mg/m². The treatment was tolerated best when given as one cycle every three weeks, and of the eight patients who were treated with this precise dose, the median number of treatment cycles was six. Complete responses were noted in two patients after five and six treatment cycles, respectively; partial responses were seen in patients treated with four and eight cycles, respectively.

When the two centre trial schedule that combined intraperitoneal rIFNα and cisplatin was applied to the co-operative group setting, (the Gynecologic Oncology Group), however, a very low response rate, only one partial response (7%), was seen.[99] This poor outcome can be compared with the other phase I-II trials of cisplatin-based or rIFNα intraperitoneal treatment in patients with persistent small-volume residual ovarian cancer (where response rates of 20–40% have been noted).[72 73 82] The difference can probably be accounted for by the fact that, in this series, most (15 of 18) evaluable patients had extensive carcinomatosis that was cisplatin-resistant and the maximum tumours were large.

Bezwoda *et al*[96] reported a clinical trial in which 35 patients with advanced ovarian cancer and ascites confined to the peritoneal cavity were treated with intraperitoneal rIFN alfa, some in combination with cisplatin; seven responses in 19 (36%) patients were seen. This observation is intriguing and suggests that this mode of immunotherapy is appropriate in some patients with solid tumours, particularly those whose tumours are localised to the peritoneal cavity—for example, primary carcinomas of the ovary. In the phase II portion of the study, 16 patients were treated randomly with cisplatin with or without IFN alfa. The combination of interferon and cisplatin produced a somewhat higher response rate than interferon alone: five of the seven patients (77%) treated with the combination responded, while two of nine (22%) treated with cisplatin alone responded. Furthermore, the responses correlated with the *in vitro* data in which the two agents acted synergistically in their antitumour effect. The authors concluded, as in previously published work, that the direct antitumour effect of cisplatin can be increased by the simultaneous exposure of the cancer cells to IFN alfa.

Intraperitoneal treatment with tumour necrosis factor

In several preclinical studies TNF has been shown to have significant antineoplastic activity against various malignant cell lines[100–102] but, in phase 1 trials in which the agent was delivered systematically there was limited clinical activity combined with considerable systemic toxicity particularly fevers, rigors, and hypotension.[105–108]

238

As with all cytokines, it was hoped that intraperitoneal TNF would produce an increased antitumour response and lower systemic side effects. In a phase I trial reported from the Memorial Sloan-Kettering Cancer Center investigators showed that recombinant human TNF could safely be given by the intraperitoneal route, and that this resulted in a pronounced pharmacokinetic advantage for intraperitoneal compared with systemic exposure.[107] After an intraperitoneal dose of TNF $50 \mu g/m^2$, peak concentrations within the peritoneal cavity ranged between 15 000 and 59 000 pg/ml, compared with unmeasurable amounts in the systemic compartment (less than 50 pg/ml). In addition, concentrations between 14 000 and 33 000 pg/ml persisted within the compartment for up to six hours after a single intraperitoneal dose of TNF. Although the TNF was not measurable in the plasma, patients experienced mild emesis, increases in temperature, and chills, even with intraperitoneal doses as low as $10 \mu g/m^2$. Only one patient developed a hypotensive episode (BP 80/50), but treatment-related abdominal pain was common. No clinical responses were observed in this phase 1 trial.[107]

In a study from the University of Heidelberg, intraperitoneal recombinant human TNF was used to control formation of malignant ascites and was shown to have a potential role in its management.[108] Thirty-two patients with symptomatic malignant ascites were treated with a weekly infusion of TNF $(80 \mu g/m^2)$ in one litre of fluid. Twenty patients had ovarian cancer and the other 12 patients had other malignancies. They received an average of 2.6 infusions of recombinant TNF. Seventeen of the 31 evaluable patients (55%) experienced complete resolution of ascites as indicated by clinical and ultrasound examination 30 days after the start of treatment, and 14 had a partial control. Interestingly, only one patient had relapsed during the follow-up of roughly eight months. As observed in the Sloan Kettering trial, the main side effects were fever, chills, abdominal pain, and emesis. On the basis of this report, further studies are indicated of intraperitoneal TNF as an agent to control malignant ascites. It is possible that this treatment can be an important option for patients with malignant ascites when alternative systemic treatments are ineffective.

The potential of TNF to increase the antitumour effect of cisplatin offers a potential strategy for immunotherapy. There is evidence that even low doses of TNF can significantly increase the antitumour properties of drugs like cisplatin, doxorubicin and cyclophosphamide.[102] If this is the case, then prolonged exposure, low-dose treatment with cytokine, given in conjunction with cytotoxic chemotherapy, could offer an advantage and minimise the toxicity of cytokine biotherapy. The intraperitoneal route offers a means by which continuous exposure can be made practical.

Intraperitoneal therapy with interferon gamma

Recombinant IFN gamma is active against malignant cell lines *in vitro*, including lines derived from patients with ovarian cancer.[109–111] In a phase

239

1 trial of intraperitoneal recombinant human IFN gamma conducted in patients with refractory ovarian cancer at the Memorial Sloan-Kettering Cancer Center, it was well tolerated when given weekly and associated with a 150–200 fold pharmacokinetic advantage compared with systemic treatment.[86] The main toxic effects were fatigue and flu-like symptoms, and there was limited local toxicity at the highest dose level tested (8 million IU/m^2), but no clinical responses were observed.

In a cooperative trial done in Europe, 40 patients with residual ovarian cancer after initial cisplatin-based chemotherapy were treated with intraperitoneal recombinant human IFN gamma at a dose of 20 million IU/m^2 twice weekly for a maximum of four months.[112] Of the 30 patients who were evaluable for response, nine (30%) achieved a surgically-defined (laparoscopy or laparotomy) complete response. Ten of 23 patients (43%) whose largest residual tumour mass measured less than 2 cm responded to this treatment. Toxicity of the therapeutic regimen included fever (90%), leucopenia (45%), increased transaminase activities (37%), abdominal discomfort (37%), and fatigue (10%). The response rate was similar to that observed with rIFN gamma. The difference between the high response rate noted in this trial and the failure of the Sloan-Kettering study to show activity for recombinant IFN gamma in a similar group of patients may have been because of the much higher dose intensity in the European study (20 million units twice a week compared with ≤8 million units weekly). Further studies will be needed to establish the potential of intraperitoneal IFN gamma for the treatment of patients with ovarian cancer.

Intraperitoneal adoptive immunotherapy and therapy with interleukin-2 (IL-2)

IL-2 given systemically with and without LAK cells or tumour-infiltrating lymphocytes, has been the subject of intense investigation in various forms of human cancer.[27–32] *In vivo* studies that evaluated the response of ovarian tumour cells implanted in nude mice[113] or the effect of LAK cells in a murine model[114] showed that treatment with LAK cells plus IL-2 can significantly prolong survival. Intraperitoneal IL-2 has been the subject of several studies,[81 115 116] and a major justification for using intraperitoneal IL-2 is the finding *in vitro* that IL-2 activity against malignant tumours is increased with increasing drug concentrations.

In a phase 1 trial reported from the National Cancer Institute, IL-2 (25 000 U/kg every eight hours) was given intraperitoneally with LAK cells after systemic priming with IL-2,[79 80] and there was a 100-fold increase in exposure of the peritoneal cavity to IL-2, compared with that when given systemically.[79] LAK activity was detectable in the peritoneal cavity during the entire period of treatment. Partial clinical responses were observed in several patients with both ovarian and colon cancers,[80] but there was considerable toxicity including fever, chills, emesis, hypotension, abdominal pain, fluid retention, bone marrow suppression, and abnormalities of liver function. Infection was common. Several patients developed extensive fibrosis in the peritoneal cavity, probably resulting from the release by IL-2 activated cells of various growth factors and cytokines capable of inducing collagen synthesis by fibroblasts. These side effects might be reduced, either by modification of LAK plus IL-2 regimens, or by the development of more targeted and less non-specific forms of adoptive immunotherapy.[30]

IL-2 has been given intraperitoneally without LAK cells.[81 115 116] There was a major pharmacokinetic advantage for intracavity treatment, with less local and systemic toxicity compared with the National Cancer Institute trial using LAK cells with IL-2. Researchers at the University of Pittsburgh did a phase I-II study in refractory ovarian cancer and used two low-dose regimens of intraperitoneal IL-2.[117] In a preliminary report, 13 patients were evaluable for response and two patients had a complete response. Systemic toxicity was mild and included fever, fatigue, myalgias, diarrhoea, emesis, and abdominal pain. These findings of antineoplastic activity and an acceptable toxicity suggest that intraperitoneal IL-2 alone should be investigated further.

Aoki *et al* reported a study that used tumour-infiltrating lymphocytes for immunotherapy in ovarian cancer.[118] In seven patients with advanced or recurrent epithelial ovarian cancer treated with the adoptive transfer of tumour-infiltrating lymphocytes after a single dose of cyclophosphamide, one complete response and four partial responses were seen. Ten additional patients were treated with a cisplatin-containing chemotherapeutic regimen as well as tumour-infiltrating lymphocytes and there were seven complete responses and two partial responses. Four of the seven patients that had complete responses had no recurrence after 15 months of follow-up. Because tumour-infiltrating lymphocytes were given without IL-2, it seems that tumour-infiltrating lymphocyte-based immunotherapy of ovarian cancer may achieve complete response rates without IL-2, so this use of tumour-infiltrating lymphocytes may be more promising for adoptive immunotherapy than LAK-based treatment.

Conclusions

Various immune mechanisms can have important roles in effective antitumour responses, including direct cytotoxicity against tumour cells, as well as other mechanisms such as action of cytotoxic or immune-enhancing lymphokines. Immunotherapy for ovarian cancer has been limited in scope and responses. Preliminary studies have indicated that immune enhancement leading to tumour rejection is most likely to occur when the various biological response modifiers are brought into direct contact with tumours, when the tumour burden is minimal (as in an adjuvant setting) or when combined with cytotoxic chemotherapy, or both.

Recent advances in biotechnology have provided large amounts of relatively pure substances that can be used in clinical trials, so newly described cytokines and monoclonal antibodies will be available in sufficient quantities to permit appropriate clinical trials of these agents. However, much additional laboratory research will be needed to develop a complete understanding of the biological activities of these substances. Adoptive immunotherapy has also created new opportunities for immunotherapy in ovarian cancer. We are clearly entering a period of extensive development and testing of these agents, both alone and in combination with adoptive immunotherapy, in the experimental treatment of ovarian cancer.

References

1 Abbas AK, Lictman AH, Pober JS. *Cellular and molecular immunology*. Philadelphia: WB Saunders Company, 1991.
2 Roitt I, Brostoff J, Male D. *Immunology*. 2nd ed. London: Gower Medical Publishing, 1989.
3 Boyer CM, Knapp RC, Bast RC. Immunology and immunotherapy. In: Berek JS, Hacker NF, eds. *Practical gynecologic oncology*. Baltimore: Williams & Wilkins, 1989: 73–108.
4 Kohler G, Milstein C. Continuous cultures of fused cells secreting antibody of predefined specificity. *Nature* 1978; **256**: 495–7.
5 Ettenson D, Sheldon K, Marks A, Houston LL, Baumal R. Comparison of growth inhibition of a human ovarian adenocarcinoma cell line by free monoclonal antibodies and their corresponding antibody-recombinant ricin A chain immunotoxins. *Anticancer Res* 1988; **8**: 833–8.
6 Ortaldo JR, Herberman RB. Heterogeneity of natural killer cells. *Ann Rev Immunol* 1984; **2**: 359.
7 Bast RC, Zbar B, Borsos T, Rapp RJ, BCG and cancer. *N Engl J Med* 1974; **290**: 1413–58.
8 Borstein RS, Mastrangelo MJ, Sulit H. Immunotherapy of melanoma with intralesional BCG. *National Cancer Institute Monographs* 1973; **39**: 213–20.
9 Di Giovine FS, Duff GW. Interleukin 1: the first interleukin. *Immunol Today* 1990; **11**: 13–20.
10 Hirano T, Akira S, Taga T, Kishimoto T. Biological and clinical aspects of interleukin 6. *Immunol Today* 1990; **11**: 443–9.
11 Watson JM, Sensintaffar JL, Berek JS, Martínez-Maza O. Epithelial ovarian cancer cells constitutively produce interleukin-6 (IL6). *Cancer Res* 1990; **50**: 6959–65.
12 Kawano M, Hirano T, Matsuda T, *et al*. Autocrine generation and requirement of BSF-2/IL-6 for human multiple myelomas. *Nature* 1988; **332**: 83–5.
13 Miles SA, Rezai AR, Salazar-Gonzalez JF, *et al*. AIDS Kaposi's sarcoma-derived cells produce and respond to interleukin-6. *Proc Natl Acad Sci USA* 1990; **87**: 4068–72.
14 Miki S, Iwano M, Miki Y, *et al*. Interleukin-6 (IL-6) functions as an in vitro autocrine growth factor in renal cell carcinomas. *FEBS Lett* 1989; **250**: 607–10.
15 Wu S, Rodabaugh K, Martínez-Maza O, *et al*. Stimulation of ovarian tumor cell proliferation with monocyte products including interleukin-1, interleukin-6 and tumor necrosis factor-alpha. *Am J Obstet Gynecol* 1992; **166**: 997–1007.
16 Martínez-Maza O, Berek JS. Interleukin 6 and cancer therapy. *In Vivo* 1991, **5**: 583.
17 Mulé JJ, McIntosh JK, Jablons DM, Rosenberg SA. Antitumor activity of recombinant interleukin 6 in mice. *J Exp Med* 1990; **171**: 629–36.
18 Fiorentino DF, Bond MW, Mosmann TR. Two types of mouse helper T cell. IV. Th2 clones secrete a factor that inhibits cytokine production by Th1 clones. *J Exp Med* 1989; **170**: 2081–95.
19 Mosmann TR, Moore KW. The role of IL-10 in crossregulation of TH1 and TH2 responses. *Immunol Today* 1991; **12**: A49.
20 Del Prete GF, De Carli M, Ricci M, Romagnani S. Helper activity for immunoglobulin synthesis of T helper type 1 (Th1) and Th2 human T cell clones: the help of Th1 clones is limited by their cytolytic capacity. *J Exp Med* 1991; **174**: 809–13.
21 Romagnani S. Human TH1 and TH2 subsets: doubt no more. *Immunol Today* 1991; **12**: 256–7.

22 Zlotnik A, Moore KW. Interleukin 10. *Cytokine* 1991; **3**: 366–71.

23 Fiorentino DF, Zlotnik A, Mosmann TR, Howard M, O'Garra A. IL-10 inhibits cytokine production by activated macrophages. *J Immunol* 1991; **147**: 3815–22.

24 Bogdan C, Vodovotz Y, Nathan C. Macrophage deactivation by IL-10. *J Exp Med* 1991; **174**: 1549–55.

25 de Waal Malefyt R, Abrams J, Bennett B, Figador CG, de Vries JE. IL-10 inhibits cytokine synthesis by human monocytes: an autoregulatory role of IL-10 produced by monocytes. *J Exp Med* 1991; **174**: 1209–20.

26 Berek JS. Epithelial ovarian cancer. In: Berek JS, Hacker NF, eds. *Practical gynecologic oncology*. Baltimore: Williams and Wilkins, 1989: 327–64.

27 Gotlieb WH, Abrams JS, Watson JM, Velu T, Berek JS, Martínez-Maza O. Presence of IL-10 in the ascites of patients with ovarian and other intra-abdominal cancers. *Cytokine* 1992; **4**: 385–90.

28 Watson JM, Gotlieb WH, Abrams JH, *et al.* Cytokine profiles in ascitic fluid from patients with ovarian cancer: relationship to levels of acute phase proteins and immunoglobulins, immunosuppression and tumor classification. *Proceedings of the Society of Gynecologic Investigation* 1993; **136**: 8.

29 Shrader JW. The panspecific hemopoitin of activated T lymphocytes (interleukin-3). *Ann Rev Immunol* 1986; **4**: 205–30.

30 Golub SH. Immunological and therapeutic effects of interferon treatment of cancer patients. *Clinical Immunology and Allergy* 1984; **4**: 377–41.

31 Rosenberg SA. Immunotherapy of cancer by systemic administration of lymphoid cells plus interleukin-2. *Journal of Biologic Response Modifiers* 1984; **3**: 501–11.

32 Rosenberg SA, Lotze MT. Cancer immunotherapy using interleukin-2 and interleukin-2 activated lymphocytes. *Annu Rev Immunol* 1986; **4**: 681–709.

33 Rosenberg SA, Lotze MT, Muul LM, *et al.* Observations on the systemic administration of autologous lymphokine-activated killer cells and recombinant interleukin-2 to patients with metastatic cancer. *N Engl J Med* 1985; **313**: 1485–92.

33 Cappuccini F, Yamamoto RS, DiSaia PJ, *et al.* Identification of tumor necrosis factor and lymphotoxin blocking factor(s) in the ascites of patients with advanced and recurrent ovarian cancer. *Lymphokine Cytokine Res* 1991; **10**: 225–9.

34 Rosenberg SA, Lotze MT, Muul LM, *et al.* A progress report on the treatment of 157 patients with advanced cancer using lymphokine-activated killer cells and interleukin-2 or high-dose interleukin-2 alone. *N Engl J Med* 1987; **316**: 889–97.

35 West WH, Tauer KW, Yannelli JR, *et al.* Constant-infusion recombinant interleukin-2 in adoptive immunotherapy of advanced cancer. *N Engl J Med* 1987; **316**: 898–905.

36 Berek JS. Intraperitoneal adoptive immunotherapy for peritoneal cancer. *J Clin Oncol* 1990; **8**: 1610–12.

37 Topalian SL, Solomon D, Avis FP, *et al.* Immunotherapy of patients with advanced cancer using tumor infiltrating lymphocytes and recombinant interleukin 2: A pilot study. *J Clin Oncol* 1988; **6**: 839–53.

38 Lotzova E. Role of human circulating and tumor-infiltrating lymphocytes in cancer defense and treatment. *Natural Immunity and Cell Growth Regulation* 1990; **9**: 253–64.

39 Garrido MA, Valdayo MJ, Winkler DR, *et al.* Targeting human T-lymphocytes with bispecific antibodies to react against human ovarian carcinoma cells growing in nu/nu mice. *Cancer Res* 1990; **50**: 4227–32.

40 Bookman MA, Bast RC Jr. The immunobiology and immunotherapy of ovarian cancer. *Semin Oncol* 1991; **18**: 270–91.

41 Zighelboim J, Nio Y, Berek JS, Bonavida B. Immunologic control of ovarian cancer. *Natural Immunity Growth Regulation* 1988; **7**: 216–25.

42 Berek JS, Martínez-Maza O, Montz FJ. The immune system and gynecologic cancer. In: Coppelson M, Tattersall M, Morrow CP, eds. *Gynecologic Oncology*. Edinburgh: Churchill Livingstone, 1992: 119.

43 Berek JS, Lichtenstein AK, Knox RM, *et al.* Synergistic effects of combination sequential immunotherapies in a murine ovarian cancer model. *Cancer Res* 1985; **45**: 4215–18.

44 Bast RC, Klug T, St. John E, *et al.* Monitoring growth of human ovarian carcinoma with a radioimmunoassay for antigen(s) defined by a murine antibody (OC125). *N Engl J Med* 1983; **309**: 883–7.

45 Slamon DJ, Godolphin W, Jones LA, *et al.* Studies of the HER-2/neu proto-oncogene in human breast and ovarian cancer. *Science* 1989; **244**: 707–12.

46 Lichtenstein A, Berenson J, Gera JF, Waldburger K, Martíez-Maza O, Berek JS. Resistance of human ovarian cancer cells to tumor necrosis factor and lymphokine-activated killer cells: correlation with expression of HER2/neu oncogenes. *Cancer Res* 1990; **50**: 7364–70.

47 Fendly BM, Kotts C, Vetterlein D, *et al*. The extracellular domain of HER2/neu is a potential immunogen for active specific immunotherapy of breast cancer. *Journal of Biologic Response Modifiers* 1990; **9**: 449–55.

48 Shepart HM, Lewis GD, Sarup JC, *et al*. Monoclonal antibody therapy of human cancer: taking the HER2 protooncogene to the clinic. *J Clin Immunol* 1991; **11**: 117–27.

49 Khoo SK, MacKay EV. Immunologic reactivity of female patients with genital cancer: Status in preinvasive, locally invasive and disseminated disease. *Am J Obstet Gynecol* 1974; **119**: 1018–25.

50 Muruhata RI, Cantrell J, Lichtenstein A, Zighelboim J. Disassociation of biological activities of Corynebacterium parvum by chemical fractionation. *Int J Immunopharmacol* 1980; **2**: 47–53.

51 Halpern B. *Corynebacterium parvum. Applications in experimental and clinical oncology.* New York: Plenum, 1975.

52 Herberman RB. *Natural cell-mediated immunity against tumors.* New York: Academic Press, 1980.

53 Scott, MT. Corynebacterium parvum as an immunotherapeutic anti-cancer agent. *Semin Oncol* 1984; **1**: 367–8.

54 Berek JS, Cantrell JL, Lichtenstein AK, *et al*. Immunotherapy with biochemically dissociated fractions of proprionebacterium acnes in a murine ovarian cancer model. *Cancer Res* 1984; **44**: 1871–5.

55 Alberts DS, Salmon ES, Moon TE. Chemoimmunotherapy for advanced ovarian cancer with Adriamycin-cyclophosphamide ± BCG: Early report of a Southwest Oncology Group study. *Recent Results Cancer Res* 1978; **68**: 160–5.

56 Bast RC, Berek JS, Obrist R, *et al*. Intraperitoneal immunotherapy of human ovarian carcinoma with Corynebacterium parvum. *Cancer Res* 1983; **43**: 1395–1401.

57 Gall SA, DiSaia PJ, Schmidt H, Mittlestaedt L, Newman P, Creasman W. Toxicity manifestation following intravenous Corynebacterium parvum administration to patients with ovarian and cervical carcinoma. *Am J Obstet Gynecol* 1978; **132**: 555–60.

58 Montovani A, Sessa C, Peri G, *et al*. Intraperitoneal administration of Corynebacterium parvum by chemical fractionation. *Int J Immunopharmacol* 1981; **2**: 437–46.

59 Rao B, Wanebo HJ, Ochoa M, Lewis JL, Oettgen HK. Intravenous C. parvum: An adjuvant to chemotherapy for resistant advanced ovarian carcinoma. *Cancer* 1977; **39**: 514–26.

60 Wanebo HJ, Ochoa M. Randomized chemoimmunotherapy trial of CAF and intravenous C. parvum for resistant ovarian cancer-preliminary results. *Proceedings of the American Association for Cancer Research* 1977; **18**: 225.

61 Creasman WT, Gall SA, Blessing JA, *et al*. Chemoimmunotherapy in the management of primary stage III ovarian cancer: A Gynecologic Oncology Group study. *Cancer Treatment Reports* 1979; **68**: 319–23.

62 Gynecologic Oncology Group (GOG) *Statistical Report* 1983.

63 Hanna MG, Key ME. Immunotherapy of metastases enhances subsequent chemotherapy. *Science* 1982; **217**: 367–9.

64 Graham JB, Graham RM. The effect of vaccine on cancer patients. *Surg Gynecol Obstet* 1962; **114**: 1.

65 Hudson CN, Levin L, McHaudy JE, *et al*. Active specific immunotherapy for ovarian cancer. *Lancet* 1976; **2**: 877–9.

66 Julliard GJF, Boyer PJ, Yamashiro CH. A phase I study of active specific intralymphatic immunotherapy (ASILI). *Cancer* 1978; **41**: 2215–25.

67 Epstein LB, Shen JT, Abele JS, Reese CC. Sensitivity of human ovarian carcinoma cells to IFN and other antitumor agents as assessed by an in vitro semi-solid agar technique. *Ann NY Acad Sci* 1980; **350**: 228–35.

68 Abdulhay G, DiSaia PJ, Blessing JA, Creasman WT. Human lymphoblastoid interferon in the treatment of advanced epithelial ovarian malignancies: A Gynecologic Oncology Group Study. *Am J Obstet Gynecol* 1985; **152**: 418–23.

69 Einhorn N, Cantrell K, Einhorn S, Strander H. Human leukocyte interferon for advanced ovarian carcinoma. *Am J Clin Oncol* 1982; **5**: 167–72.

70 Niloff TM, Knapp RC, Jones G, *et al*. Recombinant leukocyte alpha interferon in advanced ovarian carcinoma. *Cancer Treatment Reports* 1985; **69**: 895–96.

71 Thompson JA, Lee DJ, Lindgren CG, *et al.* Influence of dose and duration of infusion of interleukin-2 on toxicity and immunomodulation. *J Clin Oncol* 1988; **6**: 669–78.

72 Markman M, Howell SB. Intraperitoneal chemotherapy for ovarian cancer. In: Alberts DS, Surwit EA, eds. *Ovarian cancer.* Boston: Martinus Nijhoff, 1985; 179–212.

73 Howell SB, Kirmani S, Lucas WE, *et al.* A phase II trial of intraperitoneal cisplatin and etoposide for primary treatment of ovarian epithelial cancer. *J Clin Oncol* 1990; **8**: 137–45.

74 Berek JS, Bast RC, Hacker NF, *et al.* Lymphocyte cytotoxicity in the peritoneal cavity and blood of patients with ovarian cancer. *Obstet Gynecol* 1984; **64**: 708–14.

75 Berek JS, Knapp RC, Hacker NF, *et al.* Intraperitoneal immunotherapy of epithelial ovarian carcinoma with Corynebacterium parvum. *Am J Obstet Gynecol* 1985; **152**: 1003–10.

76 Lichtenstein AK, Spina C, Berek JS, *et al.* Intraperitoneal administration of human recombinant alpha-interferon in patients with ovarian cancer: effects on lymphocytes, phenotype, and cytotoxicity. *Cancer Res* 1988; **48**: 5853–9.

77 Boyer P, Berek JS, Zighelboim J. Lymphocyte activation by recombinant interleukin-2 in ovarian cancer patients. *Obstet Gynecol* 1989; **73**: 793–7.

78 Martínez-Maza O, Berek JS. *Immunotherapy of ovarian cancer.* In: Markman H, Hoskins WJ, eds. *Cancer of the ovary.* New York: Raven Press, 1993: 301.

79 Urba W, Clark JW, Steis RG, *et al.* Intraperitoneal lymphokine-activated killer cell/interleukin-2 therapy in patients with intra-abdominal cancer: immunologic considerations. *J Natl Cancer Inst* 1989; **81**: 602–11.

80 Steiss RG, Urba WJ, Vander Molen LA, *et al.* Intraperitoneal lymphokine-activated killer cell and interleukin-2 therapy for malignancies limited to the peritoneal cavity. *J Clin Oncol* 1990; **8**: 1618–29.

81 Chapman PB, Kolitz JE, Hakes T, *et al.* A phase I trial of intraperitoneal recombinant interleukin-2 in patients with ovarian cancer. *Invest New Drugs* 1988; **6**: 179–88.

82 Berek JS, Hacker NF, Lichtenstein AK, *et al.* Intraperitoneal recombinant alpha-interferon for salvage immunotherapy in stage III epithelial ovarian cancer: A Gynecologic Oncology Study group. *Cancer Res* 1985; **45**: 4447–53.

83 Lichtenstein A, Berek JS, Bast RC, *et al.* Activation of peritoneal lymphocyte cytotoxicity in patients with ovarian cancer by intraperitoneal treatment with Corynebacterium parvum. *Journal of Biologic Response Modifiers* 1984; **3**: 1–8.

84 Order SE, Rosenshein N, Klein JL, Leibel S, Pino Y Torres J, Ettinger D. The integration of new therapies and radiation in management of ovarian cancer. *Cancer* 1981; **48**: 590–6.

85 Ohkawa K, Ohkawa R. Locoregional immunotherapy and chemotherapy in advanced ovarian cancer cytotoxicity. *Asian Oceania J Obstet Gynaecol* 1981; **Oct**: 352–6.

86 D'Acquisto R, Markman M, Hakes T, Rubin S, Hoskins W, Lewis JL. A phase I trial of intraperitoneal recombinant gamma-interferon in advanced ovarian carcinoma. *J Clin Oncol* 1988; **6**: 689–95.

87 Freedman RS, Gutterman JU, Wharton JT, Rutledge FN. Leukocyte interferon (IFN alpha) in patients with epithelial ovarian carcinoma. *Journal of Biologic Response Modifiers* 1983; **2**: 133–8.

88 Willemse PHB, de Vries EGE, Mulder NH, Aalders JG, Bouma J, Sleijfer DTH. Intraperitoneal human recombinant interferon alpha-2b in minimal residual ovarian cancer. *Eur J Clin Oncol* 1990; **26**: 353–8.

89 Balkwill FR, Moodie EM. Positive interactions between human interferon and cyclophosphamide or adriamycin in a human tumor system. *Cancer Res* 1984; **44**: 904–8.

90 Welander C, Gaines J, Homesley H, Rudnick S. In vitro synergistic effects of recombinant human interferon alpha 2(rIFN-a2) and doxorubicin on human tumor cell lines. *Proceedings of the American Society of Clinical Oncology* 1983; **2**: 42.

91 Inou M, Tan YH. Enhancement of actinomycin-D and cis-diamminedichlorplatinum (II) induced killing of human fibroblasts by human beta interferon. *Cancer Res* 1983; **43**: 5484–8.

92 Aapro MS, Salmon SE, Alberts DC. Schedule dependent synergism of vinlastine and cloned leukocyte interferon A. *Stem Cells* 1981; **1**: 303–4.

93 Le J, Yip YK, Vilcek J. Cytologic activity of interferon gamma and its synergism with 5-FU. *Int J Cancer* 1984; **34**: 495–500.

94 Harabayshi N, Nishiyama M, Yamaguchi M. Assessment of the combined effects of mitomycin C with alpha interferon by the clonogenic assay technique. *Gan To Kagaku Ryoho* 1982; **12**: 1056–62.

 95 Carmichael J, Fergusson RJ, Wolf CR, Balkwill FR, Smyth JF. Augmentation of cytotoxicity of chemotherapy by human alpha interferons in human non-small cell lung cancer zenograsfts. *Cancer Res* 1986; **46**: 4916–20.

 96 Bezwoda WR, Golombick T, Dansey R, Keeping J. Treatment of malignant ascites due to recurrent/refractory ovarian cancer: the use of interferon-alpha or interferon-alpha plus chemotherapy. In vivo and in vitro observations. *Eur J Cancer* 1991; **27**: 1423–9.

 97 Nardi M, Cognetti F, Pollera CF, *et al*. Intraperitoneal recombinant alpha-2-interferon alternating with cisplatin as salvage therapy for minimal residual disease ovarian cancer: a phase II study. *J Clin Oncol* 1990; **8**: 1036–41.

 98 Berek JS, Welander C, Schink JC, Grossberg H, Montz FJ, Zighelboim J. A phase I-II trial of intraperitoneal cisplatin and alpha-interferon in patients with residual epithelial ovarian cancer. *Gynecol Oncol* 1991; **40**: 237–43.

 99 Markman M, Berek JS, Blessing JA, McGuire WP, Bell J, Homesley HD. Characteristics of patients with small-volume residual ovarian cancer resistant to platinum-based intraperitoneal chemotherapy: lessons learned from a phase II Gynecologic Oncology Group trial of intraperitoneal cisplatin and alpha interferon. *Gynecol Oncol* 1992; **45**: 3–8.

100 Bonavida B, Tsuchitani T, Zighelboim J, Berek JS. Synergy is documented in vitro with low-dose tumor necrosis factor, cisplatin, and doxorubicin in ovarian tumor cells. *Gynecol Oncol* 1990; **38**: 333–9.

101 Oettgen HF, Old LJ. Tumor necrosis factor. In: DeVita VT, Hellman S, Rosenberg SA, eds. *Important advances in oncology*. Philadelphia: Lippincott, 1987; 105–30.

102 Old LJ. Tumor necrosis factor. *Science* 1985; **230**: 630–2.

103 Chapman PB, Lester TJ, Casper ES. Clinical pharmacology of recombinant human tumor necrosis factor in patients with advanced cancer. *J Clinical Oncol* 1987; **5**: 1942–51.

104 Creagan ET, Kovach JS, Moertel CG, Frytaks S, Kvols LK. A phase 1 clinical trial of recombinant human tumor necrosis factor. *Cancer* 1988; **62**: 2467–71.

105 Feinberg B, Kurzrock R, Talpaz M, Blick M, Saks S, Gutterman JU. A phase 1 trial of intravenously administered recombinant tumor necrosis factor-alpha in cancer patients. *J Clin Oncol* 1988; **6**: 1328–34.

106 Spriggs DR, Sherman MK, Michie H, *et al*. Recombinant human tumor necrosis factor administered as a 24-hr intravenous infusion. A phase 1 and pharmacologic study. *J Natl Cancer Inst* 1988; **80**: 1039–44.

107 Markman M, *et al*. Phase I trial of recombinant tumor necrosis factor administered by the intraperitoneal route. *Reg Cancer Treat* 1989; **2**: 174–7.

108 Raeth U, Schmid H, Karck U, Kempeni J, Schlick E, Kaufmann M. Phase II trial of recombinant human tumor necrosis factor alpha in patients with malignant ascites from ovarian carcinomas and non-ovarian tumors with intraperitoneal spread. *Proceedings of the American Society of Clinical Oncology* 1991; **10**: 187.

109 Belardelli F, Gresser I, Maury C, Manoury M-T. Antitumor effects of interferon in mice injected with interferon-sensitive and interferon resistant Friend leukemia cells. *Int J Cancer* 1982; **30**: 813–20.

110 Rubin BY, Gupta SL. Differential efficacies of human type I and type II interferons as antiviral and antiproliferative agents. *Proc Natl Acad Sci USA* 1980; 77: 5928–32.

111 Crane JL, Glasgow LA, Kern ER, Youngner JS. Inhibition of murine osteogenic sarcomas by treatment with type I and type II interferons. *J Natl Cancer Inst* 1978; **61**: 871–4.

112 Pujade-Lauraine E, *et al*. Intraperitoneal human recombinant interferon gamma in patients with residual ovarian carcinoma at second look laparotomy. *Proceedings of the American Society of Clinical Oncology* 1990; **9**: 156.

113 Oomori K, Kikuchi Y, Miyauchi M, *et al*. Effects of lymphokine-activated killer cells and interleukin-2 on the ascites formation and the survival time of nude mice bearing human ovarian cancer cells. *J Cancer Res Clin Oncol* 1989; **115**: 217–20.

114 Ottow RT, Steller EP, Sugarbaker PH, Wesley RA, Rosenberg SA. Immunotherapy of intraperitoneal cancer with interleukin 2 and lymphokine-activated killer cells reduces tumor load and prolongs survival in murine models. *Cell Immunol* 1987; **104**: 366–76.

115 Markman M, *et al*. A pilot trial of daily intraperitoneal administration of recombinant interleukin-2. *Reg Cancer Treat* 1990; **3**: 44–6.

116 Lembersky B, Baldisseri M, Kunschner A, *et al*. Phase I-II study of intraperitoneal low dose interleukin-2 in refractory stage III ovarian cancer. *Proceedings of the American Society of Clinical Oncology* 1989; **8**: 163.

117 Mule JJ, Shu S, Rosenberg SA. The anti-tumor efficacy of lymphokineactivated killer cells and recombinant interleukin 2 in vivo. *J Immunol* 1985; **135**: 646–52.
118 Aoki Y, Takakuwa K, Kodama S, *et al*. Use of adoptive transfer of tumor-infiltrating lymphocytes alone or in combination with cisplatin-containing chemotherapy in patients with epithelial ovarian cancer. *Cancer Res* 1991; **51**: 1934–9.

17 Minimal access surgery in ovarian cancer

EARL A SURWIT, JOEL M CHILDERS

The operations that are done for women with ovarian cancer include surgical staging, debulking, and second-look laparotomies. The staging of patients with ovarian cancer requires a thorough evaluation of both the peritoneal cavity and the retroperitoneum. Can laparoscopy adequately evaluate these areas? Work by a few investigators suggests that it has potential in many of them.

Historical perspective

With the advent of modern operative laparoscopic equipment and techniques there has been a renewed interest in the role of laparoscopy in the treatment of gynaecological cancers. Before its modernisation, laparoscopy was used to stage ovarian malignancies. Early investigators who attempted to restage patients with presumed localised disease concluded that laparoscopy was useful in discovering undetected metastases.[1-5] It was possible to identify extraovarian metastases laparoscopically in intraperitoneal washings and from omental and diaphragmatic biopsy specimens. Laparoscopic pioneers also evaluated the usefulness of laparoscopy for second-look procedures, and numerous reports confirmed that with laparoscopy, surgeons were able to discover persistent disease in about 28% of the procedures.[4-11] The reported false-negative rate was approximately 36%, however, confirming that with laparoscopy surgeons were less accurate than with laparotomy.

Laparoscopy has now become modernised. New laparoscopy journals and reports on advanced procedures are abundant, yet few data are available on our ability to stage ovarian carcinoma adequately using modern operative laparoscopy. Reich *et al* first described laparoscopic staging of a patient with presumed stage I ovarian carcinoma.[12] Their patient refused laparotomy but consented to laparoscopy. She was managed with a laparoscopic-assisted vaginal hysterectomy, bilateral salpingo-oophorectomy, omentectomy, and pelvic lymphadenectomy. No extraovarian disease was discovered. Because the diaphragm and para-aortic lymph nodes were not biopsied, this staging was criticised for not meeting oncological standards for staging.[13]

248

Querleu and LeBlanc published the first report about laparoscopic staging of adnexal malignancies, in which they described an adequate staging procedure.[14] Nine patients were staged for presumed stage I invasive carcinomas and five for tumours of low malignant potential or restaged (second look) after chemotherapy. Persistent metastatic right diaphragmatic disease was identified at one second-look laparoscopy. There were no complications and the mean hospital stay was 2.8 days.

Historically, surgical staging of patients with "apparent" stages I and II ovarian carcinoma resulted in the discovery of occult metastatic disease in as many as 28% of patients.[15-22] Reports indicate that occult metastases in patients with presumed localised disease are found in several high-risk areas: omentum, diaphragmatic peritoneum, peritoneal cytology specimens, and pelvic and para-aortic nodes. Young *et al*, in a cooperative national study in which 100 patients with apparent stage I and stage II disease underwent surgical staging, upstaged 15% (8/49) of their apparent stage I cases and 43% (22/51) of their apparent stage II cases.[15] The Gynecologic Oncology Group ovarian cancer staging study upstaged 9/97 apparent stage I cases.[16] About 1% had signs of involvement in random biopsy specimens of the diaphragm, omentum, or pelvic lymph nodes. Of the 97 subjects 4% had subclinical para-aortic node involvement. Over the last 15 years we have learned that patients with ovarian carcinoma are more than likely to have involved lymph nodes, even patients with clinically localised disease. Table 17.1 summarises some studies on nodal involvement in patients with clinical stage I ovarian carcinoma. The overall rate of involvement of 11% (23/219) is consistent with the results of our study (9%).

Table 17.1 Incidence of nodal involvement in clinical stage I ovarian carcinoma

First author	Reference No	No of patients evaluated	No of patients with involved nodes (%)
Knapp	17	24	3 (12)
Delgado	18	5	1 (20)
Musumeci	19	24	5 (21)
Piver	9	1	0
Chen	20	11	2 (18)
Wu	21	9	2 (22)
Buchsbaum	16	95	4 (4)
Burghardt	22	37	9 (24)
Young	15	52*†	6 (12)
Young	15	11*§	1 (9)
Childers		11	1 (9)
Total		284	31 (12)

* Includes patients with IIa and IIb disease; † considers para-aortic nodes only; and § considers pelvic nodes only.

By contrast, second-look surgical staging identifies persistent disease much more often (in about half) and in sites other than the high-risk sites, even when high-risk sites contain no carcinoma. The rate of involvement reported by various investigators varies widely (28%–70%), and is dependent on the original stage of disease, extent of residual disease after primary cytoreductive surgery, and the histological grade of the tumour.[23-24] Current reports indicate that about half of the patients who have undergone optimal cytoreductive surgery and platinum-based adjuvant chemotherapy will have persistent disease if they are restaged at second-look laparotomy. The inadequacy of this procedure is highlighted by the fact that about a half of patients with no sign of invasion at second-look laparotomy will have recurrent disease.

Second-look laparoscopy

The accuracy of laparoscopic surgical staging of ovarian cancer was discussed by Childers *et al* (Childers JM *et al*. Laparoscopic staging of ovarian cancer. Presented at 26th annual meeting of the Society of Gynecologic Laparoscopists, San Francisco, 1995.) Our experience with staging included both patients with advanced disease who were eligible for second-look procedures and patients with presumed stage I disease who needed surgical staging. Our initial experience with second-look laparoscopy at the University of Arizona encompasses 44 procedures in 40 patients with advanced ovarian cancer, all of whom had definitive cytoreductive surgery and platinum-based chemotherapy and were considered without evidence of disease, with negative examinations and/or negative CT examinations prior to second look. The patients' ages ranged from 27 to 80 years, with a mean of 61 years, and they weighed from 47.7 to 87.0 kg, with a mean of 62.3 kg.

The second-look procedure is the same as that done through a laparotomy, with specimens for peritoneal cytology taken from the pelvis, paracolic gutters, and right diaphragm. Extensive adhesiolysis is often required before the entire peritoneal cavity can be evaluated. Any suspicious areas are then biopsied. If no obvious disease is encountered, multiple random biopsy specimens are obtained with emphasis on high-risk areas including the paracolic gutter, omentum, and diaphragm. The diaphragm can better be evaluated laparoscopically than by laparotomy, and the large peritoneal patches that can be peeled off the diaphragm are better specimens of the diaphragm than are generally obtained at laparotomy. If there is no evidence of intraperitoneal disease, sampling of pelvic and para-aortic lymph nodes is required.

In patients with ovarian cancer it is extremely important to evaluate the lymph nodes that are high in the para-aortic chain where the lymph from the ovarian cancer can drain directly. As a consequence, it becomes necessary to mobilise the duodenum, identify the inferior mesenteric artery, identify the ovarian artery, visualise the renal veins, and do a high para-aortic lymphadenectomy (figs 17.1–17.4).

In our series, persistent disease was found laparascopically in 24 of the 44 procedures. Interestingly, five of these showed microscopic disease only—that is, they had no evidence of gross disease at laparoscopy but had

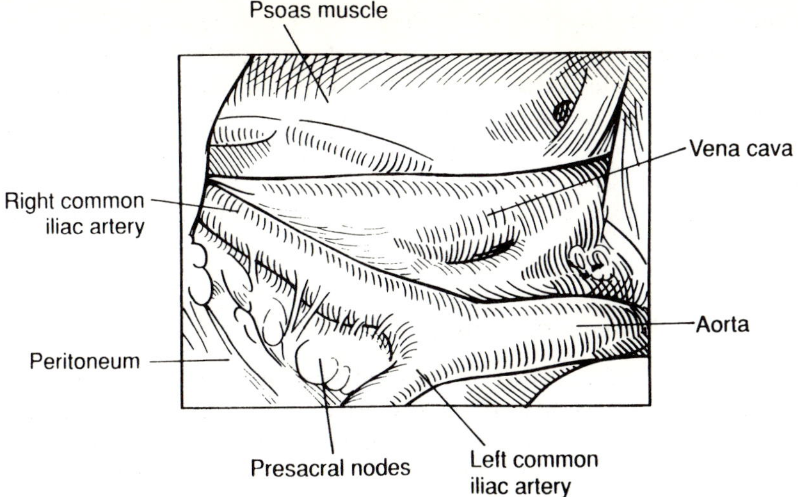

Fig. 17.1 After completion of a low right para-aortic lymphadenectomy the ureter and the ovarian vessels are retracted upwards (out of the drawing) and the entire field is inspected for adequate haemostasis (as viewed from the umbilicus).

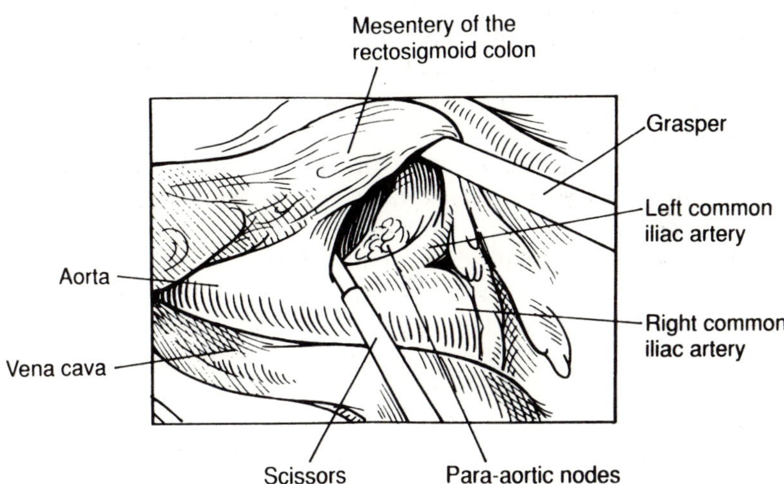

Fig. 17.2 During a low left para-aortic lymphadenectomy the mesentery of the rectosigmoid (which contains the inferior mesenteric artery) is lifted off the aorta by blunt dissection. Lateral dissection is then continued until the left ureter, ovarian vessels, and psoas muscle are identified.

signs of involvement in biopsy specimens one of which was in the omentum, two in para-aortic nodes, and one on the pelvic peritoneum. In one patient only the peritoneal washings indicated disease. This overall incidence of positive findings (56%) is consistent with that found at second-look

251

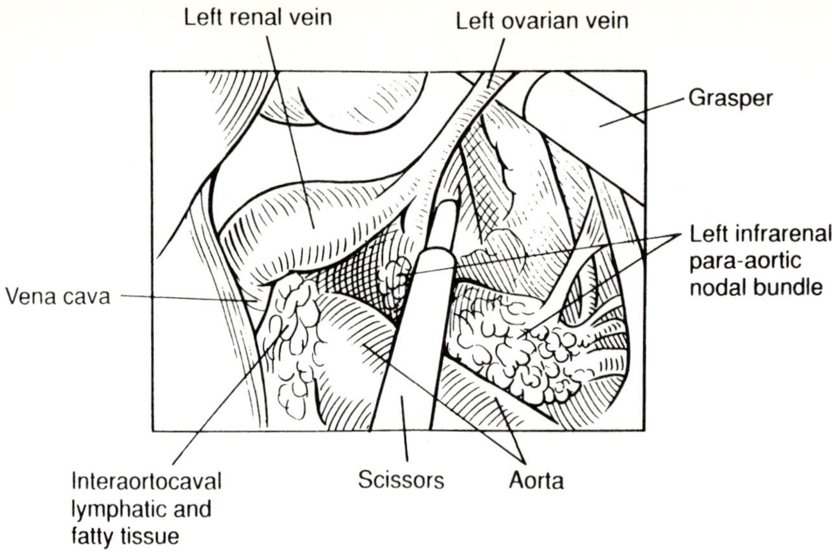

Fig. 17.3 To remove the left infrarenal para-aortic nodes to the lateral extent of the dissection the left ovarian vein must be separated from the nodal tissue.

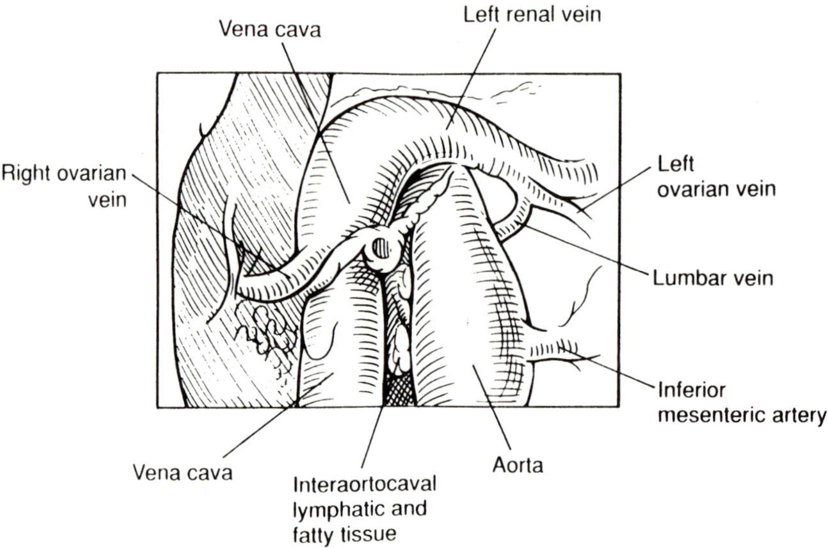

Fig. 17.4 On completion of the infrarenal para-aortic lymphadenectomy, vascular structures are easily identified. The surgeon removes any remaining tissue as desired and adequate haemostasis is assured.

laparotomies, and in our opinion clearly documents our ability to identify disease in these patients by laparoscopy.

Looking at the patients who had negative second-look procedures, an average of 14 areas had biopsy specimens taken. There were five patients in whom we felt we could not do an adequate second-look procedure. We chose not to do laparotomies in these patients but rather accept that the second-look was inadequate. In four of these we could not evaluate the peritoneal cavity because of extensive adhesions; in the other, we could not do a para-aortic lymphadenectomy. Interestingly, all five of these patients had developed recurrences at seven to 13 months.

Because all of these patients had had extensive previous surgery, they were at high risk of complications from the laparoscopy. Two patients had enterotomies repaired at laparotomy, and one patient had a hole in the vena cava repaired. One patient had an abscess that was drained under computed tomographic guidance, and one patient had a haematoma and fever that required treatment with antibiotics. In the 36 uncomplicated procedures, the mean hospital stay was 1.3 days. Now, nearly all of our second-look operations are outpatient procedures.

We think that second-look laparoscopy is an adequate staging procedure, with findings of disease in 56% of the patients. The procedure is safe, (with a 7% laparotomy rate) and it offers a shortened—or out-patient—hospital stay.

Staging early ovarian cancer

After a year and a half of doing second-look laparoscopies, we thought we could detect extraovarian disease adequately in patients with apparently localised cancer. The same high-risk areas were evaluated: washings, diaphragm, omentum, lymph nodes, and the pelvic peritoneum. We think that the para-aortic nodes are the most important area to evaluate in these patients, as they are the most likely site for metastatic disease to develop (figs 17.1–17.4). Our experience in laparoscopic para-aortic lymph-adenectomy has shown that it is a safe and adequate technique.[25]

Our experience is still small, but we have staged 11 patients with presumed early ovarian cancers, seven of which were endometrioid carcinomas, one a Sertoli-Leydig cell tumour, and three papillary serous tumours. (Childers JM *et al*. Laparoscopic staging of ovarian cancer. Presented at the 26th annual meeting of the Society of Gynecologic Laparoscopists, San Francisco, 1995.) Tumour grade varied, with five grade 1, three grade 2, and three grade 3 tumours. The ages of the patients ranged from 17 to 72 years with a mean of 65; their weights ranged from 50–91 kg with a mean of 75. Findings at laparoscopic staging showed that two patients had malignant cells in peritoneal washings, two had signs of involvement in pelvic biopsy specimens, one had fallopian tube involvement, and in one a para-aortic node showed signs of tumour. High para-aortic nodes (to the venal vein) were removed in all these cases.

Our early experience in the staging of early ovarian cancer has convinced us that it is an adequate staging procedure and that it is safe in that we encountered no major complications in the 11 patients. The procedures were associated with a short hospital stay, ranging from one to three nights with a mean of 1.7 days. An average stay of 2.2 days was noted. The five patients who underwent hysterectomy had a mean stay of 2.2 days and the six patients who did not have hysterectomies stayed a mean of 1.3 days.

Many patients with presumed early ovarian cancer are referred to oncologists without adequate surgical staging. Laparoscopy offers these patients a surgical staging procedure with a low morbidity and short hospital stay. We think this is a better option for these patients than being given chemotherapy empirically, because some patients with grade 1 or 2 disease might not require chemotherapy if the staging shows no signs of cancer. In addition, there will be occasional patients who are upstaged to either stage II or stage III disease and who should then receive multi-agent paclitaxel (taxol) and platinum-based chemotherapy.

If laparoscopic surgical staging of ovarian cancer is an adequate procedure, then it should be possible to manage patients with suspicious adnexal masses as well, and during the last four years we have attempted to do so. Many of the presumed stage I ovarian cancer patients who were staged laparoscopically came from the group with adnexal masses. Our experience in managing more than 120 patients with suspicious adnexal masses laparoscopically tells us that more than 90% of such masses are benign; most can be managed laparoscopically (>90%); the complication rate is extremely low; and only a short stay in hospital (either outpatient or one-day) is required, again accompanied by a rapid (one to two week) recovery.

Rupture of the tumour capsule

We believe that if a cancer is encountered during laparoscopic management of an adnexal mass, adequate staging can be accomplished either laparoscopically or by laparotomy. The only important issue therefore, if laparoscopy is to be used for managing suspicious adnexal masses, is that of capsular rupture and tumour spill. The only paper that we could find that states that spillage of tumour is associated with a significantly worse prognosis is by Webb *et al*[26]; they described a five year survival rate of 56% in patients whose tumours ruptured compared with 78% among those that did not (p = 0.004). However, 28 patients had adherent tumours, with a survival of 51%, and there is no mention of which groups these patients belonged to. Adherence and extracystic masses had a worse prognosis but were not correlated with rupture in the analysis.[26] In addition, there were more well-differentiated tumours in the group that did not rupture (34% compared with 26% with a survival of 85%), and that too was not taken into account. The weakness of this study is that multivariate analysis was not done. Adherence of the tumours and whether the capsule was intact are probably the two most significant factors that lead to the risk of rupture.

Two decades after the publication by Webb *et al* there are now five reports of multivariate analyses of multiple risk factors: those by Sevelda *et al*,[27] Dembo *et al*,[28] Finn *et al*,[29] Vergote *et al*,[30] and Sjövall *et al*.[31] Not one of these series showed that tumour spill was an independent risk factor in a multivariate analysis. Sjövall *et al*, in a large series of patients, showed that preoperative rupture is a risk factor. Data from the International Federation of Gynecology and Obstetrics (FIGO) have also shown that preoperative rupture has a worse prognosis than that of intraoperative rupture.

We still think that in the management of suspicious adnexal masses, it is important to avoid spill if possible. There are a number of laparoscopic techniques for doing this including placing the mass in a sac and removal or drainage through a colpotomy incision. Leakage or spill can be avoided or greatly minimised. Reported data have suggested that preoperative spill is a risk for patients, so spill followed by a delay in appropriate management of the patient with an early ovarian cancer places that patient into a preoperative spill category that may worsen prognosis. The patient's prognosis may not be worsened if spill is followed by immediate appropriate surgical or chemotherapeutic management, or both.

After evaluating published reports we think that it is justifiable to manage suspicious adnexal masses laparoscopically. The cardinal sin in the laparoscopic management of patients with ovarian carcinoma is, however, to manage the disease incompletely during the initial procedure. If one has the skills to resect the tumour completely and stage the patient laparoscopically, then one should proceed laparoscopically. If one does not have those skills, then laparotomy should be done to resect the tumour completely and stage the patient surgically at that time.

Laparoscopic cytoreductive surgery

We now turn to our experience with laparoscopic debulking in advanced ovarian cancer. (Surwitt EA, *et al*. Neoadjuvant chemotherapy for advanced ovarian cancer. Paper presented at the 26th annual meeting of the Society of Gynecologic Laparoscopists, San Francisco, 1995.) We have attempted this technique only in patients who have had a good response to neoadjuvant chemotherapy. It requires advanced laparoscopic skills, as some of the surgery (the debulking of omental disease) is in the upper abdomen. Experience in retroperitoneal surgery is also required as we must evaluate the pelvic and para-aortic nodes laparoscopically, and occasionally will have to remove disease from this area as well. As a consequence, we think that laparoscopic debulking should be done only by the surgeon with skills and experience in both laparoscopic techniques and in gynaecological oncology.

Laparoscopic debulking in advanced ovarian cancer usually requires resection of the ovarian tumour and an omentectomy. In dissection of the pelvic tumour, identification of the ureters is paramount and often the ureter has to be dissected off the tumour at the pelvic side wall. The best way to do a laparoscopic omental resection is to mobilise the splenic and hepatic flexures so that the transverse colon can be fully mobilised to the

255

midline and delivered through a small mini-laparotomy of 4–5 cm in the upper abdomen in the midline; the remainder of the omentectomy should be done with the transverse colon delivered. We have done several omentectomies completely laparoscopically, and note that it becomes exceedingly tedious and difficult to do the whole operation this way as many of the tumours are densely adherent to the transverse colon.

We have now done laparoscopic debulking for eleven patients. Although our results are early, we think that laparoscopic debulking after neoadjuvant chemotherapy is feasible for some patients with advanced ovarian cancer.

Implantation of tumour at the laparoscopic port site

Implantation of tumour in the abdominal wall after endoscopic procedures in patients with intraperitoneal carcinoma is of concern to some oncological surgeons, but our experience suggests that it should not be.[32] We reviewed 557 laparoscopic procedures done on the gynaecological oncology service between November 1990 and February 1994. In 105 malignancy was documented cytologically or histologically, 88 with intraperitoneal disease and 17 with retroperitoneal disease. Ovarian cancer with intraperitoneal malignancy comprised 80% (70 of 88) of the procedures. Histologically the ovarian carcinomas ranged from low malignant potential to poorly differentiated. Among 88 patients with intraperitoneal disease, 77 had gross disease and 11 had microscopic disease. Four hundred and thirty seven different puncture sites were made in the abdominal wall (38 Verres needle sites and 399 laparoscopic ports). In only one of the 437 (0.2%) abdominal-wall puncture sites did implantation develop, an incidence of 1.0% (one/105) per procedure. The one developed after a second-look laparoscopic procedure for ovarian carcinoma in which only microscopic disease was present. Irrigation of all abdominal puncture sites, including separate Verres needle sites, is easy and may reduce the likelihood of implantation.

We do not think that the threat of implantation in the abdominal wall should deter the ongoing investigation of laparoscopy in patients with intraperitoneal and retroperitoneal carcinoma. The inclusion of malignant conditions as a contraindication to laparoscopy would needlessly preclude many oncological patients from benefiting from the advantages of laparoscopy. Like laparotomy, laparoscopic entry into the abdominal cavity may occur in a field of intraperitoneal disease. Despite this, the incidence of implantation seems low.

Conclusion

In summary, we feel that laparoscopy in ovarian cancer begins with the management of suspicious adnexal masses, at which point we can stage adequately, as needed, those patients who have early ovarian carcinoma.

We think that we can resect minimally advanced ovarian cancer, and also resect tumours in a number of patients with advanced ovarian cancer after neoadjuvant chemotherapy.

References

1 Bagley CM, Young RC, Scheine PS, Chabner BA, DeVita VT. Ovarian carcinoma metastatic to the diaphragm—frequently undiagnosed at laparotomy. *Am J Obstet Gynecol* 1973; **116**: 397–400.

2 Rosenoff SH, Young RC, Anderson T, *et al*. Peritoneoscopy: a valuable staging tool in ovarian carcinoma. *Ann Intern Med* 1975; **83**: 37–41.

3 Piver MS, Barlow JJ, Lele B. Incidence of subclinical metastases in stage I and II ovarian carcinoma. *Obstet Gynecol* 1978; **52**: 100–4.

4 Mangioni C, Bolis G, Molteni P, Belloni C. Indications, advantages, and limitations of laparoscopy in ovarian cancer. *Gynecol Oncol* 1979; **7**: 47–55.

5 Ozols RF, Fisher RI, Anderson T, Makuch R, Young RC. Peritoneoscopy in the management of ovarian cancer. *Am J Obstet Gynecol* 1981; **140**: 611–19.

6 Spinelli R, Luini A, Pizzetti P, DePalo GM. Laparoscopy in staging and restaging of 95 patients with ovarian carcinoma. *Tumori* 1976; **62**: 493–502.

7 Smith WG, Day TG, Smith JP. The use of laparoscopy to determine the results of chemotherapy for ovarian carcinoma. *J Reprod Med* 1977; **18**: 257–60.

8 Quinn MA, Bishop GJ, Campbell JJ, Rodgerson J, Pepperell RJ. Laparoscopic follow-up of patients with ovarian carcinoma. *Br J Obstet Gynaecol* 1980; **87**: 1132–9.

9 Piver MS, Lele SB, Barlow JJ, Gamarra M. Second-look laparoscopy prior to proposed second-look laparotomy. *Obstet Gynecol* 1980; **55**: 571–3.

10 Berek JS, Griffiths T, Leventhal JM. Laparoscopy for second-look evaluation in ovarian cancer. *Obstet Gynecol* 1981; **58**: 192–8.

11 Lacey CG. Laparoscopy in gynecologic oncology. In: Morrow CP, ed. *Recent clinical developments in gynecologic oncology*. New York: Raven Press, 1983: 181–9.

12 Reich H, McGlynn F, Wilkie W. Laparoscopic management of stage I ovarian carcinoma: A case report. *J Reprod Med* 1990; **35**: 601–4.

13 Sedlacek TV. Editorial comments. *J Reprod Med* 1990; **35**: 604–5.

14 Querleu D, LeBlanc E. Laparoscopic infrarenal paraaortic lymph node dissection for restaging of carcinoma of the ovary or fallopian tube. *Cancer* 1994; **73**: 1467–71.

15 Young RC, Decker DG, Warton JT, *et al*. Staging laparotomy in early cancer. *JAMA* 1983; **250**: 3072–6.

16 Buchsbaum HJ, Brady MF, Delgado G, *et al*. Surgical staging of carcinoma of the ovaries. *Surg Gynecol Obstet* 1989; **169**: 226–32.

17 Knapp RC, Friedman EA. Aortic lymph node metastases in early ovarian cancer. *Am J Obstet Gynecol* 1974; **119**: 1013–7.

18 Delgado G, Chun B, Caglar H, Bepko F. Para-aortic lymphadenectomy in gynecologic malignancies confined to the pelvis. *Obstet Gynecol* 1977; **50**: 418–23.

19 Musumeci R, Banfi A, Bolis G, *et al*. Lymphangiography in patients with ovarian epithelial cancer. *Cancer* 1977; **40**: 1444–9.

20 Chen SS, Lee L. Incidence of para-aortic pelvic lymph node metastases in epithelial carcinoma of the ovary. *Gynecol Oncol* 1983; **16**: 95–100.

21 Wu P, Qu J, Lang JH, Huang R, Tang M, Lian L. Lymph node metastases in ovarian carcinoma: A preliminary study of 74 cases of lymphadenectomy. *Am J Obstet Gynecol* 1986; **155**: 1103–8.

22 Burghardt E, Girardi F, Lahousen M, Tamussino K, Stettner H. Patterns of pelvic and para-aortic lymph node involvement in ovarian carcinoma. *Gynecol Oncol* 1991; **40**: 103–6.

23 Ozols RF, Rubin SC, Dembo AJ, Robboy S. Epithelial ovarian cancer. In: Hoskins WJ, Perez CA, Young RC, eds. *Principles and practices of gynecologic oncology*. Philadelphia: JB Lippincott, 1992: 758–62.

24 Gershenson DM. Epithelial ovarian cancer. In: Copeland LJ, ed. *Textbook of gynecology*. Philadelphia: WB Saunders, 1993: 1069–82.

25 Childers JM, Hatch KD, Tran AN, Surwit EA. Laparoscopic para-aortic lymphadenectomy in gynecologic malignancies. *Obstet Gynecol* 1993; **82**: 741–7.

26 Webb MJ, Decker DG, Mussey E, Williams TJ. Factors influencing survival in stage I ovarian cancer. *Am J Obstet Gynecol* 1973; **116**: 222–8.

27 Sevelda P, Dittrich C, Salzer H. Prognostic value of the rupture of the capsule in stage I epithelial ovarian carcinoma. *Gynecol Oncol* 1989; **35**: 321–2.

28 Dembo AJ, Davy M, Stenwig AE, Berle EJ, Bush RS, Kjorstad K. Prognostic factors in patients with stage I epithelial ovarian cancer. *Obstet Gynecol* 1990; **75**: 263–73.

29 Finn CB, Luesley DM, Buxton EJ, *et al*. Is stage I epithelial ovarian cancer overtreated both surgically and systemically? Results of a five-year cancer registry review. *Br J Obstet Gynaecol* 1992; **99**: 54–8.

30 Vergote I, Kaern J, Abeler V, Pettersen E, DeVos L, Trope C. Analysis of prognostic factors in stage I epithelial ovarian carcinoma: Importance of degree of differentiation and deoxyribonucleic acid ploidy in predicting relapse. *Am J Obstet Gynecol* 1993; **169**: 40–52.

31 Sjövall K, Nilsson B, Einhorn N. Different types of rupture of the tumor capsule and the impact on survival in early ovarian carcinoma. *International Journal of Gynecological Cancer* 1994; **4**: 333–6.

32 Childers JM, Aqua KA, Surwit EA, Hallum AV, Hatch KD. Abdominal-wall tumor implantation after laparoscopy for malignant conditions. *Obstet Gynecol* 1994; **84**: 765–9.

18 Cancer of the fallopian tube

FRANCES EVANS, CHRISTOPH LEES,
FRANK LAWTON

Cancer of the fallopian tube presents in a manner similar to ovarian cancer, is often misdiagnosed as such preoperatively, and shares many aspects of treatment and prognosis. It is therefore appropriate to include a chapter about this unusual malignancy in this book.

Incidence

Primary cancer of the fallopian tube is rare with estimates of incidence of around 0.3% of all gynaecological cancers.[1,2] Based on data from the National Cancer Institute's cancer registries in the United States, the average annual incidence of malignant epithelial fallopian tube neoplasms between 1973 and 1984 was 3.6/million women.[3]

Case reports of fallopian tube cancer have been published since the mid-nineteenth century with about 1500 cases having been reported worldwide. The rarity of the condition means that there are few series containing large numbers of patients. For example, the Roswell Park experience details 64 cases over a period of 23 years and so there is no one centre with a large experience of treating the disease.[4] The natural history, prognostic factors, and optimum management therefore remain uncertain.

Aetiology

The aetiology of fallopian tube cancer is unknown. It presents mainly in the fourth, fifth and sixth decades of life with a mean age at presentation of 55 years, although there have been cases reported in women as young as 17 years.[5] The incidence seems to be higher in white than in negro women.

The higher incidence of infertility among women with tubal cancer than in the general population has been thought to indicate chronic tubal damage as an aetiological factor—for example, it may be secondary to pelvic tuberculosis or bacterial salpingitis. However, this may be a rare malignancy arising in the tubes, which are, not uncommonly, the site of an infective process.

As tubal epithelium responds to cyclical hormonal activity (which is thought to be an aetiological factor in ovarian, breast and endometrial malignancy) it may be that in association with nulliparity, the response plays a part in the aetiology of tubal cancer. No link, however, has been established between oral contraceptive use and fallopian tube cancer.

Pathology

The most common primary malignancy of the fallopian tube is an adenocarcinoma, with a papillary pattern being the commonest histological type in this group. These comprise some 90% of tubal cancers.[6] Many other rare tubal malignancies have been reported including lymphosarcoma, medullary carcinoma, sarcoma, transitional cell carcinoma, squamous cell, and clear cell carcinoma.

The classic macroscopic appearance of tubal cancers is a fusiform swelling of the tube; however, the tube is not invariably enlarged. On opening the tube the tumour may appear as a solid nodule, an ulcerated lesion or a friable tumour mass.

An abnormal looking fallopian tube discovered at laparoscopy or laparotomy and with an appropriate history and examination should be opened and biopsied to make the diagnosis. It is important that surrounding structures and, in particular, the ovaries are carefully examined as histologically most tubal cancers resemble ovarian cancers. The rare choriocarcinoma in the fallopian tube is confused with ectopic pregnancy both clinically and at operation. Some series have reported tubal cancer as part of a multifocal upper genital tract malignancy associated with other primary tumours of the ovary, uterus, or cervix. Other series have reported primary tumours of the breast, colorectal tumours, and melanomas in association with tubal cancer.[7] In some series, however,[8] there have been few coincidental malignancies. Carcinoma *in situ* has been reported as an incidental histological finding in "normal" fallopian tubes removed at hysterectomy,[9] but the significance of this finding is not clear. Benign tumours such as lymphangiomas or leiomyomas may also cause diagnostic confusion occasionally.

Diagnosis

The fallopian tube is a common site of secondary spread from malignancies arising from the uterus and ovaries and may also be involved in the spread of malignant disease arising in the gastrointestinal tract. To establish a diagnosis of primary tubal cancer the criteria set forward by Hu *et al* are used,[10] namely that the tumour arises from tubal epithelium, with histological examination suggesting a tubal origin and showing a transition between normal endosalpinx and malignant endosalpinx; that most of the cancer is located in the tube; and that the uterus and ovaries are either normal or contain less tumour than the tube.

Atypical epithelial hyperplasia of the tubal epithelium as occurs in association with tuberculosis and bacterial infection may in more florid examples exhibit papillary formation, nuclear pleomorphism, and mitotic activity which make it difficult to distinguish from early invasive tubal cancer.

Clinical features

The presentation of primary fallopian tube cancer is similar to that of ovarian cancer in that it is usually a disease which is asymptomatic in its early stages, has an insidious onset, and may give rise to a variety of often non-specific symptoms.

The classic association of "hydrops tubae profluens" described by Latzko in 1916 (profuse watery vaginal discharge associated with pain and an adnexal mass) is a rarely reported presentation. Abnormal vaginal bleeding or blood-stained discharge is one of the most common presenting symptoms, and may be associated with pelvic or abdominal pain and a pelvic mass. Menstrual disturbance may also be the presenting complaint, with other symptoms including weight loss, abdominal distension secondary to ascites, rectal bleeding, and night sweats.

As with other rapidly growing adenocarcinomas, paraneoplastic neurological disorders such as subacute cerebellar degeneration may be the presenting symptom before the diagnosis of a gynaecological malignancy is made. Such cases are usually associated with a high level of anti-Purkinje cell antibodies.[8] The presence of malignant glandular cells on cervical cytology associated with negative colposcopy and negative cervical and endometrial histology may lead to a suspicion of tubal cancer.

Vaginal ultrasound in such cases may be able to show tubal enlargement with laparoscopy as a further diagnostic tool.[5] As with ovarian cancer, the tumour marker CA125 may be raised in fallopian tube cancer,[8] but is of no value in distinguishing between the two diagnoses before laparotomy.

At laparotomy the presence of unilateral tubal enlargement with a normal tube on the other side may suggest the diagnosis of tubal cancer, as other diseases that cause tubal enlargement (such as hydrosalpinx) are usually bilateral.

Differential diagnosis

The main differential diagnosis of tubal cancer both at presentation and at laparotomy is ovarian cancer, but certain benign conditions must be considered including genital tuberculosis (which commonly affects the fallopian tubes and can cause a use in serum CA125 titres), benign adenomatoid tumours of the fallopian tube and salpingitis isthmica nodosa (a benign proliferation of the isthmic portion of the tube which occurs as a consequence of chronic bacterial salpingitis).

Staging

The International Federation of Gynecology and Obstetrics (FIGO) have established a staging system for fallopian tube cancer based on that for ovarian cancer (table 18.1) with stage I disease confined to the tubes, stage II disease involving pelvic organs, stage III disease involving intraperitoneal but extrapelvic structures, and stage IV disease including distant metastases, including parenchymal liver metastases. As with ovarian cancer, fallopian tube cancer is staged surgically at laparotomy which includes peritoneal washings, and palpation and biopsy of lymph nodes and peritoneal surfaces.

Table 18.1 International Federation of Gynecology and Obstetrics staging of tubal cancer

Stage	Definition
0	Carcinoma in situ (limited to tubal mucosa)
I	Growth limited to fallopian tubes
Ia	Growth limited to one tube with extension into the submucosa or muscularis, or both, but not penetrating the serosal surface; no ascites
Ib	As Ia, but growth limited to both tubes
Ic	Tumour stage Ia or Ib with extension to serosa; or ascites or peritoneal washings containing malignant cells
II	Growth involving one or both fallopian tubes with pelvic extension
IIa	Extension or metastasis, or both, to the uterus or ovaries, or both
IIb	Extension to other pelvic tissue
IIc	stage IIa or IIb with ascites containing malignant cells or positive peritoneal washings
III	Tumour involves one or both fallopian tubes with extrapelvic peritoneal implants or retroperitoneal or inguinal nodes containing tumour Tumour seems limited to true pelvis, but with histologically proved malignant extension to the small bowel or omentum
IIIa	Tumour grossly limited to the true pelvis with negative nodes, but with histologically confirmed microscopic seeding of the abdominal peritoneal surfaces
IIIb	Tumour involving one or both tubes with histologically confirmed implants of abdominal peritoneal surfaces, none exceeding 2 cm in diameter. Lymph nodes clear
IIIc	Abdominal implants greater than 2 cm in diameter or retroperitoneal or inguinal nodes, or both, containing tumour
IV	Growth involving one or both fallopian tubes with distant metastases Pleural effusion cytology must show malignant cells to be stage IV Parenchymal liver metastases equals stage IV

Erez *et al* in 1967 devised a system of staging for tubal cancer which took into account tumour invasion into the wall of the fallopian tube and is similar to Dukes' classification of colorectal carcinoma.[11] This may be appropriate in assessing early stage disease when tubal penetration is important, but in late stage disease, once lymphatic permeation and spread over peritoneal surfaces has occurred, tubal cancer most strongly resembles ovarian cancer and therefore the FIGO classification of staging is more useful.

Once the serosa of the tube has been breached the extensive lymphatic drainage of the tube facilitates spread of tumour. In one series of 15 patients, a third had para-aortic lymph node metastases, with 6% having inguinal, mediastinal, and supraclavicular deposits.[12]

Using the FIGO staging roughly two thirds of patients present with stage I or II disease with a 55% five-year survival for those with stage I disease. The disease may present at an earlier stage than ovarian cancer because it causes vaginal bleeding which leads to earlier diagnosis; it may be difficult to establish the tubal origin of a more advanced cancer.

Surgical management

Fallopian tube cancer is usually encountered as an unexpected finding at laparotomy. In one series of 71 patients the correct pre-operative diagnosis had been made in only two making preoperative management planning a rare possibility.[13] The recommended surgical management is the same as that for ovarian cancer, namely total abdominal hysterectomy with bilateral salpingo-oophorectomy and omentectomy. Examination of samples of retroperitoneal lymph nodes, peritoneal washings, and biopsy specimens all contribute to accurate staging.

In those patients with advanced stage disease, surgical debulking should be undertaken, as the amount of residual tumour is an important prognostic factor. In the series by Eddy *et al*[13] median survival in those patients with no gross residual disease was 30 months compared with 22 months for those with residual disease up to 2 cm diameter, and 17 months with those with more extensive residual tumour.

Postoperative treatment

As few reports contain large numbers of patients treated prospectively with standard regimens it is difficult to draw any firm conclusions about the most effective form of postoperative treatment. The similarity between fallopian tube cancer and serous ovarian cancer suggests that treatment that is effective in the management of ovarian cancer may also be appropriate in the management of tubal cancer.

Interestingly, patients with early stage fallopian tube cancer have poor prospects of survival; the median survival for patients with FIGO stage I disease treated by surgery alone was less than three years in one study,[13] so some form of postoperative treatment is probably appropriate for patients with almost all stages of disease.

Radiotherapy

Radiotherapy to the pelvis alone is inadequate, as shown by McMurray *et al*[14] who reported 50% recurrence in the upper abdomen with 44% of patients developing recurrence at an extraperitoneal site. Semrad *et al*[15] reported 10 out of 14 patients with evidence of relapse outside the peritoneal cavity, seven of which were in inguinal mediastinal or supraclavicular nodes and three within the pleural cavity. Peters *et al*[16] projected a 10 year survival of only 15% after pelvic irradiation for stage II disease, implying that most women would die of their disease despite radiotherapy. This means that locoregional radiotherapy has little place as first line postoperative treatment in the management of fallopian tube cancer.

Chemotherapy

Many studies have shown response rates similar to those in epithelial ovarian cancer after cisplatin-based combination chemotherapy, the most common combination being cisplatin, doxorubicin, and cyclophosphamide or bleomycin.[4 7 14 17–20] In stage II-IV disease, these drugs in combination produce response rates of more than 80%, with durations of response varying from 12 months to more than six years.

As with ovarian cancer, therefore, adequate surgical debulking before adjuvant cisplatin-based combination chemotherapy seems to increase both response rates and survival.

Theoretically hormonal therapy may have a role in the treatment of women with tubal cancer, as the tubal epithelium shows hormonal variation during the menstrual cycle, but this treatment has not been evaluated in a sufficient number of patients to allow any conclusions to be drawn.

Second-look procedures

There have been attempts to evaluate the role of "second-look" laparotomy in establishing prognosis in tubal cancer in a similar manner to ovarian cancer. Barakat *et al* retrospectively reviewed patients who had received cisplatin-based combination chemotherapy after primary surgical treatment for tubal cancer and had undergone a "second-look" laparotomy four to six weeks after completion of chemotherapy.[21] They found that neither stage nor grade of tumour at primary presentation were predictive of findings at second-look laparotomy except in those with stage I or grade I tumours, of whom all five had no evidence of residual disease at second laparotomy.

Of 11 patients who presented with advanced stage disease and had a negative second-look laparotomy, 10 were disease free at a median of 49 months, but the absence of gross residual disease after primary surgery was the strongest predictor that no macroscopic disease would be found at second-look laparotomy.

Second-look laparotomy may have a role as a prognostic indicator after appropriate primary surgery and platinum-based combination chemotherapy, but as recurrences are often extra-abdominal it has a limited usefulness. In addition, although it may assess prognosis, it does not alter it and means the patient has to undergo a further operation. Its limitations in fallopian tube cancer reflect those in epithelial ovarian cancer.

Follow-up

In those patients with tubal cancer who have a raised serum CA125 titre at presentation which falls after primary treatment, a subsequent rise may indicate recurrence of disease. CA125 titres can therefore be used in monitoring of the disease. In the absence of clinical recurrence and with no effective second line treatment, however, the clinical usefulness of this form of monitoring is limited, although it may provide useful prognostic information.

Most recurrences are extrapelvic with half or more being extraperitoneal, usually in combination with recurrent intraperitoneal disease,[5] and most occur within three years of the initial diagnosis; some have been reported up to 13 years later, however, so adequate long term follow-up is essential. As there may be an association between primary tubal cancer and other primary malignancies, for example of the breast, regular mammographic screening may be appropriate.

Prognosis

Overall the prognosis for women with tubal cancer is poor with no significant improvement in the last two decades. As with most cancer, the stage or distribution of tubal cancer at the time of presentation is the most important prognostic factor. Baeklandt *et al*[5] combined survival figures from eight relatively large studies, and calculated a five year survival rate of 60% in patients with disease confined to the tube at the time of diagnosis, with 40% five year survival in stage II disease and only 10% survival in those patients who presented with advanced stage disease.

The occlusion of the abdominal tubal ostium may delay or prevent local spread of tumour; however, the importance of this as a prognostic factor is not clear. Similarly, whether the site of the cancer within the tube is of prognostic significance is uncertain although it seems logical to assume that earlier dissemination to other pelvic organs may occur in tumours located in the distal end of the tube.

Conclusion

Primary tubal cancer is a rare disease which needs to be considered in the differential diagnosis of women with appropriate unexplained gynaecological symptoms. From the evidence accumulated over many years at different centres it is apparent that a thorough staging laparotomy with adequate surgical debulking is essential in its effective primary management. Adjuvant treatment in the form of cisplatin-based combination chemotherapy should be offered to all patients with the exception of those with stage I disease and minimal mucosal penetration, as this is the form of adjuvant treatment that is most likely to prolong the disease free interval. The use of carboplatin rather than cisplatin may reduce toxicity.

The wider use of transvaginal ultrasound in detecting tubal disease may lead to earlier diagnosis of tubal cancer and therefore a larger number of surviving women.

Further advances in the management of tubal cancer are unlikely unless international collaborative studies can be established, the feasibility of which is limited by the rarity of the disease. The use of the FIGO staging system universally may help in drawing conclusions from those cases reported, as the small number of women who present with tubal cancer means that full and accurate case reporting remains important for the future.

References

1 Green TH, Scully RE. Tumors of the Fallopian tube. *Clin Obstet Gynecol* 1952; **5**: 866–70.
2 Roberts JA, Lifshitz S. Primary adenocarcinoma of the fallopian tube. *Gynecol Oncol* 1982; **13**: 301–6.
3 Hanton EM, Malkasian GD Jr, Dahlin DC, Pratt JH. Primary carcinoma of the fallopian tube. *Am J Obstet Gynecol* 1966; **94**: 832–9.
4 Rose PG, Piver MS, Tsukada Y. Fallopian tube cancer. *Cancer* 1990; **66**: 2661–7.
5 Baeklandt M, Kockx M, Wesling F, Gerris J. Primary adenocarcinoma of the fallopian tube. Review of the literature. *International Journal of Gynecological Cancer* 1993; **3**: 65–70.
6 Park RC, Parmley TH. Fallopian tube cancer. In: McGowan L, ed. *Gynecologic oncology*. New York: Appleton Century Crofts, 1978: 274–80.
7 Podratz KC, Podczaski ES, Gaffey TA, O'Brien PC, Schray MF, Malkasian GD Jr. Primary carcinoma of the fallopian tube. *Am J Obstet Gynecol* 1986; **154**: 1319–26.
8 Pryse Davies A, Fish A, Woollas R, Oram D. Raised serum CA125 preceding the diagnosis of carcinoma of the fallopian tube. *Br J Obstet Gynaecol* 1991; **98**: 602–3.
9 Bannatyne P, Russell P. Early adenocarcinoma of the fallopian tubes. A case for multifocal tumorigenesis. *Diagnostic Gynecology and Obstetrics* 1981; **3**: 49–60.
10 Hu CY, Taylor ML, Hertig AT. Carcinoma of the fallopian tube. *Am J Obstet Gynecol* 1950; **59**: 58–67.
11 Erez S, Kaplan AL, Wall JA. Clinical staging of carcinoma of the uterine tube. *Obstet Gynecol* 1967; **30**: 547–50.
12 Tamimi HK, Figge DC. Adenocarcinoma of the uterine tube: potential for lymph node metastases. *Am J Obstet Gynecol* 1981; **141**: 132–7.
13 Eddy GL, Copeland LJ, Gerhenson DM, Atkinson EN, Wharton JT, Rutledge FN. Fallopian tube carcinoma. *Obstet Gynecol* 1984; **64**: 546–52.
14 McMurray EH, Jacobs AJ, Perez CA, Camel HM, Kao MS, Galakatos A. Carcinoma of the fallopian tube. *Cancer* 1986; **58**: 2070–5.
15 Semrad N, Watring W, Fu Y-S, Hallat J, Ryoo M, Lagasse L. Fallopian tube adenocarcinoma: common extraperitoneal recurrence. *Gynecol Oncol* 1986; **24**: 230–5.
16 Peters WA, Andersen WA, Hopkins MP, Kumar NB, Morley GW. Prognostic features of carcinoma of the fallopian tube. *Obstet Gynecol* 1988; **71**: 757–62.
17 Jacobs AJ, McMurray EH, Parham J, *et al.* Treatment of carcinoma of the fallopian tube using cisplatin, doxorubicin and cyclophosphamide. *Am J Clin Oncol* 1986; **9**: 436–9.
18 Raju KS, Barker GH, Wiltshaw E. Primary carcinoma of the fallopian tube. Report of 22 cases. *Br J Obstet Gynaecol* 1981; **88**: 1124–9.

19 Maxson WZ, Stehman FB, Ulbright TM, Sutton GP, Ehrlich CE. Primary carcinoma of the fallopian tube: evidence for activity of cisplatin combination therapy. *Gynecol Oncol* 1987; **26**: 305–13.
20 Gurney H, Murphy D, Crowther D. The management of primary fallopian tube carcinoma. *Br J Obstet Gynaecol* 1990; **97**: 822–6.
21 Barakat RR, Rubin SC, Saigo PE, *et al.* Cis-platin based combination chemotherapy in carcinoma of the fallopian tube. *Gynecol Oncol* 1991; **42**: 156–60.

19 Aspects of quality of life in women undergoing treatment for ovarian cancer

SEAN KEHOE

It is difficult to define quality of life as it means different things to different people, but a usable definition is "Quality of life refers to patients' appraisal of and satisfaction with their current level of functioning as compared to what they perceive to be possible or ideal."[1] It is important that patients judge their own quality of life, as it has been shown that not only are assessments by physicians and patients not comparable, but also that the more objective scores such as the Karnofsy may not accurately reflect how patients feel.[2]

Assessments of quality of life should include psychological, physical, social, and occupational aspects, and it is difficult to find one questionnaire that is acceptable for them all. This should not prevent us from trying, however, because it is essential to evaluate the effects of treatment on the quality as well as the duration of life, particularly in oncology.

Why measure quality of life?

Trials in patients with metastatic disease have shown that aggressive regimens result in improvements in the quality of life,[3,4] which contradicts the medical assumption that the more aggressive the treatment the worse the quality of life.

Quality of life is also important in the assessment of new drugs. If, for example, a new drug is being compared against the best standard treatment and they are found to be similar in terms of cost and toxicity, it might not be noticed that one has a considerable impact on quality of life unless specific questions are asked.

Quality of life is also an independent prognostic indicator,[5] and though it is not always possible to improve prognosis by improving quality of life it should always be borne in mind.

What to measure

The most important variables are physical concerns (symptoms), function (activity), family wellbeing, emotional wellbeing, spirituality, satisfaction

268

with treatment, financial concerns, hopes for the future, sexuality, social function, and occupational function.[6] If any of these is omitted the validity of the conclusions may be affected.

Method of measurement

Before the assessment is started the objectives must be clearly defined and the best method chosen for the target population. For example, it would be pointless to use "return to work" as an end point if the group being studied comprised retired people, or to use a test that was designed to be repeated at three monthly intervals if the life expectancy of patients with the disease being studied was two months. The reliability, validity, standardisation, and specificity of the method must be checked to make sure that it is suitable.

Reliability

This assesses the capability and consistency of the test, which should be applied to the group to be tested in two different formats ("split test" reliability) and at different time intervals ("test-retest" reliability). Test-retest reliability may not be appropriate as time can change the patients' disease state. "Alternate form" reliability will make sure that different versions of the test measure the same thing, and "internal consistency" whether a particular item significantly influences the final score. A reliability coefficient of 0.70 or more is just acceptable, but one of 0.80 is preferable.

Validity

Validity indicates the accuracy of the test. "Face" validity is a subjective measurement of whether all the necessary questions relating to the end point are asked. "Content" validity examines whether the important questions about the disease or treatment have been asked. "Criterion" validity is predictive, and shows how the test questions compare with previously validated methods. "Construct" validity assesses how well particular subgroups within the study group can be differentiated, and is associated with the sensitivity (which identifies grades of quality of life) and specificity (which identifies changes brought about by treatment or by progression of the disease).

One should approach construct validity by asking patients with a clearly defined disease to answer open ended questions. This should be done by people trained in quality testing who will identify the problems and put together the skeleton of the questionnaire together with other questions that are considered appropriate. Relatives may help. The reliability and validity must then be assured, which takes a long time with associated high costs.

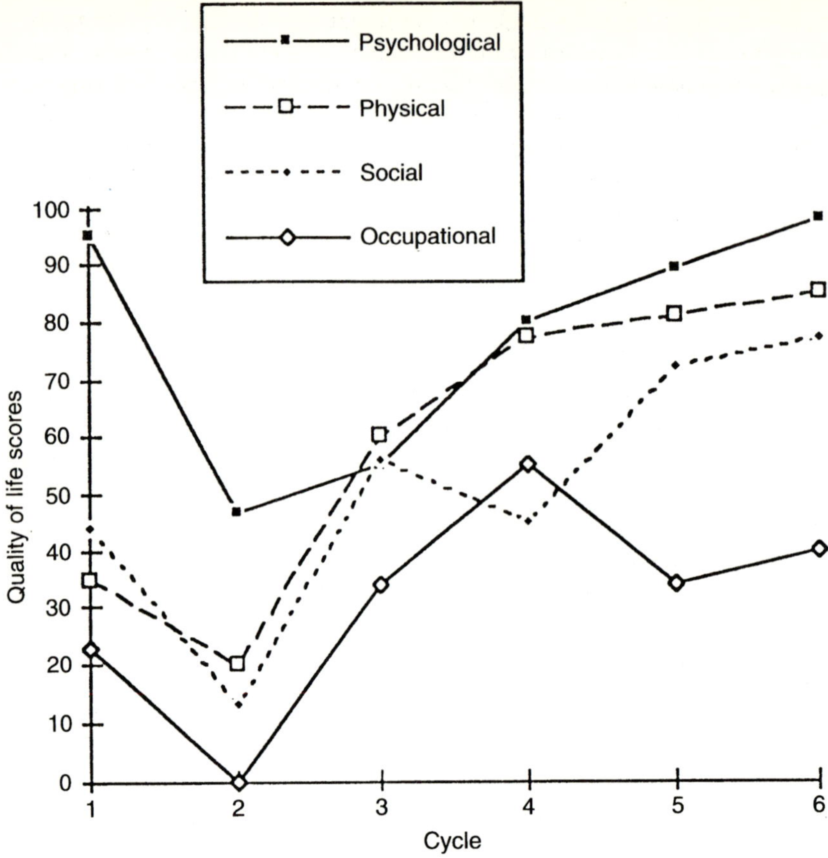

Fig. 19.1 Treatment cycles and quality of life scores.

Alternatively one can amalgamate parts of previously validated questionnaires to construct a new one. Though this is simpler it may not be as reliable as the first method, and an expert opinion should be sought before the questionnaire is introduced.

To avoid bias, staff who undertake the assessments should not be involved in the patients' treatment and any evaluations made while patients are being treated should be kept until the end of the course to avoid influencing decisions about treatment. The quality of life at any particular point of treatment could be poor, but improve after further intervention. A hypothetical example is given in fig. 19.1.

Statistical analysis

To measure quantity is easy; to measure quality difficult. "Hard" information is inevitably sought that will help carers to assess the benefits (or lack of them) associated with a particular treatment, but without

270

knowledge of the quality of life in a comparable group of unaffected people or of the standard quality of life of people with the disease being studied this is impossible to obtain. Another major problem is that of missing data, which obviously detracts from the validity of an analysis. In the absence of hard data physicians may have to accept that the best that they are going to get from assessments of quality of life is overall trends, and these can be good enough indicators of the effects of treatment.

How often to measure quality of life

There are no hard and fast rules about timing; too often is a burden to patients and prolonged intervals will tax their memory. Carey *et al* found that weekly compared well with monthly and two monthly assessments, though some patients had difficulty in recalling their previous quality of life after two months.[7] Daily questionnaires give too much detail, and patient compliance is poor.

The decision about the timing will depend on the disease and its treatment but it is necessary to apply the questionnaire before, during, and after the intervention (be it operation, chemotherapy, or radiotherapy) and then at intervals for follow up.

Types of tests available

There are two types of tests: those that use linear analogue scales and those that require categorical answers (usually from a choice of four or five). Linear analogue scales are lines about 10 cm long with one end indicating the worst response and the other indicating the best. Patients are asked to mark the point on the line that applies to their response. Unlike categorical questions, linear analogue scales are difficult to score and not as useful for identifying more subtle variations in the quality of life. Examples of some questions are shown in Table 19.1.

Measurements of function (activity)

These measurements are often used as criteria for entry into clinical trials. One of the most important is the Karnofsky scale which was first used in 1949 to assess nursing needs and involves the assessment of patients by physicians.[8] It gives a broad idea about functional state and is a crude predictor of survival. Similar systems have been developed by the World Health Organisation (five point scale) and the Eastern Cooperative Oncology Group (ECOG).[9]

Table 19.1 Types of questions about quality of life

Linear Visual Analogue
   Linear analogue self assessment
      How do you feel?
      Very sad ——————————————————————— Very happy
   Functional living index cancer
      How well do you feel today?

1	2	3	4	5	6	7
Extremely						Extremely
ill						well

Categorical method
   Rotterdam symptom checklist
      Tick the box closest to how you have felt in the last X days
      Depression:
      1 Not at all
      2 A little
      3 Somewhat
      4 Very much
   EORTC core question
      During the last week:

	Not at all	A little	Quite a bit	Very much
Did you feel depressed?	1	2	3	4

Hospital anxiety and depression scale[10]

Though this was originally developed for patients who did not have cancer it is now part of the European Organisation for Research in Treatment of Cancer (EORTC) questionnaire; its validity and specificity are good and it is easy to apply.

Neither of these two tests alone is adequate for assessing quality of life, but they are often used together with more concise measures.

Linear analogue scales

Linear analogue self assessment scale

This test contains 25 core questions for quality of life assessment. It was originally developed for use in patients with breast carcinoma,[11] and is simple, sensitive and reproducible. Though its reliability and validity are now known it does seem to detect changes over time and distinguish between treatments.

Functional living index cancer[12]

This contains 22 items and is reliable in its physical and psychological components though difficulties have been reported with compliance.[13 14] It has not been compared with other validated tests which limits its value.

Categorical tests

Rotterdam symptom checklist

These questions were carefully selected and modified from other tests and it is one of the best quality of life questionnaires. It is reliable and valid, takes only 10–15 minutes to complete and is acceptable to patients. It was recently shown to be reliable with regard to psychological and physical variables and has been used in patients with ovarian cancer.[15]

Quality of life index[16]

This is an observer rated score which takes only a minute to do. Like many such scores it correlates poorly with the patients' version.

Sickness impact profile[17]

This is reliable and valid, and its 136 items cover all the necessary components of a quality of life measurement. It has been used in various conditions and although it is a good measure of the quality of life it is too long. It has been used in patients with ovarian cancer but though it is useful extra questions about sexual function are required.[18]

EORTC quality of life questionnaire C30[19]

Much research has gone into this test to try to develop a questionnaire applicable to all cancers, which would be useful as it would permit comparison of results from different centres. The questionnaire has been validated on 177 patients with different cancers and results indicated good external validity but less good internal consistency.[20]

Quality adjusted life years (QALYs) and Time without symptoms or toxicity (TWIST)

Both these methods are used to assess cost effectiveness of treatments and in both methods duration of survival is adjusted to the degree of toxicity or disability caused. They were introduced by Bush et al,[21] and should be used only to assess cost effectiveness.[22 23]

Which test to use

These are only examples of quality of life measures, of which there are many.[24] The Medical Research Council Working Party on Quality of Life and Cancer concluded that the best was probably the Rotterdam symptom checklist with the addition of questions related to specific diseases.[25] The hospital anxiety and depression scale and linear analogue self assesment was also thought to be useful, as was the EORTC scale.[26]

Quality of life and ovarian cancer

There have been few reviews of quality of life measurements in ovarian cancer, which is of concern because three quarters of patients present with advanced disease and most die within five years.[27] Treatment is mainly palliative and so quality of life is particularly important,[28] and yet in a review of publications from 1988–94 there were only 10 about the management of ovarian cancer that measured quality of life. Many claimed that treatment improved quality of life, but gave no evidence.

Blythe and Wahl studied quality of life in relation to surgery for ovarian cancer and concluded that debulking improved quality of life.[29] Willemse *et al* used TWIST analysis in 69 patients with ovarian cancer who were undergoing intensive platinum-based treatment, 52 of whom had second-look laparotomy. They found it useful but it is more applicable to cost effectiveness studies.[30]

Guidozzi used the Spitzer questionnaire to evaluate prospectively the impact of treatment and relapse of disease in 28 women with ovarian cancer, and found that patients with residual disease had less disruption of their personal relationships during the first year than those without.[31] Walczak *et al* used the functional living index cancer to compare the effects of standard and intensified doses of platinum in patients with advanced cancer, but had a serious problem with incomplete data.[14] Oschega *et al* interviewed 30 patients with testicular or ovarian cancer a year after treatment with high dose cisplatin to find out the effect of peripheral neuropathy on their quality of life but because it was a retrospective study with a small number of patients were unable to draw any conclusions.[32]

Homsley *et al* did probably the best investigation[33]; they undertook a prospective randomised trial of cisplatin plus epirubicin or doxorubicin and measured quality of life with Priestman and Baum's visual analogue scale.[11] They concluded that as survival and quality of life scores were similar in the two groups the preferred regimen was cisplatin plus epirubicin because patients in this group had less cardiotoxicity and thrombocytopenia. Although the test used is questionable the overall approach was good.

The potential uses of quality of life measurements were described by Anderson *et al*.[34] The impact of treatment on the patients' psychological wellbeing, particularly sexual function, can be assessed with appropriate tests. Though most work has been done in patients with cervical and vulval diseases the same principles apply to ovarian cancer.

Conclusion

Measuring quality of life is difficult and costly but it must become part of the standard approach to the care of people with ovarian cancer. When new treatments are tried they often fail to show a significant improvement in five year survival but it could be that they make a difference to the quality of life. Retrospective analysis is not possible so a unique opportunity to glean information about the effects of treatment has been lost. Although the complexities of introducing quality of life measurements are daunting, neglecting to do so could have serious implications. For example the Food and Drugs Administration has stated that there are two requirements for the acceptance of new treatments for cancer; one is the effect on survival and the other quality of life.[35] No doubt this will soon be the case in Europe.

Cost effectiveness of treatment is becoming increasingly important, and the relevance of quality of life is obvious. Patients expect doctors to be able to tell them as much about the effects of treatment on quality of life as on survival. The lack of such information may be one reason why people seek "alternative" treatments.[36]

Physicians and surgeons must attempt to give the best possible care to their patients, and the measurement of the quality of life will supplement our knowledge about the efficacy of treatments, particularly when they are palliative. Improving quality of life may be less inspiring than improving survival rates, but the management of ovarian cancer can only be improved by it.

References

1 Cella DF, Cherin EA. Quality of life during and after cancer treatment. *Compr Ther* 1988; **14**: 69–75.
2 Slevin ML, Plant H, Lynch D, Drinkwater J, Gregory WM. Who should measure quality of life—the doctor or the patient? *Br J Cancer* 1988; **57**: 109–12.
3 Tannock IF. A randomised trial of two dose levels of cyclophosphamide, methotrexate and fluoracil chemotherapy for patients with metastatic breast cancer. *J Clin Oncol* 1988; **6**: 1377–87.
4 Coates A, Gebski M, Bishop J, *et al*. Improving the quality of life during chemotherapy for advanced breast cancer. *N Engl J Med* 1987; **317**: 1490–5.
5 Kassa S, Mastekassa A, Lund E. Prognostic factors for patients with inoperable non-small lung cancer, limited disease. The importance of patients subjective experience of disease and psychosocial well-being. *Radiother Oncol* 1989; **15**: 235–42.
6 Cella DF, Tuskey DS. Measuring quality of life today: Methodological aspects. *Oncology* 1990; **4**: 25–38.
7 Carey RD, Perez DJ, Campbell AV. Quality of life scales for patients with cancer: how often to administered? *NZ Med J* 1993; **106**: 250–1.
8 Karnofsky DA, Burchenal JH. The clinical evaluation of chemotherapeutic agents in cancer. In MacLeod CM, ed. *Evaluation of chemotherapeutic agents*. New York: Columbia Press, 1949: 199–205.
9 Zubrod CG, Schneiderman M, Frie E III, *et al*. Appraisal of methods for the study of chemotherapy of cancer in man: Comparative therapeutic trials of nitrogen mustard and triethylene thiophosphoramide. *J Chron Dis* 1960; **11**: 7–33.
10 Zigmond AS, Snaith RP. The hospital anxiety and depression scale. *Acta Psychiatr Scand* 1983; **67**: 361–70.
11 Priestman TJ, Baum M. Evaluation of quality of life in patients receiving treatment for advanced breast cancer. *Lancet* 1976; **1**: 899–901.

12 Schipper H, Clinch J, McMurray A, Levitt M. Measuring quality of life of cancer patients. The functional living index-cancer: development and validation. *J Clin Oncol* 1984; **2**: 472–83.

13 Ganz PA, Haskell CM, Figlin RA, *et al*. Estimating the quality of life in a clinical trial of patients with metastatic lung cancer using the Karnofsky Performance Status and Functional Living Index-Center. *Cancer* 1988; **61**: 849–56.

14 Walczak JR, Brady M, Stonebraker B. Use of functional living index cancer (FLIC) in cancer clinical trials: problems and issues. *Oncol Nurs Forum* 1989; **16** (suppl 2): 141.

15 De Haes JC, van Knippenberg FC, Neijt JP. Measuring psychological and physical distress in cancer patients: structure and application of the Rotterdam Symptom Checklist. *Br J Cancer* 1990; **62**: 1034–8.

16 Spitzer WO, Dobson AJ, Hall J, *et al*. Measuring the quality of life of cancer patients: A concise QL-Index for use by physicians. *J Chron Dis* 1981; **34**: 585–97.

17 Bergner M, Bobbitt RA, Carter WB, *et al*. The sickness impact profile: Development and final revision of a health status measure. *Med Care* 1981; **19**: 787–805.

18 Pearson KE. Perceived functional status of ovarian cancer patients. *Oncol Nurs Forum* 1989; **16** (suppl 2): 203.

19 Aarsonson NK, Bullinger N, Ahmedzai S. A modular approach to quality of life assessment in cancer clinical trials. *Recent Results Cancer Res* 1988; **111**: 231–49.

20 Ringdal GI, Ringdal K. Testing the EORTC quality of life questionnaire on cancer patients with heterogenous diagnoses. *Quality of Life Research* 1993; **2**: 129–40.

21 Bush JW, Chen M, Patrick DL. Cost-effectiveness using health status index: Analysis of the New York State PKU screening programme. In: Berg R, ed. *Health status index* Chicago: Hospital Research and Educational Trust, 1973: 172–208.

22 Gelber RD, Goldhirsch A, Cole BF. Evaluation of effectiveness: Q-Twist. *Cancer Treat Rev* 1993; **19** (suppl A): 73–84.

23 Kaplan RM. Quality of life assessment for cost/utility studies in cancer. *Cancer Treat Rev* 1993; **19** (suppl A): 85–96.

24 Fallowfield L. *The quality of life*. London: Souvenir Press, 1991.

25 Maguire P, Selby P. Assessing quality of life in cancer patients. *Br J Cancer* 1989; **60**: 437–40.

26 Filiberti A, Flechtner H, Fleishman SB, *et al*. The European Organization for Research and Treatment of Cancer QLQ-C30: a quality of life instrument for use in international clinical trials in oncology. *J Natl Cancer Inst* 1993; **85**: 365–76.

27 OPCS Mortality Statistics England and Wales (1993). Series DH 2. No 18. London: HMSO, 1993.

28 Gough IR, Dalsleish LI. What value is given to quality of life assessment by health professionals considering response to palliative chemotherapy for advanced cancer? *Cancer* 1991; **68**: 220–5.

29 Blythe JG, Wahl TP. Debulking surgery: Does it increase the quality of life? *Gynecol Oncol* 1982; **14**: 396–408.

30 Willemse PHB, Van Lith J, Mulder NH, *et al*. Risk and benefits of cisplatin in ovarian cancer. A quality-adjusted survival analysis. *Eur J Cancer* 1990; **26**: 345–52.

31 Guidozzi F. Living with ovarian cancer. *Gynecol Oncol* 1993; **50**: 202–7.

32 Oschega Y, Donohue M, Fox N. High-dose cisplatin related peripheral neuropathy. *Cancer Nurs* 1988; **11**: 23–32.

33 Holmsley H, Harry D, O'Toole R, Franklin E, Cavanagh D, Nahhas W, Hoogstraten B, Smith J. Randomized comparison of cisplatin plus epirubin or doxorubicin for advanced ovarian cancer: A multicentre trial. *Proceedings of the Annual Meeting of the American Society of Clinical Oncology* 1990; **9**: A627.

34 Anderson BL. Predicting sexual and psychological morbidity and improving the quality of life for women with gynaecological cancer. *Cancer* 1993; **71** (suppl): 1678–90.

35 Johnson JR, Temple R. Food and Drug Administration requirements for approval of new anticancer drugs. *Cancer Treatment Reports* 1985; **69**: 1155–7.

36 Cassileth BR, Lusk EJ, DuPont G, *et al*. Survival and quality of life among patients receiving unproven as compared with conventional cancer therapy. *New Engl J Med* 1991; **324**: 1180–5.

20 Living with dying from ovarian cancer: the palliative care of patients with advancing disease

WENDY MAKIN

Unlike many malignancies, the management of ovarian cancer is concerned in many patients with the treatment of advanced disease from the time of diagnosis. Over half have stage III or IV disease and only 10%–30% of these will survive more than five years.[1,2] Women who fail to respond or to progress on initial chemotherapy have a poor outlook. Those who relapse after remaining off treatment for some time may benefit from second-line chemotherapy, but overall there are a number of women who are receiving active treatment into the last months and weeks of their lives.

Another feature of advanced ovarian cancer is that major symptoms are predominantly related to progressive intra-abdominal disease rather than to distant metastatic spread. Measures which alleviate symptoms and offer support to the patient and her family should clearly be an integral part of care throughout the illness and achieving a comfortable, dignified death is the final part of this process for most women.

When does active treatment stop and last stage begin?

In many cancers, the definition of the terminal phase can be uncertain, sometimes because of the gradual nature of a slowly progressing illness and sometimes because there are unpredictable events that change the course of the disease. The patient, the family, and at times the physician may be reluctant to acknowledge the real situation for fear of "giving up hope". There must be careful objectivity about the continuation of active treatment in the face of progressing disease.

Palliative chemotherapy in ovarian cancer may usefully reduce symptoms, even if in the short term, by shrinking tumour masses or reducing the re-accumulation of ascites. The patient also feels supported by continued care from the expert professionals at the cancer treatment centre. This must be weighed against the cost in terms of discomfort from side-effects or difficult venous access, and time taken up by hospital attendance. Evaluation of palliative treatment has been poor in the past but there is now better assessment of the quality of life achieved for individual patients and of the use of health care resources in general.[3]

Indications to stop active treatment are lack of tumour response or of evidence of clinical benefit—for example, the development of inoperable bowel obstruction or organ failure. Other considerations include unacceptable toxicity in the context of deteriorating general condition, poor venous access, or chronic myelosuppression. Sometimes the patient may ask if the treatment is doing any good and this creates an opportunity for open discussion. Some people cope best by understanding what is likely to happen and it is important to them to be able to make realistic plans and prepare others. It is also important that family doctors outside the oncology unit should be involved as they will be key figures in the provision of continuing care and support.

The decision to stop, made at the right time, need not be an admission of failure. Confidence in dealing with symptoms and assurance of continuing support, even in the face of deterioration and death, enable the focus of care to be directed to other goals and this is the aim of palliative care. It does not, of course, preclude the use of active measures such as transfusion or paracentesis if these might improve quality of life.

Clinical problems in advanced ovarian cancer

Abdominal distension, whether caused by tumour or ascites, is usual and this alone produces a number of symptoms such as discomfort, anorexia, and breathlessness. Widespread intraperitoneal tumour may reduce bowel motility and lead to mechanical obstruction: the patient complains of nausea, vomiting, constipation, and pain. Other symptoms may arise from subdiaphragmatic tumour or a pelvic mass; enlarged pelvic and retroperitoneal lymph nodes may cause nerve compression and oedema.

Breathlessness is much more likely to be caused by a pleural effusion than intrapulmonary metastases, and tumour deposits adherent to the liver capsule are far commoner than hepatic secondaries from haematogenous spread, even after ultrasound scanning.[4] Bone and cerebral metastases and paraneoplastic effects such as hypercalcaemia although well documented are rare clinical problems compared with other epithelial malignancies such as breast cancer.[5–7]

The problems of advanced cancer cannot be managed effectively without appreciating their impact on the patient and family: this means being aware that there are emotional and spiritual as well as physical needs to be met. Constrained by the pressures of a busy clinical service, it is all too easy for doctors to focus on what to them seems to be the main problem, particularly if related to a decision about treatment. Other seemingly less important symptoms remain unsolicited but may greatly impair the enjoyment of life for that person. If this is the case for physical problems, it is hardly surprising that emotional distress is not tackled by the clinician until confronted by overt anger, anxiety, or just withdrawal. Appreciable depression is under-diagnosed and treated but is estimated to affect at least one in four patients with cancer at some time during their illness.[8]

An important part of care for these patients is to give them the time and opportunity to express all of their concerns. A consultation should go beyond assessment of tumour spread and response to treatment; table 20.1 suggests areas which should be covered although it may not be possible or appropriate to explore all of these in one interview.

Table 20.1 Assessment of the patient with advanced cancer

Points to consider:
- The patient's understanding of her illness and her present condition
- Psychological adjustment; the main fears and what helps her cope
- Availability of social and professional support
- Special concerns—for example, financial, dependent children
- Physical symptoms including pain
- Effect of the illness on mobility, sleep, and daily life
- Extent of disease, prognosis, and treatment options
- Hopes and expectations of the patient and family
- Information to be communicated with patient, carers, and other professionals

Management of common symptoms

Bowel problems

Chronic constipation is extremely common in advanced cancer. Many patients will be taking opioid analgesics and drugs that have an anticholinergic effect such as hyoscine, cyclizine, and tricyclic antidepressants. Other factors include reduced fluid and fibre intake, immobility, and muscle weakness. Prophylactic measures to avoid constipation are important, particularly in patients with intra-abdominal disease. An effective laxative regimen comprises a stool softener and stimulant in combination such as co-danthrusate or co-danthamer, or senna plus docusate or lactulose. The need to take these regularly should be explained to the patient and also the importance of increasing the amount of laxative as the analgesic dose is increased.

The accumulation of hard, desiccated faeces in the colon produces abdominal distension, nausea, and colic; palpable faecal masses can be mistaken for tumour. Sometimes this can be difficult to distinguish from incipient obstruction and plain abdominal *x* ray films may be useful here. The rectum should be emptied using lubricants and bisacodyl suppositories; oil enemas may be used to soften stool in the descending colon followed by phosphate enemas or a short-lasting evacuant such as oral bisacodyl 10 mg to clear the bowel. If there is extensive faecal loading and colicky pain stimulant laxatives are not well tolerated; the answer is to use a stool softener alone to begin with—for example, docusate 100–200 mg twice daily.

Bowel obstruction develops in 25%–42% of patients with ovarian cancer.[9,10] The small bowel is most often involved and there may be several sites of obstruction occurring together or sequentially. This may be the terminal event for a number of women; strangulated obstruction is unlikely as the intestinal loops tend to become fixed, but chronic dilatation leads to ischaemia of the bowel wall and septicaemia.

Surgery to relieve the obstruction undoubtedly offers the best relief of symptoms and the chance of prolonging survival if done promptly. In one series the mean survival was halved if the operation had been delayed for a trial of conservative treatment.[10] Not surprisingly, survival times are also better if effective postoperative chemotherapy is given, as is the case for patients who obstruct early on in their illness. Surgical intervention also ensures that cases of obstruction caused by adhesions rather than tumour are not missed.[11]

The decision to advise laparotomy becomes more difficult in patients with documented recurrence for whom there is no further systemic treatment to be offered. While surgery should always be considered, it must be weighed against the considerable morbidity and mortality likely in patients with advanced disease.[12] There is also the likelihood of reobstruction in women with widespread intraperitoneal disease.

Perhaps it is easier to suggest the relative contraindications to surgery in advanced disease:

- Reluctance of patient to undergo further procedures
- Extensive intraperitoneal tumour
- Ascites requiring frequent drainage
- Involvement of multiple bowel loops
- Reobstruction within weeks of previous palliative surgery
- Poor general condition with serious problems other than obstruction.

What of the patients for whom surgery is not an option? At the worst they may face a miserable terminal illness of discomfort and starvation while they are attached to a drip stand and with a nasogastric tube constantly in place—unfortunately these appendages do not guarantee good control of their symptoms but are evidence that "you could not possibly manage at home." At best they may receive comfort and the support to die at home if this is their choice. These are the aims of "conservative" management of intestinal obstruction. This would not be expected to prolong survival and it is also unlikely that the obstruction would reverse. Even with suction and intravenous hydration, few patients show any prolonged improvement and reobstruction is usual.[13] Sucess should be measured by the control of distressing symptoms.

Realistic objectives are the control of pain and nausea and reduction in the frequency of vomiting to once or twice daily. The patient is able to take small amounts of fluids and low residue foods and in most, intubation and intravenous fluid replacement is avoided. As absorption of oral medication is uncertain, other routes (rectal, transdermal, and subcutaneous) should be used.

Symptom control in obstructed patients

Control of colic—stimulant laxatives should be discontinued and an antispasmodic such as hyoscine butylbromide 80–120 mg/24 hours given by subcutaneous infusion. This also reduces the volume of gastrointestinal secretions. Hyoscine hydrobromide, 800–2400 µg/24 hours is an alternative which also has a central anti-emetic and sedative effect.

Control of background pain—morphine or diamorphine can be given by subcutaneous infusion. The total daily oral dose of morphine should be divided by three to obtain the equivalent dose of diamorphine that is required.

Control of nausea and vomiting—cyclizine has a potent effect on the vomiting centre and also reduces secretions. It may be given rectally, 50 mg eight hourly or by subcutaneous infusion 150–250 mg/24 hours. Haloperidol, 1.5–5 mg daily by subcutaneous injection, will counteract nausea that is caused by uraemia or bacterial toxaemia in longstanding obstruction. Methotrimeprazine is a more sedative antiemetic than cyclizine which may be a useful alternative in the terminal stages.

Clinically there may be partial obstruction with occasional passage of flatus or stool. It is worth continuing with a stool softener but it is better to avoid stimulant laxatives which would cause more colic and necessitate the use of antispasmodics which further reduce bowel motility. In these patients the use of antiemetics which increase gastric emptying and forward peristalsis in the proximal bowel will reduce the volume and frequency of vomits. Metoclopramide can be given by subcutaneous infusion, 30–90 mg/24 hours or domperidone suppositories 60 mg three times daily. These drugs are likely to aggravate colic if the obstruction becomes complete.

A trial of high-dose steroids such as dexamethasone 16 mg daily may also be considered with the object of reducing oedema around the tumour deposits that are adherent to the bowel wall. It is difficult to evaluate the usefulness of this regimen in incomplete obstruction but improvement may follow within a few days. If not, the steroids should be discontinued.

An indwelling nasogastric tube is unpleasant, particularly for prolonged periods of time, and can usually be avoided except in cases of very high obstruction where aspiration is preferable to frequent large vomits. Alternatively, drainage of secretions through a venting gastrostomy that has been formed by an endoscopic technique may improve the quality of life for a small number of patients.[14][15]

If patients are encouraged to drink and suck ice, and to take antiemetics, they may be able to absorb enough fluid to prevent dehydration. The discomfort of a dry mouth must be distinguished from thirst, and unfortunately drugs with anticholinergic effects will make this worse; good mouth care is therefore essential. Profound dehydration produces hypotension and may cause unpleasant dizziness and nausea. If it is not appropriate to use sedation to maintain comfort, fluid replacement will make the patient feel much better very quickly. Indefinite attachment to a drip is cumbersome so the use of intermittent intravenous fluids or hypodermocyclis may be needed occasionally.[16]

Intestinal obstruction is a frightening development for all patients, particularly when surgical intervention is not possible. A simple explanation of the above measures together with reassurance that they can take in some liquids and calories to "get by" is an important part of helping them live with the problem.

Diarrhoea and fistulas

Frequent loose stools may be a problem after resection or by-pass of small bowel. Spurious diarrhoea secondary to constipation must of course be excluded. Enterocutaneous fistulas are distressing complications and palliative surgery should be considered to improve quality of life. Skin protection from excoriation by small bowel contents is essential: adhesive carmellose wafers can be placed around the fistula and a stoma bag attached if possible. Loperamide 2–4 mg six hourly reduces stool volume and frequency. Ocreotide, a synthetic analogue of somatostatin, can dramatically switch off gastrointestinal secretion; a dose of 150–600 µg/24 hours infusion by subcutaneous injection has been used effectively in patients with fistulas and bowel obstruction but the cost precludes long term use.[17 18]

Nausea and vomiting

Nausea and vomiting is a common problem, particularly for those with intra-abdominal disease. Apart from the wretchedness of the symptoms, chronic vomiting leads to electrolyte and water depletion, and weakness and cachexia are accelerated. Not surprisingly patients become depressed and apathetic; they lose interest in life and only sleep may bring them some relief. Absorption of medication becomes erratic and often compliance with taking tablets or medicines is reduced, so that control of pain and other symptoms is impaired.

Successful management of nausea and vomiting depends on identification of the underlying cause, often more than one, and choice of the appropriate

Table 20.2 Causes of nausea and vomiting in ovarian cancer

Causes	Antiemetic	Routes
Stimuli to chemoreceptor trigger zone		
Drugs: cytotoxics	Ondanstron	Orally,
opioids	Haloperidol	subcutaneously
		intravenously
Metabolic: raised calcium/urea	Ondansetron	Orally,
concentrations		subcutaneously,
		intravenously
Impaired GI motility		
Gastric stasis	Domperidone	Orally, rectally
Squashed stomach	or Metoclopramide	Orally,
Autonomic dysfunction	or Cisapride	subcutaneously,
		intravenously
	Cisapride	Orally, rectally
Stimuli to vomiting centre		
By visceral autonomic nerves:	Cyclizine	Orally,
gastritis, obstruction, hepatomegaly,		subcutaneously,
constipation		intravenously
Central: cerebral metastases		
	or Methotrimeprazine	Orally,
		subcutaneously,
		rectally

antiemetic and the route to ensure adequate absorption (table 20.2). A combination of drugs may be needed in some patients. The role of 5 hydroxytriptamine$_3$-serotonin antagonists such as ondansetron for other than chemotherapy-induced emesis is not yet clear; they may prove to be useful in counteracting opioid-induced nausea and vomiting but are expensive drugs. Delivery by subcutaneous infusion has been described.[19]

In ovarian cancer, recognisable syndromes occur quite often. One example is the combination of symptoms in association with partial and complete bowel obstruction, which is discussed elsewhere. Another is the "squashed stomach syndrome", in which large intra-abdominal tumour masses or ascites cause compression and impede gastric emptying.[20] The clinical features are loss of appetite and early satiety, intermittent nausea with occasional large vomits, and oesophageal reflux with hiccups. It is useful to recognise these patterns as they can be effectively alleviated by a combined approach: regular domperidone to increase gastric emptying, dimethicone suspension as an antacid and antiflatulent, plus cyclizine if there is persistent nausea. This regimen often relieves the hiccups as well, but if they are still troublesome baclofen 5–10 mg or nifedipine 20 mg three times daily may curtail them.[21]

Anorexia, cachexia and generalised weakness

The patient with an uncomfortable, distended abdomen or who is constipated or nauseated will understandably be reluctant to eat. Candida infections can be responsible for a dry, sore mouth and should be treated promptly. Anxiety, depression, or poorly controlled pain are all potent appetite suppressants, but anorexia and cachexia are a consequence of altered metabolism in malignancy and this is difficult to reverse. Dexamethasone 2–4 mg daily may be tried to improve appetite but should be discontinued if there is no benefit within two weeks. Progesterones such as megestrol acetate 800 mg–1600 mg daily have lead to weight gain but the duration of this benefit is uncertain.[22]

Eating is a social activity and mealtimes may be dreaded by the patient who is harangued by an anxious family. They must be encouraged to move from formal meals to a succession of attractively presented, small snacks; often intake is best early in the day. Nutritional supplements are available but it is better to incorporate these in preparation of dishes to boost calories and protein content rather than as replacement foods.

Progressive weight loss leads to reduction in mobility and independence and drastically alters personal appearance, and many patients experience general lassitude. They and their families need help and advice to readapt to the changes imposed by the disease and assessment by physiotherapists and occupational therapists is invaluable at this stage. While any treatable causes of weakness such as hypokalaemia or anaemia should be sought, a wheelchair can transform lifestyle if the patient can manage to go only short distances. Pressure-relieving cushions and mattresses and hygiene aids may be required. New clothes that are loose and casual lessen the impact of cachexia and can be a big morale boost.

Problems with fluid accumulation: oedema, ascites and effusions

Swollen legs

Soft peripheral oedema is commonly associated with a falling albumin concentration. Initially appearance causes the most concern, but as fluid accumulates mobility is affected and leakage through the skin can be troublesome. Low pressure (10–20 mmHg) stockings, put on before getting up, will contain the swelling initially and are preferable to diuretics, but these may be needed if there is coexistent ascites or pulmonary oedema.

Lymphoedema develops as a result of lymphatic obstruction—for example, secondary to a pelvic mass—and may be compounded by venous occlusion. It is associated with extracellular deposition of protein and does not respond to treatment with diuretics. However, inaction leads inevitably to progressive increase in size and weight of the limb; episodes of cellulitis may occur and brawny secondary skin changes may develop. Treatment comprises a combination of truncal massage, use of properly fitted compression hosiery or bandaging plus exercise and skin care.[23] This is effective in reducing the volume of the swollen limb and maintaining the improvement, but requires continuing treatment and motivation.

284

Ascites

Ovarian cancer is the commonest malignancy associated with ascites, which is present in over 60% of patients at the time of death.[24] Intraperitoneal tumour deposits occlude lymphatics, impair venous drainage, and alter membrane permeability; liver metastases may be an additional factor in a few patients. Viscous ascites may occasionally develop with mucinous adenocarcinoma (pseudomyxoma peritonei). Ascites causes considerable impairment in the quality of life through the discomfort of distension and breathlessness as a result of diaphragmatic splinting and compression of the stomach. In addition this contributes to the general weakness and debility of the patient.

An attempt should be made to drain ascites if the abdomen is tense and uncomfortable, but caution is needed if there are large tumour masses or possibly dilated bowel. If in doubt an abdominal *x* ray film should be obtained first. Ultrasound-guided paracentesis may be needed after unsuccessful attempts, as the peritoneal cavity may become multiloculated. The viscous material associated with pseudomyxoma peritonei does not drain easily and gentle suction may be needed.

Often the benefit of paracentesis is short-lasting; if there is much peripheral oedema ascites may re-accumulate within days to the disappointment of the patient. Repeated attempts deplete protein levels further, although salt-poor albumin infusions may be used in an attempt to counteract this. Insertion of a peritoneovenous shunt to enable recirculation of ascitic fluid is another option.[25 26] This may be considered for those patients who may be expected to survive for several months but who do not have the problems of a multiloculated abdomen, liver involvement, or coexistent pleural effusion. There is evidence from necropsy examinations that such shunts can permit dissemination of malignant cells—for example, to the lungs[27]—but this is unlikely to produce clinically important metastases in the remaining time.

Diuretics have a slower effect, but at least avoid the problems of protein depletion. A regimen of oral frusemide 40 mg–80 mg plus spironolactone 100–200 mg may be tried, with monitoring of electrolytes and output. An infusion of frusemide, 100 mg over 24 hours, may initiate the diuresis.[28] Spironolactone counteracts the effect of secondary aldosteronism, but can cause nausea and vomiting. In all patients it is important to relieve the background abdominal discomfort with analgesics and to treat the symptoms of gastric compression.

Effusions and breathlessness

Pleural effusions may arise through metastatic involvement of the pleura, lungs, or mediastinal nodes but may also be secondary to ascites through communicating lymphatic channels.[29] If this produces troublesome breathlessness, simple needle aspiration of 500–1500 ml will produce relief.

If recurrent pleural effusion is the main problem, more effective intercostal drainage followed by instillation of tetracycline 500 mg or bleomycin 60 mg should be considered to prevent reaccumulation.

With advancing illness, there may be no further interventional measures to help the breathless patient, but there are still ways of reducing the discomfort. Good mouthcare, a fan by the bedside, and company can all help. It is important to ensure a comfortable position, often fairly upright, by day and night. A wheelchair may be needed for both indoor and outdoor use. Oxygen is indicated if there is hypoxia, as with pulmonary emboli or lymphangitis, but better psychological support is achieved by simple explanation and reassurance. Many breathless patients fear that they will choke or suddenly not be able to get their breath, and are relieved to know this is not likely to happen. The distress may be helped by regular oral morphine, starting with 2.5 mg four hourly, and the use of nebulised opioids is currently under evaluation. Benzodiazepines—for example, diazepam 5 mg 12 hourly and nabilone, 100–500 μg eight hourly, are other options.

Steroids can help the dyspnoea and cough associated with widespread intrapulmonary metastases and lymphangitis. It is logical to aim for a beneficial effect quickly by using a trial of dexamethasone 16 mg daily. This has no mineralocorticoid effect and so should not aggravate the condition by causing fluid retention. If there is improvement within a few days the dose should be gradually reduced by about 2 mg every second or third day to a level which maintains the benefit.

A dry persistent cough may be helped by methadone linctus or 2–5 ml of nebulised 0.25% bupivicaine. A productive cough with troublesome secretions can be distressing. If this is the result of infection, appropriate oral antibiotics should be given unless death is likely within a few days. This should be viewed as good symptom control, not an attempt to prolong a terminal illness (which in reality, antibiotics rarely do). If it is not appropriate to use antibiotics, hyoscine hydrobromide can be used to dry secretions and it also has a sedative effect. If pulmonary oedema is caused by heart failure an injection of frusemide should be given even in the last hours; this may help the distress of the family although the patient may be unaware.

Pain

People with advanced cancer are likely to have several different sites and types of pain; a fundamental step in pain control is to identify these and the probable underlying causes (table 20.3). In addition to careful clinical assessment, investigations are justifiable even in advanced disease if they will assist pain management and this is preferable to an empirical "morphine for everything" approach. Metastatic bone pain is fortunately quite rare in ovarian cancer and pain in the abdomen and pelvis, often visceral in origin, predominates. Not all pain will be directly caused by tumour and other straight forward causes must not be overlooked: examples are

286

Table 20.3 Examples of pain encountered in advanced ovarian cancer

Problem	Cause	Possible Treatment
Sore mouth	Candida infection	Antifungal drug Benzydamine mouthwash
Right upper quadrant pain	Subdiaphragmatic tumour Stretched liver capsule	Primary analgesics Non-steroidal anti-inflammatory drugs Steroids Coeliac block
Colic	Constipation Bowel obstruction	Laxatives Antispasmodics
Tenesmoid rectal pain	Tumour in pouch of Douglas	Primary analgesics Steroids Nifedipine Sympathectomy
Backache plus pain in legs on movement	Para-aortic nodes Nerve compression	Primary analgesics Steroids Transcutaneous electrical nerve stimulation Spinal analgesia
Burning pain and numbness in foot	Sacral nerve infiltration by pelvic tumour	Primary analgesics Amitriptyline Anticonvulsants Transcutaneous electrical nerve stimulation

Table 20.4 Reasons for poor pain control

- Inadequate assessment of pain and underlying causes
- Failure to prescribe strong enough analgesic for severity of pain
- Suboptimal dose, frequency or provision for breakthrough pain
- Disregard of opioid side-effects
- Fears of respiratory depression or addiction
- Failure to consider adjuvant analgesics
- No recognition of pain that is not responding to morphine
- Failure to address psychological impact of pain
- Failure to review frequently

musculoskeletal discomfort in the cachectic patient, or drug-induced dyspepsia.

Most cancer-related pain is the result of tumour invasion and compression of pain-sensitive structures and should respond to primary analgesics. These range in potency from paracetamol to strong opioids. Appropriate use of this "analgesic staircase" according to the severity of symptoms should control pain in over 70% of cancer patients.[30] Sadly cancer pain is still often poorly controlled and this is often because basic principles are overlooked (table 20.4). Most patients will eventually need a regular strong opioid and morphine remains the drug of choice; it is so familiar that doctors assume they know how to use it well. However, problems often arise through failure to deal with side-effects, incorrect titration of dose to achieve pain control, and errors in changing between routes of administration of different strong opioids. Hanks and Hoskins published

Table 20.5 Drugs suitable for use in subcutaneous infusions

Analgesics	Diamorphine
	Morphine
	Hyoscine butylbromide
Steroids	Dexamethasone (incompatible with midazolam)
Antiemetics	Cylizine (irritant at high concentrations)
	Haloperidol
	Methotrimeprazine
	Hyoscine hydrobromide
Anticonvulsants	Phenytoin (incompatible with other drugs)
Sedatives	Midazolam
	Methotrimeprazine

Avoid combinations of more than two drugs; if in doubt about compatibility or use of this route for other drugs seek the advice of a pharmacist. Some solutions are too irritant; examples to avoid are diazepam and prochlorperazine.

an excellent review of opioid analgesics and their use.[31] An avoidable bad experience may deter a patient from accepting adequate pain relief in the future. Addiction and psychological dependence are not a problem in the management of chronic cancer pain and should not be used as an excuse for suboptimal pain control.[32]

The oral route may not be suitable for some patients, particularly if they are obstructed. Morphine and oxycodone suppositories are available, and other strong analgesics—for example, buprenorphine and phenazocine can be used sublingually, but these routes may not be practical at higher doses. Continuous subcutaneous infusions delivered by a syringe driver are often used; the painkiller can be combined with other drugs (table 20.5). Diamorphine is eight times more soluble than morphine which is advantageous for parenteral use as smaller volumes are needed; this becomes a consideration in patients who require infusions of several grams of morphine daily. It has no therapeutic advantage as it is directly metabolised to morphine. Transdermal delivery systems are a useful option; if the daily morphine requirement is stable this can be converted to an equivalent dose of fentanyl worn as patches on the skin.[33]

Doctors should be aware that morphine will not relieve all cancer pain and examples of pains that respond partially or poorly to opioids include colic, pain from pressure sores, and neuropathic pain. Inappropriate escalation of the opioid dose can produce unacceptable sedation, confusion, and myoclonic jerks in a patient who is still distressed by pain. It is also difficult to eradicate movement-related pain although the patient may be comfortable at rest. In these circumstances other approaches to pain control must be considered.

Adjuvant analgesics

A range of other drugs contribute to pain control which should be combined with the primary analgesics when indicated. Non-steroidal anti-inflammatory agents (NSAIDs) can help not just bone pain, but soft tissue inflammation such as a tender liver capsule. Drugs which reduce spasm in skeletal and smooth muscle are used frequently—an obvious example being hyoscine butylbromide for bowel colic. Steroids can sometimes dramatically reduce pain when it is caused by pressure of a tumour mass, particularly on nerves. A trial of dexamethasone 16 mg daily for five days may be used, reducing slowly after the initial response or discontinuing promptly if there is no benefit.

It is important to recognise neuropathic pain, which may have a burning, stabbing, or shooting quality and is felt in an area of altered sensation. This may follow damage to nerves by infiltrating tumour or as a side-effect of cytotoxic drugs such as cisplatin. Drugs useful in neuropathic pain include the tricyclic antidepressants and anticonvulsants. Amitriptyline 10–20 mg at night may be increased gradually up to 75 mg. Any response will be apparent within a few days and is separate from any effect on mood. Carbamazepine or sodium valproate are also effective for neuropathic pain and can be increased slowly up to the therapeutic anticonvulsant dose; they may be combined with a tricyclic antidepressant. Transcutaneous nerve stimulation (TENS) can be extremely effective for this type of pain.

Other aspects of pain control

No patient should be left with uncontrolled, severe pain. A few patients will have symptoms that are difficult to manage and they should be referred to a palliative care physician or an anaesthetist who specialises in pain management. Interventional procedures are necessary in about 10% of patients. A coeliac block may alleviate upper abdominal pain and colic; a lumbar sympathectomy can help intractable rectal pain. As well as neurolytic techniques, spinal analgesia may be an option. Advice from colleagues should be sought early to make the remaining time as comfortable as possible.

Chronic pain leads to depression which needs to be diagnosed and actively treated. The hospital anxiety and depression scale is a useful tool in assessment of cancer patients.[34] Honesty about what can be achieved is important; in spite of our best efforts patients still live with day to day discomfort. It is well worth looking at practical measures to find the most comfortable position with attention to pressure areas. Massage and relaxation techniques can help many people to live with pain and of course a good night's sleep is important. Above all patients need our continuing empathy, time, and support.

The final stage

Attempts to predict the prognosis are often inaccurate even with very ill patients admitted to a hospice, and usually the time remaining is over-estimated.[35] A useful guide is the rate of change; if there is perceptible deterioration day to day, time is likely to be short. The family should be prepared and given the opportunity to ask questions; sometimes there is great anxiety because the patient is no longer eating or drinking and it is helpful to emphasise that the focus of care is on comfort. At this stage the patient will not be hungry and attention to a dry mouth achieves more than rehydration.

All medication should be reviewed and simplified to the essential drugs for control of active symptoms. Analgesics should be continued even if the patient is unconcious. Oral delivery may no longer be possible and the rectal or subcutaneous routes are used.

Terminal restlessness causes distress to staff and family. This is often the result of causes other than pain, but if uncertain an extra injection of diamorphine should be given and the response assessed. Sometimes a full bladder or rectum is the problem and a change of position should also be tried. If the patient continues to remain unsettled, a subcutaneous infusion of a sedative such as methotrimeprazine 50–300 mg/24 hours or midazolam 15–250 mg/24 hours may be used.[36] Hyoscine hydrobromide may be indicated for secretions and is also sedative, but can cause paradoxical agitation in a few patients.

At no other time is support for the woman and carers more important together with attention to privacy, dignity, and cultural beliefs. The sadness of death can be combined with relief at the end of suffering and great professional satisfaction for those who look after the patient to the end.

References

1 Richardson GS, Scully RE, Nikrui N, Nelson JH Jr. Common epithelial cancer of the ovary [2]. *N Engl J Med* 1985; **312**: 415–24.

2 Neijt JP. Treatment of advanced ovarian cancer: 10 years of experience. *Ann Oncol* 1992; **3**: 17–27.

3 Dodwell DJ, Rathmell AJ, Ash DV. Assessment of palliative chemotherapy: a step beyond response. *Clin Oncol* 1993; **5**: 114–7.

4 Paling MR, Shawker TH, Dwyer A. Ultrasonic evaluation of therapeutic response in tumors: its values and implication. *Journal of Clinical Ultrasound* 1981; **9**: 435–41.

5 Dauplat J, Hacker NF, Nieberg RK, Berek JS, Rose TP, Sagae S. Distant metastases in epithelial ovarian carcinoma. *Cancer* 1987; **60**: 1561–6.

6 Ferenczy A, Okagi T, Richart RM. Para-endocrine hypercalcemia in ovarian neoplasms. Report of mesonephroma with hypercalcemia and review of the literature. *Cancer* 1971; **27**: 427–33.

7 Harbert CJ, Rocha L, Smith FP, Delgado G. The efficacy of radionuclide liver and bone scans in the evaluation of gynecological cancers. *Cancer* 1982; **49**: 1040–2.

8 Massie MJ, Holland JC. Depression and the cancer patient. *J Clin Psychiatry* 1990; **51** (suppl 7): 12–17.

9 Tunca JC, Buchler DA, Eberhard AM, Ruzicka FF, Crowley JJ, Carr WF. The management of ovarian cancer caused bowel obstruction. *Gynecol Oncol* 1981; **12**: 186–92.

10 Beattie GJ, Leonard RCF, Smyth JF. Bowel obstruction in ovarian carcinoma: a retrospective study and review of the literature. *Palliative Medicine* 1989; **3**: 275–80.

11 Walsh HPJ, Schofield PF. Is laparotomy for small bowel obstruction justified in patients with previously treated malignancy? *Br J Surg* 1984; **71**: 933–5.

12 Krebs HB, Gopelrud DR. Surgical management of bowel obstruction in advanced ovarian carcinoma. *Obstet Gynecol* 1983; **61**: 327–30.

13 Chan A, Woodruff RKJ. Intestinal obstruction in patients with widespread intra-abdominal malignancy. *Journal of Pain Symptom Management* 1992; 7: 339–42.

14 Malone JM Jr, Koonce T, Larson DM, Freedman RS, Carrasco CH, Saul PB. Palliation of small bowel obstruction by percutaneous gastrostomy in patients with progressive ovarian cancer. *Obstet Gynecol* 1986; **68**: 431–3.

15 Ashby MA, Game PA, Devitt P, *et al.* Percutaneous gastrostomy as a venting procedure in palliative care. *Palliative Medicine* 1993; **5**: 147–50.

16 Bruera E, Legris MA, Kuehn N, Miller MJ. Hypodermoscyclis for the administration of fluids and narcotic analgesics in patients with advanced cancer. *Journal of Pain Symptom Management* 1990; **5**: 218–20.

17 Nubiola P, Badia JM, Martinez-Rodenas F, *et al.* Treatment of 27 post-operative enterocutaneous fistulas with the long half-life somatostatin analogue SMS 201-995. *Ann Surg* 1989; **210**: 56–8.

18 Mercadante S, Spoldi E, Caraceni A, Maddaloni S, Simonetti MT. Octreotide in relieving gastrointestinal symptoms due to bowel obstruction. *Palliative Medicine* 1993; 7: 295–9.

19 Mulvenna PM, Regnard CF. Subcutaneous ondansetron. *Lancet* 1992; **339**: 1059.

20 Twycross RG, Lack S. *Control of alimentary symptoms in far advanced cancer.* Edinburgh: Churchill Livingstone, 1986.

21 Anonymous. *Drug Ther Bull* 1990; **28**: 36.

22 Aisner J, Tchekmadyian NS, Tait N, Parres H, Novak M. Studies of high-dose megesterol acetate: potential applications in cachexia. *Semin Oncol* 1988; **15**: 68–75.

23 Twycross RG, Badger C. Oedema in advanced disease—a flow diagram. *Palliative Medicine* 1989; **3**: 213–15.

24 Lifshitz S. Ascites: pathophysiology and control measures. *Int J Radiat Oncol Biol Phys* 1982; **8**: 1423–6.

25 Downing R, Black J, Windsor CWP. Palliation of malignant ascites by the Denver peritoneovenous shunt. *Ann R Coll Surg Engl* 1984; **66**: 340–3.

26 Edney JA, Hill A, Armstrong D. Peritoneovenous shunts palliate malignant ascites. *Am J Surg* 1989; **158**: 598–601.

27 Tarin D, Price JE, Kettlewell MGW, Souter RG, Vass ACR, Crosley B. Clinicopathological observations on metastasis in man studied in patients treated with peritoneovenous shunts. *BMJ* 1984; **288**: 749–51.

28 Amiel SA, Blackburn AM, Rubens RD. Intravenous infusion of frusemide as treatment for ascites in malignant disease. *BMJ* 1984; **288**: 1041.

29 Friedman MA, Slater E. Malignant pleural effusions. *Cancer Treat Rev* 1978; **5**: 49–66.

30 *Cancer pain relief.* Geneva: WHO, 1986.

31 Hanks GW, Hoskins PJ. Opioid analgesics in the management of pain in patients with cancer: a review. *Palliative Medicine* 1987; **1**: 1–25.

32 Schug SA, Zech D, Grond S, Jung H, Mevsert T, Stobbe B. A long-term survey of morphine in cancer pain patients. *Journal of Pain Symptom Management* 1992; 7: 259–64.

33 Ripamonti C, Bruera C. Transdermal and inhalatory routes of opioid administration: the potential application in cancer pain. *Palliative Medicine* 1996; **6**: 98–104.

34 Zigmond AS, Snaith RP. The hospital anxiety and depression scale. *Acta Psychiatr Scand* 1983; **67**: 361–70.

35 Heyse-Moore LH, Johnson Bell VE. Can doctors accurately predict the life expectancy of patients with terminal cancer? *Palliative Medicine* 1987; **1**: 165–6.

36 McNamara P, Minton P, Twycross RG. Use of midazolam in palliative care. *Palliative Medicine* 1991; **5**: 244–9.

Index